Perspectives in Animal Ecology and Reproduction

Perspectives in Animal Ecology and Reproduction

— *Volume 3* —

Editors
Dr. V.K. Gupta
Dr. Anil K. Verma

2006
DAYA PUBLISHING HOUSE
Delhi - 110 035

ISBN 81–7035–424–2

Published by : **Daya Publishing House**
1123/74, Deva Ram Park
Tri Nagar, Delhi - 110 035
Phone: 27383999
Fax: (011) 23244987
e-mail : dayabooks@vsnl.com
website : www.dayabooks.com

Showroom : 4760-61/23, Ansari Road, Darya Ganj,
New Delhi - 110 002
Phone: 23245578, 23244987

Laser Typesetting : **Classic Computer Services**
Delhi - 110 035

Printed at : **Chawla Offset Printers**
Delhi - 110 052

PRINTED IN INDIA

Dedicated to

Dr. G.N. QAZI

Director

Regional Research Laboratory (CSIR)

Jammu – 180 001, India

Editorial Board

— Chief-Editors —

— *Editors* —

Prof. Clement A. Tisdell
Professor Emeritus,
School of Economics,
The University of Queensland, Brisbane,
Australia
E-mail: c.tisdell@economics.uq.edu.au

Prof. Vanitha Kumari
Head,
Department of Zoology,
Bharathiar University,
Coimbatore, India
E-mail: regr@bharathiaruni.org

Prof. Tej Kumar Sherstha
Head,
Central Department of Zoology,
Tribhuvan University,
Kathmandu, Nepal
E-mail: drtks@ccsl.com.np

Dr. Sunil Kumar
Deputy Director (Sr.Grade),
Division of Reproductive Toxicology & Histochemistry,
National Institute of Occupational Health,
Meghani Nagar, Ahmedabad-380016, India
E-mail: sunilnioh@yahoo.com

Dr. Justin Gerlach
Chief Scientist,
The Nature Protection of Seychelles,
133 Cherry Hinton Road,
Cambridge CBI 7BX,UK
E-mail: justgerlach@aol.com

An Appreciation

Dr. G.N. Qazi (born, August 14, 1946) got his M.Sc. Bio-chemistry and Ph.D. Microbiology from M.S. University, Baroda, India. He was Post-Doctoral Research Fellow at Federal Republic of Germany where he worked on Biochemical engineering at University of Dortmund and in Molecular biology at Max Plank Institute of Biochemistry in Munich. He is Scientist of eminence and Director, Regional Research Laboratory (CSIR), Jammu. Earlier he headed the Biotechnology Department at R.R.L., Jammu and brought international recognition to the department by giving it a lead in fermentation technology and genetic engineering. His work on gene cloning and genetic manipulation is widely acclaimed and recognized internationally.

Dr. Qazi has been the recipient of Best Invention Award Exhibition Society, Hyderabad (2004); CSIR Technology Award (2001) and Vasvik Industrial Research Award in Chemical Sciences and Technology (2000). He is founder Course Director, Department of Biotechnology, University of Jammu, Jammu. He is also Chairman, Committee for Bio- technology Policy, for J&K state; Co-chairman, All India Bio-tech Association, North Chapter; and Fellow member of Bio-tech Research Society of India. He has been associated with several national and international organizations and has International Scientific Collaboration with Federal Republic of Germany, USA, Malaysia, Thailand, Poland, Italy and U.K. to name a few.

Dr. Qazi has about 100 research papers published in the national and international journals of repute and has participated and chaired numerous sessions of national and international conferences. He has to his credit about 70 patents and has delivered a number of invited lectures in India and abroad. His areas of interest are plant biotechnology, mutagenesis, fermentation technology, drug development and crop-improvement through biofertilizers and bio-control agents.

Over and above his personal attributes, the type of dynamic leadership he has provided to the Regional Research Laboratory's scientific staff in developing an interface with Indian Pharmaceutical sector has resulted in marketing of one antihepatotoxic/ hepatoprotective drug speaks volumes about his leadership quality. I had a chance of working very closely with him during three and half years of my research tenure in R.R.L., Jammu. I have found in him a sharp learner, astute leader and a reliable colleague. He has taken his parent institute to a place where its research contributions are recognized and appreciated especially in herbal drug development and fermentation technology. I wish him further success in this self-carved out ventures.

Professor O.P Suri
School of Biotechnology & Dean Sciences,
Shri Mata Vaishno Devi University, Jammu
Formerly, Director Grade Scientist & Chairman
Natural Products Chemistry & Instrumentation Division,
Regional Research Laboratory (CSIR), Jammu

Foreword

It is indeed a matter of great pride for me to know that Dr.V.K.Gupta & Dr. Anil K. Verma having made best use of their experience in the area of biology have come up with the 3rd volume of the book "Perspectives in Animal Ecology and Reproduction". The updated volume of the book shall provide recent information along with future directions to all those who select biology as their future.

Today, there is an explosion in the field of biology. Everyday new discoveries are being reported by workers actively engaged in research work in the area of life sciences. While going through the contents of the book it seems that both the authors have kept in mind the latest developments in biology and have put in their sincere and untiring efforts in compiling the original research papers/review articles on the reproductive and ecological aspects of diverse forms of life. In the present day world since the environment is degrading fast, it becomes imperative for all of us to know the effects of environmental degradation on human health. The book will provide an opportunity to its readers to understand the relationships of the animals with the environment in which they live and also the impact of various toxicants and radiations on the animal reproduction.

Material contained in the book in itself shall be a treasure of knowledge to the students of biological sciences which would prove very useful preparing them for various competitive examinations / seminars / public discussions etc.

I congratulate both the authors & appreciate their efforts in writing this book that shall be of tremendous use both in the classroom & outside the classroom.

My best wishes for the authors.

Dr. Subash Gupta
Professor of Zoology & Principal Investigator,
Human Genetic Research cum Counseling Centre,
University of Jammu, New Campus, Jammu-180 001

Preface

The science of biology today is not the same science of fifty years or even ten years ago. Today's accelerated pace of research, aided by new techniques and instruments imparts to biology a rapidly changing character as discoveries go to pile up one on top of the other. Each new and important discovery is not just a mere addition to our knowledge, it also throws our establish beliefs into questions, and forces us certainly to reappraise and often to reshape the foundations upon which biology rests. An adequate presentation of the dynamic state of modern biology is, therefore, a formidable task, a challenge worthy of our best teachers, scientists and researchers.

The subject of animal ecology and reproduction is likely to dominate all areas of biology in early part of the new millennium. Animal ecology and reproduction, is one of the most important and interesting aspects of the study of animal sciences. Therefore, the need for periodic updating of books particularly meant for the students and researchers of ecology and reproduction, is more acute today than ever before.

The present knowledge of ecology and reproductive strategies in animal world has already assumed gigantic proportions and something, as discussed earlier, keeps on being added every day, which has encouraged us, in our endeavour to compile a series of book on these aspects. This volume of the series (Vol. III) comprises twenty one research articles, fourteen compiled under the section-animal ecology and seven under-animal reproduction section. The topics under each unit have been contributed by the leading experts in the field with exhaustive, relevant and up-to-date information. The book aims at providing the much needed information to teachers, researchers and others interested in environmental aspects and reproductive characteristics of various animal groups from diverse and varied ecological habitations.

The Editors express their gratefulness to the contributors who have shared valuable thoughts through their scholarly contributions and timely submission of manuscripts for this volume. Though meticulous care has been taken in editing the material, some mistakes might have crept in. We, therefore, close with the request to all who may use this book to be free to pass on to us their comments or suggestions, to enable us to improve our future volumes.

V.K. Gupta

Anil K. Verma

Contents

Part I
Animal Ecology

Chapter 1

Studies on the Environmental Correlates with the Infra Population of *Hemicriconemoides mangiferae* Infesting *Mangifera indica* in Western Uttar Pradesh

Anita* and *A.K. Chaubey*

Nematology Laboratory, Department of Zoology, C.C.S. University, Meerut –250 005

ABSTRACT

The living organisms have several levels of organizations in the environment and the physiology of the organism is concerned with the responses of the individual's environment. The plant parasitic nematodes are petite, un-segmented round worms that feed on or inside the roots of the host plant with the help of stylet. Plant parasitic nematodes are unseen, unheard hidden enemies of the farmers and are considered as the worst enemies of a wide host range of plants. Nematodes have a sensitive relationship with their environment and they respond to the spatial and temporal dynamics of resources and stressors in soil habitats. These attributes individually and collectively influence the structure and function of nematode biodiversity.

* Corresponding author: E-mail: kumar_ac2001@rediffmail.com

The biotic and abiotic factors influence the various stages and activities of nematodes and their relationship with the plants.

The study revealed the strong negative correlations at almost all the depths and distances during 2000–2001 and 2001–2002. Contrary to above few positive correlations were also recorded at 20.0–40.0 cm depth ($r = 0.304$; $P < 0.335$) at 30 cm distance and at 40.0–60.0 cm depth ($r = 0.223$; $P > 0.484$) at 60 cm distance around the root zones of *M. indica* during 2001–2002 whereas the uniform negative correlations were recorded with organic carbon to that of the *H. mangiferae* during 2000–2001 and 2001–2002 under the same host plant.

The functional linear regression depicting correlations in between nitrogen and *H. mangiferae* population indicated the high degree positive correlations at almost all the depths and distances during 2000–2001 and 2001–2002 (Table 1.3). However, a negative correlation was observed at 40.0–60.0 cm depth ($r = -0.295$; $P < 0.350$) at 30 cm horizontal distance from the stem of *M. indica* during 2000–2001. The functional linear regression trends showed significant negative correlations with phosphorous to that of the total *H. mangiferae* at almost all the depths and distances during both the years of the study

Keywords: *Population, Hemicriconemoides mangiferae, Mangifera indica, Environmental factors.*

Introduction

The living organisms have several levels of organizations in the environment. These organizations are the individuals, population, guild, community, ecosystem and ecosphere (biosphere). In fact at individual level the physiology of the organism is concerned with the responses of the individual's environment. The most important living organism that feed externally around the roots and internally inside the roots of plants are known as plant parasitic nematodes. The soil or plant nematodes are petite, unsegmented round worms that are different from the flatworms (cestode and trematode). All plant parasitic nematodes share one distinguishing feature–a stylet that the nematodes uses to puncture cells and feed. Plant parasitic nematodes are considered worst enemies of a wide host range of plants. Due to their microscopic nature, they cause severe losses to important crops (Chiarappa, 1971).

Nematodes have a sensitive relationship with their environment and they respond to the spatial and temporal dynamics of resources and stressors in soil habitats. Resources are the sources of nutrition and characteristic of the habitat that enable nematodes to survive and reproduce whereas stressors tie nematode biodiversity to fundamental attributes of ecosystems, the plants and micro-organisms in a habitat upon which nematodes feed, natural and anthropogenic disturbances, physical and chemical properties of soil and climate. These attributes individually and collectively influence the structure and function of nematode biodiversity. The composition of the soil nematode community depends on the vegetation present, soil type, season, moisture level, amount of organic matter, and many other factors because they are responsive to so many different factors (biotic and abiotic). It is believed that nematodes may be used as bio-indicators of the condition of the soil environment (Chaubey and Malhotra, 1988). The biotic and abiotic factors influence the various stages and activities of nematodes and their relationship with the plants. Among the edaphic factors like soil texture, temperature, moisture, pH, organic carbon and too many chemical constituents of the soil play a pivotal role on influencing the population as well as distribution of plant parasitic nematodes (Norton, 1978).

Western Uttar Pradesh is not only known for its agricultural crops but also known for some economically important fruit plants. Some of the most important fruit plants grown in this area are– mango, papaya, lemon, guava, grape, peach, etc. Among these mango (*Mangifera indica*) is the major fruit that are extensively used in the fruit processing industry and used in diet in this region.

The present study was undertaken to correlate some of the edaphic factors (temperature, organic carbon, nitrogen and phosphorus) with the population of *Hemicriconemoides mangiferae* infesting *M. indica* under field conditions in Western Uttar Pradesh.

Materials and Methods

The present study was carried out in some districts (Baghpat, Bulandshahr, Ghaziabad, Meerut, Muzaffarnagar and Saharanpur) of Western Uttar Pradesh. Soil Samples were collected randomly, from September 2000 to August 2002, for the distribution of the dominant ectoparasitic Tylenchids–*H. mangiferae*. Each soil sample (250 cc) was consisting of the mixture of 9 sub-samples. Sampling was done at monthly interval with the help of classical cylindrical, tube type auger from the three vertical depths (0.0-20.0, 20.0–40.0 and 40.0–60.0 cm) and at three horizontal distances (30, 60 and 90 cm) from the stem of *M. indica*. Samples were stored in well-labelled polythene bags to prevent the evaporation and loss of moisture. Soil samples were processed for nematode extraction by Cobb's Sieving and Decanting method (Cobb, 1918). The counting of the parasite was done under 10 × 10X magnification of stereoscopic binocular microscopic.

Determination of Edaphic Factors

Temperature

Soil temperature was measured on the spot at each depth and distance with the help of dial thermometer.

Organic Carbon

Organic Carbon was determined by adopting the method of Rapid Titration (Walkley and Black, 1934).

Calculations

1. Wt of soil taken–Wg
2. Blank Reading–Volume of N/2 FAS required for reduction 10 ml of $K_2Cr_2O_7$ solution = x ml
3. Experimental Reading–Volume of N/2 FAS required for reducing the excess of dichromate = y ml

   % of Carbon in Soil (Z) = (x −y) × 0.003 × 100/2 × Wg

   (1 ml of 1 N $K_2Cr_2O_7$ = 0.003 g Carbon)

Nitrogen

To determine the nitrogen in soil, the same procedure of Organic Carbon was adopted by applying the formula:

Nitrogen = Organic Carbon × 96.9 + 141.8 ppm

Phosphorus

Phosphorus was determined by the method given by Olsen, Cole, Watanabe and Dean (1954).

Calculations

1. Weight of the soil sample = 2.5 g

2. Volume of N/2 $NaHCO_3$ = 20 ml
3. Volume of the filtrate = 5 ml
4. Final volume after the development of the colour (dilution = 20 × 5 = 100 times)
5. Transmittance of the test solution = T (for comparing with standard curve)
6. Concentration of P read from the standard curve = X ppm
7. ppm of available P in the soil or kg of available P/Acre = X × 100

Computer software SPPS was applied to analyse the data.

Results and Discussion

Temperature was found to be detrimental to that of the total nemic population of *H. mangiferae* during the 24 months of study. The study revealed the strong negative correlations at almost all the depths and distances during 2000–2001 and 2001–2002 (Table 1.1). Contrary to above, positive correlations were recorded at 20.0–40.0cm depth ($r = 0.304$; $P < 0.335$) at 30 cm distance and at 40.0–60.0 cm depth ($r = 0.223$; $P > 0.484$) at 60 cm distance around the root zones of *M. indica* during 2001–2002. Study on organic carbon was found to have substantial detrimental effect with the same parasite. The uniform negative correlations were recorded with organic carbon to that of the *H. mangiferae* during 2000–2001 and 2001–2002 under the same host plant (Table 1.2). However some insignificant positive correlations were observed at 0.0–20.0 ($r = 0.127$; $P > 0.694$) and 20.0–40.0 cm depths ($r = 0.172$; $P > 0.593$) at 60 cm distance during 2000–2001.The insignificant positive correlations was also observed at 0.0–20.0 cm depth ($r = 0.263$; $P > 0.407$) at 30 cm distance around the root zones of *M. indica* during 2001–2002. The functional linear regression depicting correlations in between nitrogen and *H. mangiferae* population indicated the high degree positive correlations at almost all the depths and distances during 2000–2001 and 2001–2002 (Table 1.3). However, a negative correlation was observed at 40.0–60.0 cm depth ($r = -0.295$; $P < 0.350$) at 30 cm horizontal distance from the stem of *M. indica* during 2000–2001. The functional linear regression trends showed significant negative correlations with phosphorous to that of the total *H. mangiferae* at almost all the depths and distances during both the years of the study (Table 1.4). However, insignificant positive correlations were recorded at 0.0–20.0 ($r = 0.183; P > 0.568$), 20.0–40.0 ($r = 0.059; P > 0.856$) and 40.0–60.0 cm depths ($r = 0.061; P > 0.851$) at 90 cm distance from the collar region of the host plant during 2000–2001. The observations also indicated insignificant positive correlations at 0.0–20.0 ($r = 0.239; P > 0.454$), at 20.0–40.0 ($r = 0.255; P > 0.421$) and at 40.0–60.0 cm depths ($r = 0.021; P > 0.947$) at 30 cm distance; at 20.0–40.0 cm depth ($r = 0.203; P > 0.525$) at 60 cm distance and also at the same depth ($r = 0.198; P > 0.536$) at 90 cm distance from the stem of the host plant during 2001–2002.

Table 1.1: Functional Linear Regression Depicting Correlation of Soil Temperature with *H. mangiferae* Population Around Root Zones of *M. indica*

Distance/Depth	*2000–2001*	*2001–2002*
H = 30 cm		
V= 0.0–20.0 cm	Y = 179.95–4.816X	Y = 16.92–0.139X
	r = –0.786; P < 0.002	r = –0.160; P > 0.618
V = 20.0–40.0 cm	Y = 142.41–3.838X	Y = 3.50+0.264X
	r = –0.799; P < 0.001	r = 0.304; P < 0.335
V = 40.0–60.0 cm	Y= 121.32–3.235X	Y= 17.75–0.148X
	r = –0.813; P < 0.001	r = –0.107; P > 0.740
H = 60 cm		
V = 0.0–20.0 cm	Y = 175.64–4.571X	Y = 28.38–0.148X
	r = –0.763; P < 0.003	r = –0.038; P > 0.907
V= 20.0–40.0 cm	Y = 118.83–2.498X	Y = 26.83–0.214X
	r = –0.412; P < 0.181	r = –0.093; P > 0.772
V= 40.0–60.0 cm	Y = 145.97–3.842X	Y = 5.21+0.325X
	r = –0.615; P < 0.031	r = 0.223; P > 0.484
H = 90 cm		
V = 0.0–20.0 cm	Y = 134.03–3.155X	Y = 97.92–2.623X
	r = –0.321; P < 0.306	r = –0.574; P < 0.049
V = 20.0–40.0 cm	Y = 52.44–0.593X	Y = 42.75–0.820X
	r = –0.195; P > 0.543	r = –0.426; P > 0.165
V = 40.0–60.0 cm	Y = 80.44–1.867X	Y = 17.39–0.077X
	r = –0.542; P < 0.066	r = –0.041; P > 0.899

H: Horizontal distance from the stem of the host plant.

V: Vertical depth.

Table 1.2: Functional Linear Regression Depicting Correlation of Soil Organic Carbon with *H. mangiferae* Population Around Root Zones of *M. indica*

Distance/Depth	*2000–2001*	*2001–2002*
H = 30 cm		
V = 0.0–20.0 cm	Y = 58.78–46.547X r = –0.162; P > 0.614	Y = 10.26+11.852X r = 0.263; P > 0.407
V = 20.0–40.0 cm	Y = 75.57–160.888X r = –0.550; P < 0.062	Y = 15.40–24.086X r = –0.423; P < 0.168
V = 40.0–60.0 cm	Y = 36.75–11.852X r = –0.086; P > 0.789	Y = 16.52–13.521X r = –0.202; P > 0.527
H = 60 cm		
V = 0.0–20.0 cm	Y = 41.30+45.874X r = 0.127; P > 0.694	Y = 33.37–38.505X r = –0.185; P > 0.564
V = 20.0–40.0 cm	Y = 31.46+107.406X r = 0.172; P > 0.593	Y = 31.96–39.014X r = –0.424; P < 0.167
V = 40.0–60.0 cm	Y = 55.72–64.980X r = –0.145; P > 0.652	Y = 18.54–17.351X r = –0.209; P > 0.514
H = 90 cm		
V = 0.0–20.0 cm	Y = 77.56–172.469X r = –0.271; P > 0.392	Y = 41.81–66.813X r = –0.332; P < 0.289
V = 20.0–40.0 cm	Y = 52.54–114.474X r = –0.488; P < 0.104	Y = 23.43–12.410X r = –0.131; P > 0.683
V = 40.0–60.0 cm	Y = 39.47–50.760X r = –0.398; P > 0.198	Y = 18.53–11.407X r = –0.124; P > 0.701

H: Horizontal distance from the stem of the host plant.

V: Vertical depth.

The multiple regression analysis of major edaphic factors with *H. mangiferae* population indicated that temperature showed positive correlations at 20.0–40.0 cm depth (r = 0.272; P > 0.442) at 30 cm distance and at 0.0–20.0 cm depth (r = 0.253; P > 0.482) at 60 cm distance around the root zones of *M. indica* during 2001–2002 (Table 1.5). Multifactorial etiology of macronutrients indicated that organic carbon was most effective at 40.0–60.0 cm vertical depth (r = 0.799; P < 0.224) at 30 cm distance and at 20.0–40.0 cm depth (r = 0.645; P < 0.051) at 60 cm distance during 2001–2002 (Table 1.6). The significant positive correlations of nitrogen to that of total *H. mangiferae* was recorded at 20.0–40.0 cm (r = 0.948; P < 0.020) and at 40.0–60.0 cm depth (r = 0.914; P < 0.050) at 30 cm distance, at 0.0–20.0 cm depth (r = 0.993; P < 0.032) and at 40.0–60.0 cm depth (r = 0.719; P < 0.375) at 90 cm distance during 2000–2001 (Table 1.6). Nitrogen was also most effective at 40.0–60.0 cm depth (r = 0.580; P < 0.063) at 90 cm distance during 2001–2002 (Table 1.6). The multiple regression analysis of phosphorus to that of *H. mangiferae* indicated positive correlation only at 20.0–40.0 cm depth (r = 0.930; P < 0.034) during 2001–2002 (Table 1.6).

Table 1.3: Functional Linear Regression Depicting Correlation of Soil Nitrogen with *H. mangiferae* Population Around Root Zones of *M. indica*

Distance/Depth	*2000–2001*	*2001–2002*
H = 30 cm		
V = 0.0–20.0 cm	Y = –105.84+0.989X	Y = 11.76+0.008X
	r = 0.244; P > 0.444	r = 0.018; P > 0.956
V = 20.0–40.0 cm	Y = –110.42+1.004X	Y = –9.89+0.125X
	r = 0.413; P < 0.179	r = 0.397; P < 0.194
V = 40.0–60.0 cm	Y = 152.49–0.802X	Y = –46.13+0.363X
	r = –0.295; P > 0.350	r = 0.538; P < 0.069
H = 60 cm		
V = 0.0–20.0 cm	Y = 4.15+0.297X	Y = 9.69+0.098X
	r = 0.076; P > 0.814	r = 0.159; P > 0.621
V = 20.0–40.0 cm	Y = –172.32+1.424X	Y = –21.42+0.263X
	r = 0.540; P < 0.067	r = 0.270; P > 0.394
V = 40.0–60.0 cm	Y = –251.66+ 1.929X	Y = –20.24+0.212X
	r = 0.582; P < 0.045	r = 0.372; P < 0.231
H = 90 cm		
V = 0.0–20.0 cm	Y = –464.11+3.287X	Y = –129.31+0.971X
	r = 0.761; P < 0.003	r = 0.564; P < 0.053
V = 20.0–40.0 cm	Y = –98.13+0.846X	Y = –34.56+0.344X
	r = 0.423; P < 0.168	r = 0.445; P < 0.144
V = 40.0–60.0 cm	Y = –87.58+0.769X	Y = –58.43+0.482X
	r = 0.323; P < 0.303	r = 0.519; P < 0.081

H: Horizontal distance from the stem of the host plant.

V: Vertical depth.

Table 1.4: Functional Linear Regression Depicting Correlation of Soil Phosphorus with *H. mangiferae* Population Around Root Zones of *M. indica*

Distance/Depth	*2000–2001*	*2001–2002*
H = 30 cm		
V = 0.0–20.0 cm	Y = 146.16–5.859X	Y = 8.17+0.329X
	r = –0.534; P < 0.071	r = 0.239; P > 0.454
V = 20.0–40.0 cm	Y = 108.81–4.090X	Y = 6.75+0.219X
	r = –0.588; P < 0.042	r = 0.255; P > 0.421
V = 40.0–60.0 cm	Y = 123.99–6.046X	Y = 13.12+0.045X
	r = –0.827; P < 0.001	r = 0.021; P > 0.947
H = 60 cm		
V = 0.0–20.0 cm	Y = 157.05–5.931X	Y = 59.58–1.990X
	r = –0.803; P < 0.001	r = –0.298; P > 0.345
V = 20.0–40.0 cm	Y = 98.80–2.803X	Y = 12.85+0.500X
	r = –0.470; P < 0 121	r = 0.203; P > 0.525
V = 40.0–60.0 cm	Y = 135.57–5.557X	Y = 18.84–0.333X
	r = –0.700; P < 0.010	r = –0.176; P > 0.582
H = 90 cm		
V = 0.0–20.0 cm	Y = –7.85+3.134X	Y = 35.15–0.457X
	r = 0.183; P > 0.568	r = –0.053; P > 0.871
V = 20.0–40.0 cm	Y = 32.97+0.189X	Y = 12.60+0.546X
	r = 0.059; P > 0.856	r = 0.198; P > 0.536
V = 40.0–60.0 cm	Y = 25.10+0.309X	Y = 30.38–1.034X
	r = 0.061; P > 0.851	r = –0.500; P < 0.095

H: Horizontal distance from the stem of the host plant.

V: Vertical depth.

Amongst the soil edaphic factors, the soil temperature is the most important parameter in the ecological study of phytonematodes. Soil temperature affects activities of organisms, it greatly influences and is regulated by other parameters. Different organisms are affected differently, whereas a certain temperature itself may limit activity in conjuction with other parameters. It may have a modified effect. Temperature is important in the formation of soil and the life their

Table 1.5: Multiple Regression Analysis Depicting Correlation of Major Edaphic Factors with *H. mangiferae* Population Around Root Zones of *M. indica*

Distance/Depth		
	2000–2001	
H = 30 cm		
V = 0.0–20.0 cm	$Y = 140.5287+8.2569X_1-3.8550X_2+4.8471X_3$	
	$r = 0.637$; $P < 0.036$	Moisture (+ ve)
V = 20.0–40.0 cm	$Y = -78.8637+7.3345X_1-3.2498X_2+2.7331X_3$	
	$r = 0.702$; $P < 0.017$	pH (+ve)
V = 40.0–60.0 cm	$Y = -82.3990+ 1.1442X_1-2.7499X_2+2.5726X_3$	
	$r = 0.820$; $P < 0.002$	pH (+ve)
H = 60 cm		
V = 0.0–20.0 cm	$Y = 317.7399+1.1493X_1-3.2295X_2-2.6080X_3$	
	$r = 0.644$; $P < 0.033$	Moisture (+ve)
V = 20.0–40.0 cm	$Y = 362.6470+1.6035X_1-1.1842X_2-4.1635X_3$	
	$r = 0.281$; $P > 0.424$	Moisture (+ve)
V = 40.0–60.0 cm	$Y = -31.7812+1.0875X_1-2.5663X_2+1.7514X_3$	
	$r = 0.467$; $P < 0.150$	Moisture (+ve)
H = 90 cm		
V = 0.0–20.0 cm	$Y = 415.8816+3.7283X_1+4.4929X_2-7.4137X_3$	
	$r = 0.260$; $P > 0.467$	Moisture (+ve)
V = 20.0–40.0 cm	$Y = 41.6036+1.0399X_1+9.4903X_2-6.5626X_3$	
	$r = 0.636$; $P < 0.036$	Moisture (+ve)
V = 40.0–60.0 cm	$Y = -47.6266+5.6484X_1-1.6101X_2+1.5634X_3$	
	$r = 0.472$; $P<0.145$	pH (+ve)
	2001–2002	
H = 30 cm		
V = 0.0–20.0 cm	$Y = 105.3627+8.8187X_1-1.7514X_2-1.1389X_3$	
	$r = 0.107$; $P > 0.811$	Moisture (+ve)
V = 20.0–40.0 cm	$Y = -167.3940+1.1260X_1+5.2972X_2+2.1122X_3$	
	$r = 0.272$; $P > 0.442$	Temperature (+ve)
V = 40.0–60.0 cm	$Y = 154.2833+5.3564X_1+3.148X_2-2.0089X_3$	
	$r = 0.235$; $P > 0.520$	Moisture (+ve)

Contd...

Table 1.5–Contd...

Distance/Depth		
H = 60 cm		
V = 0.0–20.0 cm	$Y = -473.835 + 2.4752X_1 + 4.3074X_2 + 4.6316X_3$	
	$r = 0.253$; $P > 0.482$	Temperature (+ve)
V = 20.0–40.0 cm	$Y = 25.0540 + 3.0401X_1 + 7.8981X_2 - 1.2467X_3$	
	$r = 0.032$; $P > 0.964$	Moisture (+ve)
V = 40.0–60.0 cm	$Y = -13.2346 + 7.4650X_1 + 1.0802X_2 - 1.3734X_3$	
	$r = 0.670$; $P < 0.025$	Moisture (+ve)
H = 90 cm		
V = 0.0–20.0 cm	$Y = 616.8955 + 1.2007X_1 - 1.4571X_2 - 7.3565X_3$	
	$r = 0.469$; $P < 0.148$	Moisture (+ve)
V = 20.0–40.0 cm	$Y = -179.4556 + 3.6362X_1 - 1.2841X_2 + 3.1364X_3$	
	$r = 0.332$; $P < 0.033$	pH (+ve)
V = 40.0–60.0 cm	$Y = 29.4524 - 5.4938X_1 - 7.3612X_2 + 1.3758X_3$	
	$r = 0.062$; $P > 0.910$	pH (+ve)

H: Horizontal distance from the stem of the host plant; V: Vertical depth; X_1 = Moisture; X_2 = Temperature; X_3 = pH.

in. Internally the biological activity, moisture, texture, structure, salt concentration and radiation from the soil affects soil temperature directly or indirectly. In the present investigation the population of *H. mangiferae* revealed inverse relation at almost all the depths and distances. The influence of temperature on nematode reproduction may in turn affect plant growth and productivity. The inverse relationship with nemic population indicated that the food depletion during peak summer may reduce the population below economic threshold levels at faster rate (Sehgal, Chawla, Singh and Balakrishnan, 1990). The results of the present investigations can be correlated with the work of Thorne (1961); Ross (1962); Norton (1978); Malek (1980); Zirakparvar, Norto and Cox (1980); Rai (1985). Chaubey and Dwivedi (1993) provide evidence of temperature dependence parasite rejection by the host plant at high altitude. The temperature range from 20–26°C apparently showed an adverse effect on the growth of phytonematodes and their developmental stages. The nematode development rate is also influenced by the soil temperature (Verain *et al.*, 1978; Trudgill, 1995). The influence of

Table 1.6: Multiple Regression Analysis Depicting Correlation of Macronutrients with *H. mangiferae* Population Around Root Zones of *M. indica*

Distance/Depth		
	2000–2001	
H = 30 cm		
V = 0.0–20.0 cm	$Y = -864.3233 - 4.0020X_1 + 2.1999X_2 + 3.2277X_3 - 5.2945X_4 - 3.3976X_5 + 1.4636X_6 + 2.1714X_7$	
	$r = 0.644$; $P < 0.051$	Sulphur (+ve)
V = 20.0–40.0 cm	$Y = 2696.8563 + 1.6763X_1 + 6.8660X_2 + 1.8093X_3 - 4.6059X_4 + 3.6910X_5 + 2.7990X_6 - 9.3808X_7$	
	$r = 0.948$; $P < 0.020$	Nitrogen (+ve)
V = 40.0–60.0 cm	$Y = -71.5174 - 5.9418X_1 - 4.4671X_2 + 8.3090X_3 - 6.5307X_4 + 4.4762X_5 + 8.0538X_6 + 2.8877X_7$	
	$r = 0.914$; $P < 0.050$	Nitrogen (+ve)
H = 60 cm		
V = 0.0–20.0 cm	$Y = 937.3642 + 5.0793X_1 + 1.3768X_2 + 5.8174X_3 - 4.4895X_4 - 1.6325X_5 + 9.0614X_6 - 2.9324X_7$	
	$r = 0.896$; $P < 0.071$	Potash (+ve)
V = 20.0–40.0 cm	$Y = -947.5223 - 7.7000X_1 + 1.4531X_2 - 4.3932X_3 - 9.0372X_4 - 4.711X_5 - 7.9283X_6 + 4.2617X_7$	
	$r = 0.907$; $P < 0.059$	Potash (+ve)
V = 40.0–60.0 cm	$Y = 514.6571 + 3.2049X_1 + 9.6802X_2 + 8.7555X_3 - 5.9863X_4 - 1.3661X_5 + 1.2994X_6 - 1.7549X_7$	
	$r = 0.921$; $P < 0.043$	Potash (+ve)

Contd...

Table 1.6–Contd...

Distance/Depth	
H = 90 cm	
V = 0.0–20.0 cm	$Y = -3634.2847 - 1.2614X_1 + 3.0811X_2 + 9.3440X_3 + 1.0703X_4 + 2.6624X_5 + 4.4600X_6 + 7.1165X_7$
	$r = 0.933;\ P < 0.032$ Nitrogen (+ve)
V = 20.0–40.0 cm	$Y = -283.0490 - 7.0036X_1 + 3.7081X_2 + 9.1565X_3 + 5.7002X_4 + 5.7008X_5 + 8.7032X_6 + 3.2423X_7$
	$r = 0.913;\ P < 0.052$ Potash (+ve)
V = 40.0–60.0 cm	$Y = -3373.6569 - 1.6004X_1 - 5.6304X_2 + 9.81\ 09X_3 + 3.7355X_4 + 3.0965X_5 + 2.0864X_6 + 8.8703X_7$
	$r = 0.719;\ P < 0.375$ Nitrogen (+ve)
	2001–2002
H = 30 cm	
V = 0.0–20.0 cm	$Y = 15.6903 + 2.617X_1 + 7.1700X_2 + 1.3786X_3 + 9.0706X_4 - 4.2224X_5 - 1.1687X_6 - 2.1351X_7$
	$r = 0.417;\ P < 0.085$ Sulphur (+ve)
V = 20.0–40.0 cm	$Y = -5.8118 - 7.2053X_1 + 3.6518X_2 + 5.3899X_3 + 6.7892X_4 - 1.1685X_5 + 4.9421X_6 + 1.5888X_7$
	$r = 0.930;\ P < 0.034$ Phosphorus (+ve)
V = 40.0–60.0 cm	$Y = 228.2205 + 2.0516X_1 - 3.4232X_2 + 1.4240X_3 - 7.6689X_4 - 3.1509X_5 - 5.3062X_6 - 1.5412X_7$
	$r = 0.799;\ P < 0.224$ Organic Carbon (+ve)

Contd...

Table 1.6–Contd...

Distance/Depth		
H = 60 cm		
V = 0.0–20.0 cm	$Y = 1002.2056 + 5.8872X_1 - 1.2224X_2 + 1.3398X_3 + 8.9773X_4 + 4.0603X_5 - 2.1887X_6 - 3.3404X_7$	
	$r = 0.623$; $P < 0.055$	Sulphur (+ve)
V = 20.0–40.0 cm	$Y = 1757.0848 + 8.7569X_1 - 2.3126X_2 + 4.2039X_3 + 1.5982X_4 - 9.1342X_5 + 1.4045X_6 - 4.8973X_7$	
	$r = 0.645$; $P < 0.051$	Organic Carbon (+ve)
V = 40.0–60.0 cm	$Y = 296.3807 + 1.4503X_1 + 1.7454X_2 + 2.2843X_3 + 6.7034X_4 - 3.4795X_5 + 8.5015X_6 - 8.3755X_7$	
	$r = 0.780$; $P < 0.258$	Suphur (+ve)
H = 90 cm		
V = 0.0–20.0 cm	$Y = -3147.9977 - 1.4825X_1 + 2.8431X_2 + 1.2181X_3 + 6.6873X_4 - 3.1393X_5 + 1.2583X_6 + 8.2103X_7$	
	$r = 0.609$; $P < 0.058$	Organic Matter (+ve)
V = 20.0–40.0 cm	$Y = -1559.2573 - 7.7260X_1 + 7.8020X_2 - 3.1428X_3 + 1.0886X_4 + 4.0276X_5 + 2.6952X_6 + 4.3049X_7$	
	$r = 0.555$; $P < 0.067$	Organic Matter (+ve)
V = 40.0–60.0 cm	$Y = 97.9033 + 8.8457X_1 + 1.2174X_2 + 1.0553X_3 - 8.0260X_4 + 2.4054X_5 + 4.5549X_6 - 4.8053X_7$	
	$r = 0.580$; $P < 0.063$	Nitrogen (+ve)

H: Horizontal distance from the stem of the host plant; V: Vertical depth; X_1: O.C.; X_2: K; X_3: S; X_4: P; X_5: Na; X_6: N; X_7: O.M.

temperature and the root exudates of specific host hatching is also studied by Castillo and Vovlas (2002). Walia and Mathur (1997) observed that plant parasitic nematodes dominated with the rising of the temperature. Significant negative correlations were observed between the soil temperature and nematode population (Muzna *et al.*, 1998; Anita and Chaubey, 2003). The role of temperature on the population of predatory nematode *Mylonchulus minor* has also been studied by Choudhury and Sivakumar (1999), where they have reported the negative effect of temperature on this nemic population at 0.0–10.0 cm depth.

Organic carbon plays an important role in influencing the population of soil mites (Reddy, 1981) and plant parasitic nematodes (Norton, 1978). The observations can be correlated with the findings of Verma and Jain (2001) in which they reported that nematode galling and reproduction decreased with increased level of organic carbon in cotton. Some chemical compounds as phenols, tannins, ammonia etc. are known to be released during the process of decomposition and these may have accounted for nematicidal action (Fayad and Sweelam, 1989; Chaubey and Dwivedi, 1993). The organic manures improve soil conditions for more rapid growth. According to Addabbo (1995) this enhances the utilization of soil nutrients and masks the effect of nematode damage and such observations were obtained in the present investigation.

Nitrogen in the soil is an important nutrient element that affects the growth of the plant as well as minimizes and maximizes the living organisms in the soil environment or associated with the plants. It is essential for carbohydrate use within plants and stimulates root growth, development as well as the uptake of other nutrients. Most of the soil is in the organic form, proteins and other nitrogenous compounds are associated with humus and with silicate clays, which protect them from rapid microbial breakdown. The study related to the population of *H. mangiferae*, where the regression analysis and multivariate analysis depicted positive correlations with nitrogen can be correlated with the work of Kamra, Chaudhary, Biswas and Garg (2004), where they reported that the nematode numbers increased in the presence of nitrogen and bacteria. Cole (1955) has reported the similar observations related to the increased nematode population. Heald and Burton (1968) found that organic nitrogen in the form activated sewage sludge reduced number of *P. longicaudatus*, *H. graminis* and *T. semipenetrans* on turf (Sinha and Neog, 2003).

The availability of phosphorus in soil affects the proper growth of the plant as well as the infestation levels of different soil organisms. Since the availability and chemistry of phosphorus in soil is complex. The phosphorus content of the soil directly or indirectly influences all the plant parasitic nematodes. The present study revealed that the population of *H. mangiferae* around the roots of mango plantations was suppressed by the action of phosphorus. Negative correlations of the study can be correlated with the study of Sinha and Neog (2003) where they reported the maximum reduction of nematode population with the depletion of N.P.K. The findings of the present study can also be correlated with the study of Bamel, Verma and Gupta (2003), where they studied the effect of different fertilizers source on *M. incognita* infesting okra *Abelmoschus esculentus*. They reported that murate of potash, potassium sulphate, and diammonium phosphate showed significant increase in the doses of fertilizers. Recently Sharma (2001) have reported that the phosphorus and potassium levels in the soil individually showed their significant negative impacts on *M. incognita* infesting French bean (*Phaseolus vulgaris*). In the author's opinion the negative correlation at almost all the depths and distances might be due to the effect of low reproduction that is highly induced by phosphorus in terms of low population that ultimately make the plant healthy. This might also be due to host specificity with the plant parasitic nematodes itself.

Acknowledgement

Authors are thankful to Head of Department of Zoology, CCS University, Meerut for providing laboratory facility.

References

Addabbo, T.D. 1995. The nematicidal effect of organic amendments: A review of the literature, 1982–1994. *Nematol. Medit.*, 23: 299–305.

Anita and Chaubey, A.K. 2003. Influence of soil temperature and moisture on population dynamics of major ectoparasitic phytonematodes infesting mango. *Ann. Pl. Protec. Sc.*, 11(1): 143–183.

Bamel, V., Verma, K.K. and Gupta, D.C. 2003. Effect of different fertilizer source on *Meloidogyne incognita* infesting okra (*Abelmoschus esculentus*). *Indian J. Nematol.*, 33(2): 132–135.

Castillo, P. and Vovlas, N. 2002. Factors affecting egg hatch of *Heterodera meditterranea* and differential responses of olive cultivars to infestation. *J. Nematol.*, 34(2): 146–150.

Chaubey, A.K. and Malhotra, S.K. 1988. An account of possible role of phytonematodes as indicator of draught condition. *Proc. Nat. Symp. Thr. Hab.*, 9.

Chaubey, A.K. and Dwivedi, B.K. 1993. Population dynamics of *Hoplolaimus indicus* around *Pinus roxburgii* under influence of soil temperature and organic matter. *Curr. Nematol.*, 4(1): 29–40.

Chiarappa, L. ed. 1971. Crop loss assessment methods supplement 3. FAO/CAB, Slough, England, pp. 123.

Choudhury, B.N. and Sivakumar, C.V. 1999. Seasonal fluctuations of a commonly occurring predatory mononch, *Mylonchulus minor* (Nematoda: Mononchida). *Indian J. Nematol.*, 29(1): 69–73.

Cobb, N.A. 1918. Estimating the nema population of soil USDA Bur. Plant Ind., Agr. Tech. Civ., 1: 1–48.

Cole, G.A. 1955. An ecological study of the microbenthic fauna of two Minnesota Lakes. The *American Midland Naturalist*, 53: 213–230.

Fayad, M.N. and Sweelam, M.E. 1989. Effects of phosphorus fertilization, organic matter and root-knot nematode, *Meloidogyne javanica* infection and nutrient content on growth of tomato (*Lycopersicon esculentum*). *Indian J. Agric. Sci.*, 59: 455–458.

Heald, C.M. and Burton, G.W. 1968. Effect of organic and inorganic nitrogen on nematode populations on turf. *Plant Dis. Rep.*, 52: 46–48.

Kamra, A., Chaudhary, A., Biswas, D.R. and Garg, U. 2004. Nitrogen mineralization in soil microcosms as affected by nematode bacterial interactions. *Ann. Pl. Proc. Sci.*, 12 (1): 147–151.

Malek, R.B. 1980. Population response to temperature in the sub-family Tylenchorynchinae. *J. Nematol.*, 12: 1–6.

Muzna, S., Hinai, A. and Mani, A. 1998. Seasonal population changes and management of *Tylenchulus semipenetrans* using amendments and Fenamiphos. *Nematol. Medit.*, 26: 179–181.

Norton, D.C. 1978. *Ecology of Plant Parasitic Nematodes.* John Wiley and Sons, N.Y., pp. 268.

Olsen, S.R., Cole, C.V., Watanabe, F.S. and Dean, L.A. 1954. Estimation of available phosphorus in soils by extraction with sodium bicarbonate. *Cire. U.S. Dept. Agric.*, 939.

Rai, B.B. 1985. Plant parasitic nematodes associated with maize under varying rotational practices, population dynamics of the more predominant species and their effect on growth under laboratory conditions. D. Phil. Thesis, University of Allahabad.

Reddy, M.V. 1981. Role of acari in edaphic ecosystem. In: *Contributions to Acarology*, p. 158–166.

Ross, J.P., 1962. Crop rotation effects on soybean cyst nematode population and soybean fields. *Phytopathology*, 52: 815–818.

Sehgal, M., Chawla, M.L., Singh, S. and Balakrishnan, K. 1990. Studies on interactive effect of moisture and temperature on *Tylenchus semipanetrans. Curr. Nematol.*, 1(1): 35–36.

Sharma, G.C. 2001. Impact of different soil components on the infestation potential of *Meloidogyne incognita* in French bean (*Phaseolous vulgaris*). *Indian J. Agric. Sci.*, 71(8): 535–538.

Sinha, A.K. and Neog, P.P. 2003. Effect of different levels of N.P.K. fertilizers against citrus nematode (*Tylenchulus semipeneterans*) on Khasi mandarin. *Indian J. Nematol.*, 33(1): 61–88.

Thorne, G. 1961. *Principles of Nematology*. McGraw-Hill Book Company, New York., pp. 553.

Trudgill, D.L. 1995. Host and plant temperature effects on nematode development rates and nematode ecology. *Nematologica*, 41: 398–404.

Verma, K.K. and Jain, R.K. 2001. Influence of soil moisture levels on root-knot nematode (*Meloidogyne incognita*) infesting cotton (*Gossypium hirsutum*). *Indian J. Nematol.*, 31(1): 8–11.

Vrain, T.C., Barker, K.R. and Holtzman, G.I. 1978. Influence of low temperature on rate of development of *Meloidogyne incognita* and *M. hapla* larvae. *J. Nematol.* 10: 166–171.

Walia, K. and Mathur, S. 1997. Influence of edaphic factors on soil acarines and plant parasitic nematodes and their interaction. *Indian J. Nematol.*, 27(1): 46–52.

Walkley, A.J. and Black, I.A. 1934. Estimation of soil organic carbon by the chromic acid titration method. *Soil Sci.*, 37: 29–38.

Zirakparvar, M.E., Norton, D.C. and Cox, C.P. 1980. Population increase of *Pratylenchus hexincisus* on corn as related to soil temperature and type. *J. Nematol.*, 12: 313–318.

Chapter 2

Ecological Adaptive Features of Hunter Reduviids (Insecta : Heteroptera : Reduviidae Latreille 1807) and their Biological Control Potential

*K. Sahayaraj**

Crop Protection Research Centre, Department of Zoology, St. Xavier's College (Autonomous), Palayamkottai – 627 002, Tamil Nadu

ABSTRACT

Hunter reduviids are considered to be an important biological control agent for many pests of economic importance throughout the world. They have developed many ecologically significant adaptive features. They included camouflaging behaviour, parental care, congregational feeding, death feigning and rolling behaviour, head nodding and stridulation, extension of rostrum, spitting of saliva, stinging, mimicry and protective colouration and shapes and emission of pungent odour. In addition the hunter reduviids are divided into two categories based upon the presence and absence of tibial pad such as tibialorate reduviids and non-tibialorate reduviids. Their adaptive features during the feeding behaviour are also

* E-mail: ttn_ksraj@sancharnet.in

discussed. All the above said behaviours are helping the hunter reduviids for enhancing their biological control efficiency.

Keywords: *Reduviid insects, Structural adaptations, Camouflaging, Predatory behaviour.*

Introduction

The existing flora and fauna of peninsular India are considered to represent the original Gondawana stock that experienced humid, tropical rain forest conditions. They continue to present such characters whenever such conditions prevailed. Insects are reliable indicators of microvariations in ecological conditions often presenting intraspecific variations in their micromorphology that invariably elude even the critical examination of a stereotyped museum taxonomist. Investigations on the bio-ecology, eco-physiology and ethnology of reduviids occupying different habitats presenting diverse microhabitat condition have been found to provide unlimited scope for a most sophisticated, biochemist and most conservative taxonomist alike, in their academic persuits.

Reduviidae is one of the largest families of Heteroptera with more than 6050 species described by 1999. Although cosmopolitan, reduviids are most abundant in tropical and sub-tropical climates. Important morphological characters of these "assassin bugs" include their rather elongated shape; head usually with a transverse groove between the eyes; ocelli usually present; rostrum 3-segmented, usually curved, and fitting into a groove in the pro-sternum. The front femora are usually thickened. These brownish to black, medium or large sized bugs may inflict painful bites if handled carelessly. Most of the reduviidae are predaceous on a variety of stages and groups of insects that frequent plants in a variety of terrestrial habitats. Their prey consists mainly of aphids, leafhoppers and caterpillars. They are commonly called hunter bugs or predatory reduviids. In *Amalus* sp. and *Beharus lunatus* L. and S., the fore tibiae and the reduced tarsi are covered with fresh resin, by which the prey, consisting mainly of ants of the genus *Dolichoderus*, is captured. All active stages of *Zelus peregrinus* Kirk. are predaceous, subsisting largely on aphids in fruit trees and leafhoppers in cane fields. Nymphs and adults of *Peregrinator biannulipes* Montr. associated with powder-post beetles, *Dinoderus minutus* F. in bamboo in Puerto

Rico. Both nymphs and adults attack the beetles when they emerge from their galleries. In Australia, *Pristhesancus papuensis* Stal. is an important natural enemy of plant bugs, but occasionally attacks honeybees. However, they are not completely beneficial because extensive attack on coccinellid larvae has been observed. Few species are feeding on millepeds. *Physorhynchus linnaei* Stal. apterous females, which are only 1/6th in. long, attack the giant millepede. This host, which is almost 6 in. long, is completely and permanently paralyzed, and the beak is inserted for feeding into the ventral area near the caudal end of the body.

Reduviidae of Southern India especially in Tamil Nadu, constitute the fore most order of predatory insects. The bio-ecological and eco-physiological characters of Reduviids have been found to be peculiar not by the point of view of their ecotypic specialisation but also in their behavioural manifestation (Ambrose, 1980; Vennison,1987; Kumaraswami, 1991; Sahayaraj, 1991; Kumar, 1993; Das, 1996 and Claver, 2002). Their nutritional and reproductive adaptations have reached a high position of precision and efficiency. Many of the reduviids are found to be potential predators of a number of insect pests. Investigations have been carried out to estimate the biological resources and to enlist more species of reduviids in the biological warfare against insect pests by many workers. In India Antony *et al.* (1979) evaluated the biological control potential of *Orchytes rhinoceros* in coconut palm field. Latter, the biological control potential of *Ectomocoris tibialis* Distant (Sahayaraj and Ambrose, 1997c), *Catamiarus brevipennis* Servil (Sahayaraj, 1991), *Rhynocoris kumarii* Ambrose and Livingston and *Rhynocoris marginatus* (Fab.) (Kumaraswami, 1991; Claver and Ambrose, 2001b and 2001 c) have been investigated in cotton ecosystem. Maran *et al.* (2002) also reported the suppression efficiency of *R. marginatus* against *S. litura* in cotton. Similar biocontrol potential evaluations were carried out in groundnut (Sahayaraj, 1997; Sahayaraj and Martin, 2003; Sahayaraj, 2004), lady's finger (Sahayaraj and Ambrose, 1997a), pigeon pea (Bhatnagar *et al.*, 1983; Ambrose and Claver, 1995b). Richman *et al.* (1980) evaluated the biological control potential of *Arilus cristatus* (Linn.) against *Pseudoplusia includens* Walker on soyabean by field cage release studies in USA. Furthermore, one species each has been used in biological control of the rhinoceros beetle and cotton pests in the South Pacific and Australia respectively. Hunter bugs developed many adaptive features either to capture their preys or to survive in the habitat it has dwelled. The behaviours which are being elaborated are camouflaging, parental sharing of brood care by reduviids, death feigning, rolling along with prey, structural modification of legs and rostrum, head nodding and stridulation behaviour, emission of pungent odour, mimicry, protective colouration and shape, spitting of saliva, extension of rostrum and stinging. Hence, in the present article the usefulness of adaptive behaviours of reduviids during their development and especially during predation is highlighted.

Camouflaging

Camouflaging in assassin bugs is a phenomenon by which the nymphal instars use their hind legs as shovels to throw fine particles mainly arenaceous matter on the back of their body. Camouflaging phenomenon is found only in the nymphal instars and it is called nymphal camouflaging. Nymphal camouflaging was also recorded by Aluarez (1947), Villier (1948), Immel (1955), Odhiambo (1958 a, b), Zeledon *et al.* (1969, 1970, 1973 a, b) and Luis (1974). Nymphal camouflaging among oriental assassin bugs was reported among the members of *Acanthaspidinae* (Livingstone and Ambrose, 1978 a, b, 1979, 1984; Ambrose, 1983, 1986; Vennison and Ambrose, 1986). In oriental assassin bugs, *viz.*, *Acanthaspis pedestris*, *Acanthaspis siva*, *Acanthaspis quinquespinosa* and *Edocla slateri* camouflaging condition occurs in all the five stages of nymphal instars (Livingstone and Ambrose, 1978 a, b, 1979, 1986; Ambrose, 1983, 1986 and Vennison and Ambrose, 1986).

The nymphal instars use their hind legs as shovels to take and throw arenaceous matter on the back of their bodies. So the camouflaging phenomenon mainly depends upon the number of hind leg movements that take place at the time of camouflaging. In *Acanthaspis pedestris* the total number of hind leg movement was higher in the first instar and gradually diminishes in the older instars (Livingstone and Ambrose, 1986). The initiation of hind leg movement in the act of camouflaging was almost instantaneous after eclosion and ecdysis. The initiation act of camouflaging after ecdysis in older instars is not so instantaneous (Livingstone and Ambrose, 1986 and Ambrose, 1986). But in oriental assassin bugs, both the legs were found to be employed equally in camouflaging (Livingstone and Ambrose, 1984). In between the hind leg movements of camouflaging the reduviids are found to take rest and the resting period is called as intermittent period. In *A. pedestris* (Livingstone and Ambrose, 1986) the intermittent period was relatively less in the first instar and more than double in older instars. More number of hind leg movements with less intermittent period in the first instars indicate their better camouflaging efficiency.

Types of Camouflaging

Livingstone and Ambrose (1986) classified the camouflaging into two types natural camouflaging and corpse camouflaging according to the nature of matter the nymphs use in camouflaging.

Natural Camouflaging

In natural camouflaging the nymphs use sand particles, foliage, fragements, exuvia, egg-shells, molluscan shells and other debris and rarely the empty cases of the prey as camouflaging material. In all the five nymphal stages of *A. pedestris* and the first and second instars of *A. siva, A. quinquespinosa* and *Edocla slateri* this types of camouflaging was observed (Livingstone and Ambrose, 1986). Natural camouflaging was also recorded by Breddin (1904), Subbiah and Mahadevan (1957) and Sahayaraj (1991).

Corpse Camouflaging

In this type of camouflaging the empty cases of their prey are used as camouflaging material in addition to the sand particles, foliage fragments, excuvia, egg shells, molluscan shells and debris. Older instars (II–V) of *A. siva, A. quinquespinosa* and *E. slateri* exhibit this type of camouflaging (Livingstone and Ambrose, 1986). The same phenomenon was also recorded in *A. siva* (Subbian, 1957) and *S. variegata* (Beck, 1980) and in *S. dimidiata* (Zeledon *et al.*, 1969).

Ethology of Camouflaging

Chronological analysis of hind leg movements in all the nymphal instars of *A. pedestris* indicates that the total number of hind leg movement per minute is highest in the first instar (166.3 ± 51.28) and gradually diminishes in the older instar. It also indicates that the initiation of hind leg movement is almost instantaneous (0.17 ± 0.12 min) following eclosion where as in the older instars it does not immediately follow ecdysis (1.45 ± 0.23 min to 5.37 ± 0.83 min). Further the period of rest recorded in between two successive efforts of camouflaging is relatively shorter (0.66 ± 0.05 minute) in the first instar and longer (nearly 10 times) in the older instars (Table 2.1). It appears that this tendency of hind leg movement is an instantaneous pre-adaption that gradually fades as the nymphs advance towards adulthood. This is supported by the fact that the very first act of nymphal instar immediately after eclosion is camouflaging. Similar observation was also reported by Livingstone and Ambrose (1986).

Table 2.1: Body Weight, Particle Weight (mg) and Particle Carrying Capacity in the Nymphal of *Acanthaspis pedestris*

Instars	*Body Weight*	*Particle Weight*	*Carrying Capacity*
I	2.53 ± 0.60	5.6 ± 5.1	3.51 ± 1.52
II	4.02 ± 1.77*	6.41 ± 1.42*	1.74 ± 0.47*
III	17.20 ± 1.66*	27.55 ± 6.96*	1.58 ± 0.29
IV	18.20 ± 1.38*	26.90 ± 6.06	1.43 ± 0.31
V	24.77 ± 6.87*	24.77 ± 6.87	0.98 ± 0.03

* Significance shown at 1% level.

Particle Carrying Capacity

The particle carrying capacity was studied by Ambrose (1980) in few reduviids. Our studies on *A. pedestris* revealed that all the five stages of nymphal instars of *A. pedestris* shows that the higher carrying capacity is recorded in the first instar (3.51 ± 1.52) and lowest in the fifth instar (0.98 ± 0.030) (Table 2.1). Livingstone and Ambrose (1978 a, b and 1986) worked out the effect of crowding on the carrying capacity of nymphal instars of *A. pedestris.* Table 2.2 shows the effect of crowding on body weight, particle weight and the particle carrying of the nymphal instars of *A. pedestris.* Our study reveals that crowding causes a decrease in the body weigh of the nymphal instars. Similar decrease in the body weight and body size as a function of crowding was reported earlier (Ambrose *et al.*, 1988). As the body weight and body size is a function of crowding, the particle weight decreases as a function of crowding, the particle weight carried by the nymphal instars for camouflaging is also reduced due to crowding. This indicates that a balance exist between the body weight and the particle weight. Livingstone and Ambrose (1978 and 1986) reported that immediately after each feeding the nymphal instars adjust the weight of their carrying particles in accordance with the change in their body weight. The present observation empirically proves that statement. Crowding also causes a decrease in the carrying capacity of the nymphal instars. The present observation contradict with the earlier observation of Livingstone and Ambrose (1978 and 1986) that crowding up to a level of 3 individuals causes an increase in carrying capacity.

Table 2.2: Effect of Crowding on Body Weight (BW), Particle Weight (PW) (in mg) and the Particle Carrying Capacity (CC) of the First, Second, Third, Fourth and Fifth Nymphal Instars of *Acanthasapis pedestris*

Life Stages	*No. of Individuals*	*Body Weight*	*Particle Weight*	*Carrying Capacity*
I	1	2.53 ± 0.60	5.6 ± 0.51	3.51 ± 1.52*
	2	1.45 ± 0.37	2.04 ± 0.06	1.54 ± 0.64
	4	1.33 ± 0.36*	1.34 ± 0.36	1.15 ± 0.57**
II	1	4.02 ± 1.77	6.41 ± 1.42	1.74 ± 0.47
	2	2.66 ± 0.85	2.21 ± 1.19	1.01 ± 0.54
	4	3.21 ± 1.18*	0.67 ± 0.24	0.26 ± 0.16
III	1	17.2 ± 1.66	27.53 ± 6.96	1.58 ± 0.29
	2	9.58 ± 0.4	8.4 ± 2.84	1.04 ± 0.46
	4	6.72 ± 0.93	8.08 ± 4.53*	0.67 ± 0.49
IV	1	18.66 ± 1.38	26.90 ± 6.06	1.43 ± 0.33
	2	16.26 ± 0.91	15.61 ± 1.18	0.96 ± 0.96
	4	15.39 ± 0.65*	9.83 ± 1.44	0.63 ± 0.06
V	1	24.77 ± 6.87	24.45 ± 0.69	0.98 ± 0.03
	2	23.76 ± 6.87*	19.57 ± 0.66	0.88 ± 0.28**
	4	12.44 ± 0.39	10.55 ± 0.34	0.56 ± 0.02***

Significance shown at (P): *: 0.1; **: 0.05; ***: 0.025.

Structural Adaptations

Microscopical studies reveal that the cuticle is serrated and fine silvery filaments are present intertwined with the camouflaging material. These two structures are used to hold the camouflaging materials on the dorsum of the abdomen. Livingstone and Ambrose (1986) reported the presence of special secretions. The hypodermis has special secretary glands. In *S. variegata* the body is covered with glue secreting setae (McMahan, 1982) but in adult the setae are absent. In *Chrysopa glossonae* the body is covered with glandular hairs, which secrete a sticky fluid (Eisner *et al.*, 1978). Table 2.2 represents the body weight, particle weight and particle carrying capacity of *A. pedestris*. Results revealed that crowding affects the body weight, particle weight and particle carrying capacity of *A. pedestris.*

Effect of Crowding and Camouflaging

Camouflaging has been influenced by number of factors such as crowding, age and cannibalism. The crowding influences the camouflaging phenomenon. Crowding of nymphal instars influenced the leg movement. In *A. pedestris* and *A. siva* (Livingstone and Ambrose, 1986; Ambrose, 1986) the hind leg movement increases with the increasing number of individuals. In *A. siva* due to over crowding the nymphs runs helter and skelter to escape from the cannibalistic tendency of its co-instars attempt rather than trying to camouflage (Ambrose, 1986).

Cannibalism and Camouflaging

Cannibalism is the phenomenon by which the animal feed on its own species. In *A. siva* cannibalism was higher in decamouflaged (artificially removed the camouflaging material) nymphal instars and than in the camouflaged instars (Ambrose, 1986). Similar observation was found in *A. pedestris* (Livingstone and Ambrose, 1986). Thus camouflaging gives a primary protection against the highly developed cannibalistic tendency and favour survival by escaping from other predators and natural enemies.

Camouflaging and Predation

Miller (1956) reported that camouflaging phenomenon is an aid in the capture of prey and also protective device against enemies. In *A. pedestris, A. siva* and *A. quinquespinosa* (Livingstone and Ambrose, 1986) and in *A. siva* (Ambrose, 1986) camouflaged nymphs captured their prey quicker than the decamouflaged nymphs. Moreover, the camouflaged nympns are protected by the camouflaged material. So these nymphs approach the prey faster than the decamouflaged nymphs. Similar observations were recorded by McMahan (1982) in *Salyavetu variegata* and Eishner *et al.* (1978) in *Chrysopa glossonae.*

Parental Sharing of Brood Care of Reduviids

Shared brood care or parenting care are rare behaviours in insects. While a number of species exhibit biparental care, including wood roaches, passalid beetles, and some bark and dung beetles, only buying beetle males (*Nicrophorus orbicollis*) participate in all activities, remain with the brood throughout development, and can take over all parenting responsibilities if the female disappears. In general, extended parenting by both sexes is linked to insects that use rich but ephemeral resources to provision their young. Maternal care and protection of offspring have been well documented in insects; some examples include lace bugs, reduviids, and of course, many hymenoptera. Rarer is unassisted care of the young by the father, as in belostomatids and the reduviids *Rhinocoris* and *Zelus* (Tallamy and Wood, 1986). Among insects, cooperation between the sexes in brood care is very unusual. In reduviids, parental care was observed either by male or by female toward the eggs or nymphal instars. Egg care or guarding behaviour has been observed in *Rhynocoris carmelita* Stal. (Odhiambo, 1959), *Pisilus tipuliformis* (Fab.) (Parker, 1965 a, b), *R. punctatus* (Nyirra, 1970) and *Occamus typicus* Distant (Sundararaju, 1984). *Rhynocoris marginatus* female also take care of their eggs as well as the newly hatched young ones.

Adaptation in Predatory Behaviour

All assassin bugs are predatory except *Lophocephala guerini* (Laporte) (Ambrose and Livingstone, 1979 a). Most of them are entomosuccivorous except the members of the sub-family Echtrichodiinae which feed on millipedes (Ambrose, 1980; Haridass and Anathakrishnan, 1980a and Haridass, 1985 a) and the members of Triatominae which are haematophagous. The relative development of tibial pads or fossula spongiosae in the fore and mid tibiae was considered as the most important structural adaptation in relation to their feeding behaviour and their predatory efficiency (Gillet and Wiggleswoth, 1932; Miller, 1938, 1939, 1942, 1953 and 1956 b; Edwards, 1962; Ambrose, 1980; Livingstone and Ambrose, 1978 a and 1984; Haridass and Ananthakrishnan, 1980b and Vennison, 1989). Out of the 64 species of reduviids so far examined, 27 species were having tibial pads either in their fore or/and mid tibiae whereas all the others do not have tibial pads and some species have tibial combs.

General Predatory Acts of Tibialorate Reduviids

The generalized sequential pattern of predatory behaviour was as follows: arousal – approach – pouncing – rostral probing – piercing – paralyzing the victim – sucking the contents of the prey-post predatory cleaning. However it varied from tibialorate to non-tibialorate reduviid. The predatory behaviour of reduviids with tibial pad is summarized as follows: arousal – approach – antennal manoeuvering – pouncing (capturing) stylet penetration – paralyzing the victim – prey transportation – feeding – dropping the empty case – post predatory cleaning (Figure 2.1).

Rostral Probing and Stylet Penetration

Arousal and approaching acts of both tibialorate and non-tibialorate reduviids are same. Once the prey was firmly held by the predator, it extended its rostrum to probe the various parts of the prey to locate suitable piercing sites, such as, antennal base, legs joints, junction between head and thorax and abdomen. Once the probing was over, with precision and deftmarkmanship the stylets were planted deep into the apparently predetermined sites and the toxic saliva was injected into the prey. Invariably all these events culminated into the apparently predetermined sites and the toxic saliva was injected into the prey. Invariably all these events culminated into the total paralysis of the prey. Generally paralyzing acts varied with different species. Among the species it differs from nymphs to adults.

Prey Transportation

Transporting the immobilised prey to a safer place for consumption was yet another distinct behaviour observed. By inserting the stylets at suitable places, if the prey was smaller, the predator dragged the prey forward and if it was larger it dragged backward.

Feeding

Feeding commenced while the prey continued its struggling to escape and the stylets were frequently withdrawn and reinserted at different angles. During the act of sucking, most of the reduviids (80 per cent) rotated their prey and selected the various soft cuticle areas of the prey whereas such behaviour was absent in remaining 20 per cent of the reduviids. The predators never left the prey half emptied. After sucking the contents of the prey including the brain and

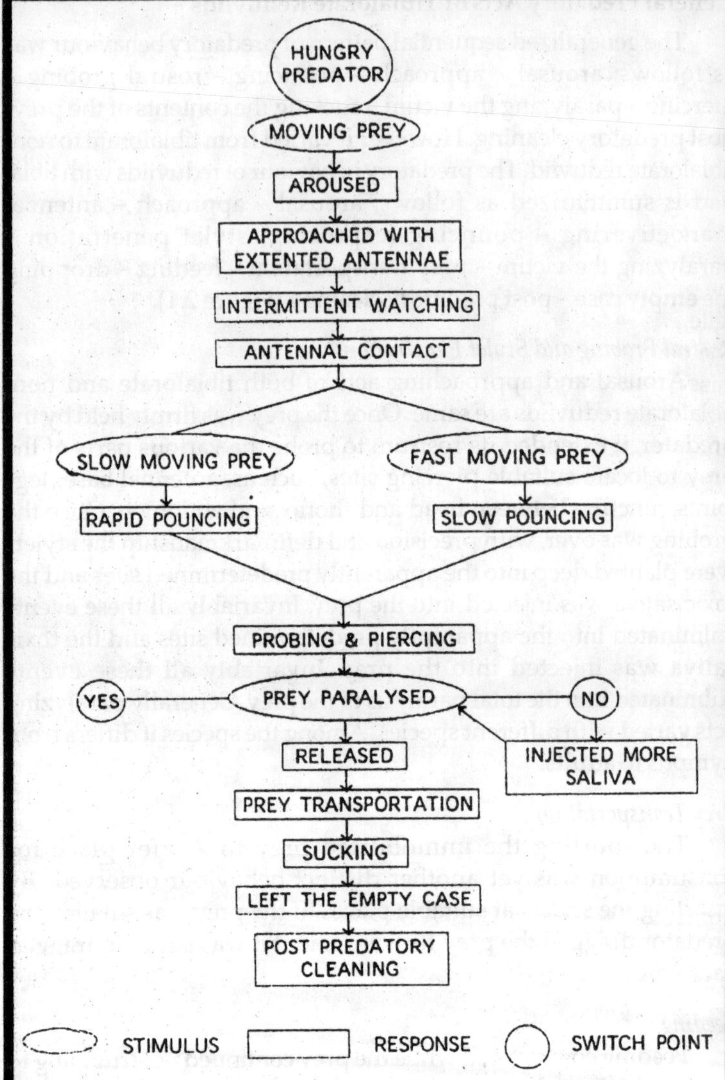

Figure 2.1: Flow Chart Depicting the General Sequential Events in the Predatory Behaviour of Tibialorate Reduviids

digestive system, the predator dropped only the empty cases of the prey. Females consumed more preys in shorter time than the males.

General Adaptive Feeding Behaviour of Non-tibialorate Reduviids

The predator behaviour sequence of this non-tibial pad reduviid can be summarized as follows: arousal –orientation –antennal extension and pinning –paralysis of the prey –prey transportation –prey sucking –dropping the empty case –post predatory behaviour.

Orientation

Once the arousal was accomplished by the non-tibialorate reduviid, it oriented towards the prey and remained motionless until the prey approached the predator. If the prey was larger and more agile, it moved away from the predator after a few minutes of waiting. The predators especially the nymphal instars became restless and raised their legs. The adults in addition to this behaviour raised their wings and they moved swiftly here and there. If the prey was small and less active the predator approached the prey with extended rostrum and antennae.

Pouncing (Capturing)

After reaching the closer proximity of the prey the predator pounced over the prey with the help of its fore leg.

Pinning

Once the prey was firmly held by the predator, it pinned the prey with the extended rostrum and inserted its rostrum and injected the toxic saliva into the prey. Invariably all the events culminated into the total paralysis of the prey.

Prey Transportation

After the prey was immobilized, the predator released its grip and transported the paralysed prey to a safe and secluded place (container lid) for feeding. The paralysed prey was kept at the tip of the rostrum and dragged forward when the prey was smaller and dragged backward when it was larger.

Feeding

Once the prey was transported to a safe and secluded place the predator started feeding. The stylets were penetrated deep into the prey and subsequently the internal contents of the prey were sucked. During the act of sucking, predators pierced and sucked various places but never rotated the prey. After sucking the contents of the prey including the digestive system, the predator dropped the empty cases of the prey (Figure 2.2). The efficiency of predators varied with

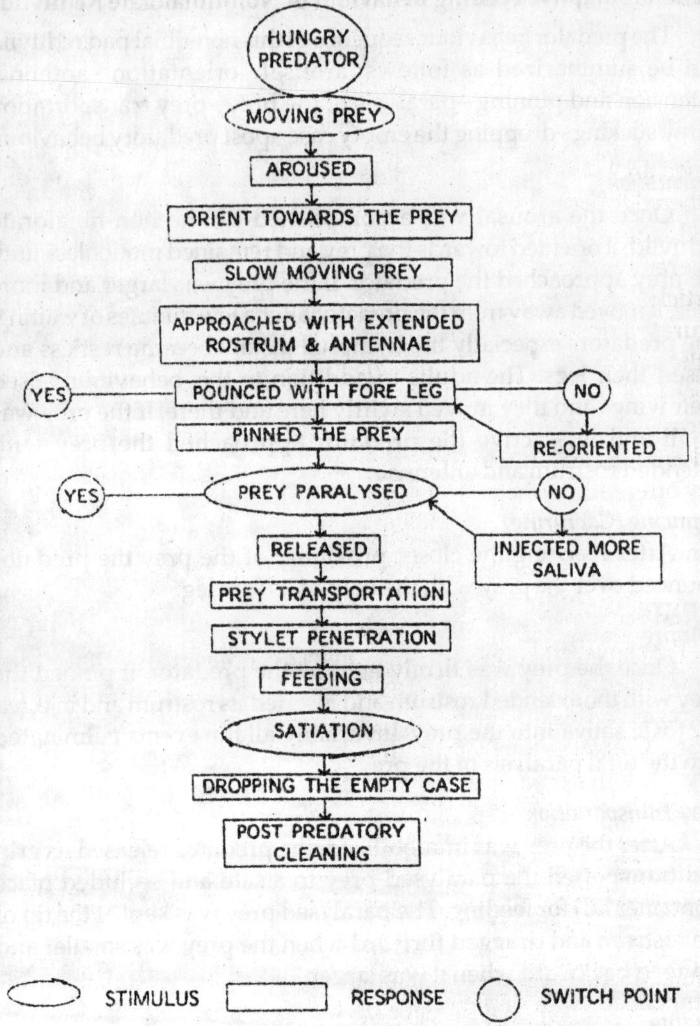

Figure 2.2: Flow Chart Depicting the General Sequential Events in the Predatory Behaviour of Non-tibialorate Reduviids

in the life stages of the reduviid. Generally, among all the nymphal stages, the V-nymphal instar was more efficient in predation. The females were found to be better predators than the males.

Post-predatory Behaviour

As the final act of predation, the empty case of victim was dropped and the satiation was manifested by antennal and rostral cleaning by the fore tibial pads in the case of tibial pad group and tibial combs in the case of non-tibial pad group. Those reduviids which are having "corpse camouflaging" they very systematically gathered the empty cases of the prey by the co-operative efforts of the hind legs and threw them over their dorsum which was already covered with debris. Rearrangement and adjustment of these particles were made by their hind legs in order to maintain the accurate balance. Grooming of antennae and rostrum was observed in the satiated bugs. The hemelytra was also cleaned by the hind legs. The predation frequency and relative duration between successive steps in the predatory behaviour varied between sexes and among the nymphal instars and for the different types of the prey offered. Starvation induced to perform rostral probing on the empty cases of the prey. Mating partners of any species were not found to predate while they were in copulating.

Congregational Feeding

Whenever the reduviid present in solitary condition they feed the available prey individually. In *Endochus umbrinus* the congregational feeding was found more common among the nymphal instars. Congregational feeding behaviour was present up to III nymphal instars. Cannibalism was common among the adults and nymphal instars of *Alleocranum quadrusignatum, Coranus* sp. nov. and *E. umbrinus.* The degrees of this cannibalistic intensity was found to vary considerably among the species. The nymphal instars of *E. umbrinus* were highly cannibalistic. But this intensity was very low in tibialorate reduviids. Adults of *A. guadrisignatum, E. umbinus* and *Coranus nodulosus* turned cannibalistic when they were subjected to starvation. Cannibalism was totally absent in the adults of *Petalochirus brachialis.* Congregational feeding observed in the present investigation is similar to that of Ambrose (1980) in the early instars of *A. pedestris, A. siva, Ectamocoris tibialis, R. marginatus, R. fuscipe, C. vitellinus* and adults of *Haematorrhophus* sp. Haridass (1985 a) in Echtrichodiinae and Harpactorinae and Vennison (1989) in the nymphal instars of *Sycanus reclinatus, Euacorus plagiatus* and *Sphedanolestes signatus.* The nymphal instars of non-tibialorate reduviid *E. umbrinus* was highly cannibalistic but this intensity was very low in tibialorate reduviids. Richman (1951), Ambrose (1980) and Vennison (1989) stated that the cannibalism was quite common among the nymphal instars and the adults of reduviids. Unusually, *A. pedestris* adult's feds *H. armigera* fifth instar larvae in congregationally.

Death Feigning Behaviour

Death feigning behaviour is behaviour when any natural enemy approaches the animal it seems to be a dead individual. This behaviour was reported in *Acanthaspis petax* (Odhiambo, 1958 a) first time. Ambrose (1980) and Vennison (1989) reported similar death feigning in other reduviids. Nymphal instars *Alleocranum quadrusignatum* were found to feign death when they were confronted with larger prey (Sahayaraj, 1991). Only after several minutes (15 to 25 minutes) they resumed their normal activities. The nymphal instars feign death at the approach of impending threat of enemies and when they confront larger and powerful prey (Edwards, 1962 a, b, 1966). This is recorded in the nymphal instars of *Neohaematorrhopus therasii* (Ectrichodiinae), *Coranus vitellinus, Irantha armipes, Polididus armatissimus* (Harpactorinae) and *Ccatamiarus brevipennis* (Peirathinae) (Ambrose, 1980; Vennison, 1988; Rukmani, 1992; George and Ambrose, 1995). Such behaviour is well pronounced in Harpactorinae than other oriental reduviids. *N. therasii* adults also feign death very often.

Rolling Behaviour

The nymphal instars, by withdrawing their cephalic and thoracic appendages roll their body into ball-like structures and lie motionless to protect themselves from their enemies or to deceive their prey fauna especially those that are larger than themselves (Readio, 1927). This behaviour is observed in *N. therasii, C. nodulosus, Coranus soosaii, C. vitellinus, E. umbrinus, Euagoras plagiatus* and *Lanca* sp. (Ambrose, 1980; Vennison, 1988; Kumaraswami, 1991; Sahayaraj, 1991; George and Ambrose, 1995). It is observed only among harpactorines and ectrichodiines. It also was observed in the nymphal instars of *Coranus* sp. and *E. umbrinus* and adults of *Coranus* sp. Ambrose (1980) reported similar behaviour in *C. vitellinus, R. fuscipes, S. aterrimus* and *N. therasii* and Vennison (1989) in *S. signatus, S. reclinatus, E. plagiatus, C. soosaii* and *P. armatissimus*. When the prey was much larger than the predator, nymphs of *Coranus nodulosus* and adults of *Endocus umbrinus* and *Coranus* sp. were found to roll their bodies and fold their legs (Sahayaraj, 1991). In my personal observation most of the biological control agents of the reduviids roll when they encountered large sized preys. *Acanthaspis pedestris* also rolls when they were provided with large size *Helicoverpa armigera*.

Adhesive Behaviour

Predatory efficiency and its concurrent prey selection have a very close relationship with structural adaptation for both prey capture and sucking. *P. brachialis* and *A. quadrusignatum* have well-developed tibial pads on both the fore and mid tibiae which help to have a powerful grip over the preys. The long rostrum of the non-tibial pad reduviid enabled the predator to straighten it out when needed for easy pinning and deep penetration of the prey. Furthermore, it helped the predator to lift the prey and to transport the prey by rostral dragging to secluded place for feeding. Arthropods make use of adhesive mechanisms in markedly different contexts, such as walking vertically or upside-down on various natural surfaces, resisting external detachment forces caused by wind gusts or attacking predators, attaching to mating partners during copulation, and holding onto the substratum in contexts such as phoresy or parasitism. The reduviid bug *Zelus leucogrammus* (Perty) has, on its legs, a sticky secretion that is produced by epidermal glands and kept in place on the cuticle by a dense arrangement of hairs (trichomes). The hairs themselves have a brush-like surface to maximize their retention potential. The fluid is hygroscopic and water soluble. Insect prey (flies) is not poisoned by the secretion but is mechanically immobilized by the large amount of fluid applied by the bug (Barth, 1953).

Other assassin bugs employ ventral adhesive pads at the distal end of their tibia to hold struggling prey (Miller, 1942; Edwards, 1960 and Edwards, 1962). Homology is assumed because of morphological and positional similarity (these so-called fossulae spongiosae occur only on the pro- and mesotibia). Ambrose (1999) summarizes the discussion on the fossula spongiosa up to his time and attributes a primarily predatory function to it. The Apiomerinae have taken this structure to the extreme: the adults are able to catch their prey from the air, because their legs are covered by 'a sticky resinous matter', hence the name resin bugs (Miller, 1942). Furthermore, the female bugs collect sticky fluid from gland hairs of plants (genus *Heterotheca*) and spread it over their eggs after oviposition. Hatching larvae obtain the sticky secretion for their leg traps from this slime (Eisner, 1988). In the sub-family Harpactorinae, a sticky fluid on their legs stems from glandular setae; however, Miller (1942) does not explicitly state that it is used for predation.

Head Nodding and Stridulation

Nymphs and adults nod their head and rub their rostrum against the transversely striated prosternal groove and produce a "crich-crich' sound at the time of danger and disturbance (Butler, 1912; Readio, 1927; Miller, 1956; Mazzotic, 1970; Schoefield, 1977, 1979; Ambrose, 1980; Vennison, 1988). According to Readio (1927) reduviids produce sound to warn or fright enemies rather than for sexual attraction. Eluding females also produce sound as a protest to males attempt for copulation. Such defensive behaviour is well pronounced in peiratines with acutely curved rostrum. Non-tibiaroliate reduviids with slightly curved or straight rostrum produce an obscure sound. Non-defensive noises emitted by reduviids by labial-prosternal mechanism and some by unknown means were not recorded for oriental reduviids (Geogala and Cokl, 1983; Gogala *et al.*, 1984).

Emission of Pungent Odour

In general animal have developed some kind of mechanism to protect themselves from their natural enemies. One such protection is to produce or emitted unpleasant odour chemicals when any natural enemies is approaching the animal. In reduviids such a kind of chemicals are produced from the dorsolateral abdominal scent gland or Brindley's gland. The idea has been proposed by Mazzotic (1970). He further added that this behaviour is well pronounced in reduviines and harpactorines than other reduviids.

Extension of Rostrum

Assassin bugs, especially under disturbed condition or restlessness, extend their rostrum intermittently to warn and attack the enemies and prey. This behaviour is seen during their failed attempts at prey capturing and mating. George and Ambrose (1995) compared this behaviour to aggressive mimicry.

Spitting of Saliva

Spitting behaviour (a drop of saliva was seen at the tip of rostrum) was seen only in some species. Spitting of saliva is another kind of behaviour observed in reduviids. This behaviour is considered either as offensive or defensive behaviour (Edwards, 1961, 1962 a). Adults and nymphal stages of *R. fuscipes, R. kumarii, R. marginatus, Sycanus reclinatus, Sycanus versicolor, E. tibialis, Ectomocoris vishnu* and *Ectomocoris xavierei* very often spit drops of watery saliva under threat of capture or when confronting larger and powerful enemies or prey (Edwards, 1961, 1962 a, b; Ambrose, 1980; Vennison, 1980; Vennison and Ambrose, 1990 e, 1991 c). In addition I observed such a kind of behaviour in both *R. longifron* and *C. brevipennis* adults in many occasions.

Stinging

Stinging is a common behaviour of hymenopteran insects. Many reduviids sting at the least provocation or when they fail to drive away or threaten their enemies by other means. They sting very ferociously and their bites cause painful swelling sometimes with raised body temperature which lasts from several hours to days (Lewis, 1958; Readio, 1927; Little, 1967). Almost all the vigorous reduviids will sting when they are in starved conditions. But this behaviour is well pronounced in harpactorines, peiratines and reduviines, for example *R. fuscipes, R. kumarii, R. mariginatus, C. brevipennis, E. tibialis, L. affinis, A. quinqespinosa* and *A. siva* (Ambrose, 1980). If the reduviid is in fully satiated level it won't sting anyone particularly the human beings. I have an experience of getting sting from different reduviids when I handled them. Among the reduviids sting the sting of *Acanthaspis siva* was more painful. I felt the pain for three days.

Mimicry, Protective Colouration and Shape

A combination of Batesian and Mullerian mimicry occurs in the milk weed and related bugs of the family Lygaeidae and similar appearing species in Rhopalidae, Largidae, Pyrrhocoridae and Reduviidae (Schuh and Slater, 1995). Three species of Phonoctonus in West Africa, each of which was found most frequently feeding on the pyrrhocorid that it most closely resembled (Stride, 1956 a, b). But they were also found feeding on less resembled pyrrhocorids. According to Stride, *Phonoctonus* sp. was extremely variable in colour, one colour morphologically resembling a brown species of pyrrhocorid and the other a red and white species. When nymphs of Phonoctonus were fed upon both brown and red and white pyrrhocorids the resulting adults were always red and white. Here the reduviid predator was protected from vertebrate predation by developing a close resemblance to its aposematically coloured prey. The red and black striped African assassin bug Rhynocoris Hahn resembled the lygaeid Spilostethus Stal., living in the same area. The neotropical assassin bug *Zellus fabricius* strikingly mimics the aposematically coloured species of *Oncopeltus*. But except such mimicked resemblance, there is no experimental evidence to suggest that assassin bugs are exhibiting aggressive mimicry of the phytophagous heteropterans. Mclver and Stonedahl (1993) reported that reduviid *Sinea diadema* (Fabricius) preferred nonmimetic mired Lopidea than the mymecomorphic mired *Coquillettia insignis* Uhler.

Predatory reduviids, especially the *Rhaphidosoma* species (Harpactorinae) and members and Stenopodainae living among grasses and sedges, develop elongate slender bodies that closely resemble the elongate slender grass and sedge stems and leaves among which they live. Members of the coranus species, *Coranus nodulosus* living in agro-ecosystems like cotton where *Oxycarenus minipennis* are abundant. Both these bugs are more or less similar in colour as well as in shape. We observed *Petalochirus brachialis* live on the barks of *Techtona grandis* tree. We cannot distinguish this reduviid when it is present on this tree bark.

Conclusion and Recommendations

All the above mentioned natural behaviours of the reduviids are very much helpful to be efficient biological control agents in the nature. However, very limited work has been available and/or studied. Because, rearing of this bug is a tactful and painful job. So I have been taking efforts to rear these bugs with artificial diet or fictitious hosts. Such types of studies are more useful to understand the ecological behaviour in biological control potential of these bugs in a meaningful manner. The same point has been emphasized by Sahayaraj (2004).

Acknowledgement

The author grateful to the Council of Science and Technology (CSIR), New Delhi for the financial assistance to carry out research projects in reduviids. He also thank Rev. Fr. A. Antonysamy, S.J., Principal and Prof. M. Thomas Punithan, Head, Department of Zoology for encouragements and facilities.

References

Alvarez, 1947. Information epidemologises dela enfermedaed de chagasen la localidad Guayagyil 1942–46. Gac Med: Guayagyil, p. 1331–1348.

Ambrose, D P. 1980. Bioecology, ecophysiology and ethology of Reduviids (Heteroptera) of the scrub jungles of Tamil Nadu, India. *Ph.D. Thesis*, University of Madras, Madras, India, pp. 1–229.

Ambrose, D.P. 1983. Diecoloty of an alate assassin bug, *Acanthaspis quinquespinosa* (r.) Fabr. (Heteroptera: Reduviidae). pp. 107–115. In: *Proceedings of Insect, Ecology and Resources Management*, (Ed.) Goel, S.C.

Ambrose, D.P. 1986. Impact of nymphal camouflaging of predation and cannibalism in the bug *Acanthaspis siva. Environ. and Ecol.*, 4(2): 197–200.

Ambrose, D.P. 1999. *Assassin Bugs*. Science Publishers, New Hampshire, U.S.A. pp. 337.

Ambrose, D.P. and Claver, M.A. 1995b. Functional response of *Rhynocoris fuscipes* Fabricius (Heteroptera: Reduviidae) to *Riptortus clavatus* Thunberg. (Heteroptera: Alydidae). *J. Biol. Control*, 9(2): 74–77.

Ambrose, D.P. and Livingstone, D. 1979 a. On the bioecology of *Lophocephala guerini* Laporte. (Reduviidae: Harpactorinae) a coprophagous reduviid from Pal ghat gap. *J. Nat. Hist.*, 13: 581-588.

Ambrose, O.P., Rani, R.S. and Vennison, S.J. 1988. Effect of crowding on the camouflaging behaviour, size development, oviposition pattern and hatchability of off springs of assassin bug *Acanthaspis pedestris* Stal a potential predator of insect pests. *Monitore Zoologica Italino*, In Press.

Antony, M., Daniel, J., Kurian, C. and Pillai, G.B. 1979. Attempts in introduction and colonization of the exotic reduviid predator *Platymeris laevicollis*. Distant for the biological suppression of the coconut rhinocerous beetle, Oryctes rhinocerous. *Proc. Plant. Crops. Symp.*, II: 445–454.

Barth, R., 1953. Die Fangdrüsen an den Beinen von *Zelus* (*Diplocodus*) *leucogrammus* (Perty 1834). *Zoologische Jahrbücher, Abteilung für Anatomie und Ontogenie der Tiere* 73, pp. 325–336.

Bhatnagar, V.S., Sithanantham, S.S., Pawar, C.S. Jadhav, D., Rao, V.R. and Reed, W. 1983. Conservation and augmentation of natural enemies with reference to integrated pest managements in chickpea (*Cicer arietinum* L.) and pigeonpea (*Cajanus cajan* L.) *Mill* sp. In: *Proc. Int. Workshop on Integrated Pest Control for Grain Legumes*, (Eds.) Goiani a, oia's Brazil, pp. 157–180.

Beck, B.B. 1980. *Animal Tool Behaviour: The use and manufacture of tools by animals*. Garland STPM Press, New York, pp. 307.

Breddin, G. 1904. Rhynchotersaus Amerisendund Termitenbauten. *Ann. Soc. Belg. Ent. Belg.*, 48: 407–416.

Brumpt, E. 1941. Importance du cannibatism *et al.*, coprophagic chez led *Reduviidea hemitophages* (*Rhodnius, Triatoma*) pour la conservationdes *Trypanosoma pathogensen* dechors de et hote vertebre. *Ball. Soc. Path. Exot.*, 7: 702–705.

Bulter, E.A. 1912. Stridulation in British Reduviidae. *Ent. Mo. Mag.* (2)43(18): 65.

Claver, M.A. and Ambrose, D.P. 2001b. Evaluation of *Rhynocoris kumarii* Ambrose and Livingstone (Hemiptera: Reduviidae) as a potential predator of some lepidopteran pests of cotton. *Journal of Biological Control*, 15(1): 15–20.

Claver, M.A. and Ambrose, D.P. 2001c. Impact of augmentative release of *Rhynocoris kumarii* Ambrose and Livingstone (Heteroptera: Reduviidae) on *Dysdercus cingulatus* (Fabricius) (Hemiptera: Pyrrhocoridae) population and damage on cotton. *Journal of Biological Control*, 15(2): 119–125.

Claver, M.A. and Ambrose, D.P. 2002. Functional response of the predator, *Rhynocoris fuscipes* (Heteroptera: Reduviidae) to three pests of pigeionpea (*Cajanus cajan*). *Shashpa*, 9(1): 47–51.

Das, S.S.M. 1996. Biology and behaviour of chosen predatory hemipterans. *Ph.D. Thesis*, Madurai Kamaraj University, Madurai, India.

Edwards, J.S. 1961. Action and composition of saliva of an assassin bug, *Platymeris rhadamanthus* Gerst. (Hemiptera: Reduviidae). *J. Exp. Biol.*, 38: 61–77.

Edwards, J.S. 1962. Observations on the development and predatory habit of two reduviids (Heteroptera). *Rhynocoris carmelita* Stal. and *Platymeris rhadamanthus* Gerst. *Proc. R. Ent. Soc. Lond.*, 32: 89–98.

Edwards, J.S. 1962a. Spitting as a defensive mechanism in a predatory reduviid. *XI Int. Cong. Entl. Vienna*, 4: 259–263.

Edwards, J.S. 1962b. Observations on the development and predatory habits of two reduviids (Heteroptera), *Rhynocoris carmelita* Stal and *Platymeris rhadamanthus* Gerst. *Proc. R. Ent. Soc. Lond.*, 37(A): 89–98.

Edwards, J.S. 1966. Observation on the life history and predatory behaviour of *Zelus exsanguis* (Heteroptera: Reduviidae). *Ibid*, 41(A): 21–24.

Eisner, T.H.K., Eisner, M., Robson, D.S. 1978. Wolf in sheep's clothing strategy of a predaceous insect larva. *Science*, 199: 770–794.

George, P.J.E. and Ambrose, D.P. 1995. Defensive and offensive behaviours of assassin bugs. *Insect Environment*, 1(4): 21.

Gillet, J.D. and Wigglesworth, U.B. 1932. The climbing organ of an insect, *Rhodnius prolixus* (Hemiptera: Reduviidae). *Proc. Royal. Soc., London*, 3(B): 364–376.

Gogala, M. and Cokl, A. 1983. The acoustic behaviour of the bug *Phymata crassipes* (F) (Heteroptera). *Rev. Can. Biol. Exper.*, 42: 249–256.

Gogala, M., Virant, M. and Blejec, A. 1984. Mocking bug, *Phymata crassies* (Heteroptera). *Acoustic Letter*, 8: 44–51.

Haridass, E.T. 1985a. Feeding and ovipositional behaviour in some reduviids (Insecta: Heteroptera). *Proc. Indian Acad. Sci. (Anim. Sci.)*, 94(3): 239–247.

Haridass, E.T. and Ananthakrishnan, T.N. 1980a. Models for the predatory behaviour of some reduviids from Southern India (Insecta : Heteroptera : Reduviidae). *Proc. Indian Acad. Sci. (Anim. Sci.)*, 89(4): 389–402.

Haridass, E.T. and Ananthakrishnan, T.N. 1980b. Functional morphology of the *fossoula spongiosa* in some reduviids (Insecta : Heteroptera : Reduviidae). *Ibid*, 89(5): 457–466.

Hase, A. 1940. Veber *Triatoma dimidiate* (Hemiptera: Triatomidae).). *Teliz Parasitenk*, 11: 419–429.

Immel, R. 1955. Zurbiologic and physiologic von. Reduvius personatu. *L. Mouphol, Oekol Tiere*, 44: 163–195.

Kumar, S.P. 1993. Biology and behaviour of chosen assassin bugs, (Insecta : Heteroptera : Reduviidae). *Ph.D. Thesis*, Madurai Kamaraj University, Madurai, India.

Kumaraswami, N.S. 1991. Bioecology and ethology of chosen predatory bugs and their potential in biological control. *Ph.D. Thesis*, Madurai Kamaraj University, Madurai, India.

Lewis, D.J. 1958. Hemiptera of medical interest in the Sudan Republic. *Proc. R. Ent. Soc. Lond.*, 33: 44.

Little, V.A. 1967. *General and Applied Entomology*. Oxford and IBH Publishing Co., New Delhi, pp. 191.

Livingstone, D. and Ambrose, D.P. 1978a. Feeding behaviour and predatory efficiency of some reduviids from the palghat gap, India. *J. Madras Univ.*, 41B(2): 1–26.

Livingstone, D. and Ambrose, D.P. 1978b. Bioecology, ecophysiology and ethology of reduviids of the scrub jungles of palghat gap Part VIII. Bioecology of *Acanthaspis pedestris* Stal. (Reduviidae : *Acanthaspidinae*) a micropterous, entomophagous species. *J. Madras University*, 41(3): 97–118.

Livingstone, D. and Ambrose, D.P. 1979. Feeding behaviour and predatory efficiency of some reduviids from the palghat gap, India. *J. Madras Univ.*, 41: 1–25.

Livingstone, D. and Ambrose, D.P. 1984. Adaptive modifications of Reduviidae of the scrub jungle of palghat gap, India, an evolutionars approach. *J. Bombay Nat. Hist. Society*, 81(3): 583–595.

Livingstone, D. and Ambrose, D.P. 1986. Nymphal camouflaging in Assassin bugs (Insecta: Heteroptra : Reduviidae). *Proceedings of the III Oriental. Entomology Symposium*, pp. 95–101.

Louis, D. 1974. Biology of Reduviidae of cocoa farm in Ghana. *Amer. Midl. Naturalist.*, 91: 68–89.

Maran, S.P.M., Babu, A. and Ignacimuthu, S. 2002. Functional response of *Rhynocoris marginatus* (Fab.) (Heteroptera : Reduviidae) on *Spodoptera litura* (Fab.) (Lepidoptera: Noctuidae). *Proc. of the Vistas of Entomological Research for the New Millennium*, (Eds.) K.P. Sanjayan, V. Mahalingam and M.C. Muralirangan. G.S. Gill Research Institute, Guru Nanak College Chennai, pp. 112–115.

Mazzotic, L. 1970. Duencia de emaniones odoriferas perceptibles and *Dipetalogster maximus* Algunos Commetrio sobre esta especies. *Rev. Inst. Med., Trop S. Paulo*, 12: 320–324.

Mc Mahan, E.A. 1982. Bait and capture strategies of termite eating Assassin bug. *Inse. Sociaux*, 29: 346–357.

Mclver, J.D. and Stonedahl, G.M. 1993. Myrmecomorphy: Morphological and behaviour mimicry of ants. *Ann. Rev. Entomol.*, 38: 351–379.

Miller, N.C.E. 1938. Function of *fossula spongiosa* in Reduviidae. *Nature*, pp. 749–750.

Miller, N.C.E. 1939. The *fossula spongiosa* in Reduviidae. *Nat Lond. Mag.*, 18: 471.

Miller, N.C.E. 1942. On the structure of the legs in Reduviidae (Rhynchota). *Proc. R. Ent. Soc. Lond.*, 19(A): 49–58.

Miller, N.C.E. 1953. Notes on the biology of the Reduviidae of Southern Rhodesia. *Trans. Zool. Soc. Lond.*, 27: 541–562.

Miller, N.C.E. 1956. *The Biology of Heteroptera.* Leonard Hill Books Ltd., London,, pp. 79–116.

Miller, N.C.E. 1956b. *The Biology of Heteroptera.* Leonard Hill Books Ltd., London, pp. 162.

Nyirra, Z.M. 1970. The biology and behaviour of *Rhynocoris albopunctatus* (Hemiptera : Reduviidae). *Ann. Ent. Soc. Amer.*; 63: 1224–1227.

Odhiambo, T.R. 1958a. Some observations on the natural history of *Acanthaspis petax* Stal. (Hemiptera : Reduviidae) living in termite mounds in Uganga. *Proc. R. Ent. Soc. Lond.*, 33: 167–175.

Odhiambo, T.R. 1958b. The camouflaging habits of *Acanthaspis petax* Stal (Hemiptera : Reduviidae) in Uganda. *Ent. Mon. Mag.*, 94: 47.

Odhiambo, T.R. 1959. An account of parental care in *Rhynocoris albospilus* Signoret (Hemiptera : Heteroptera : Reduviidae) with notes on its history. *Proc. R. Ent. Soc. Lond.*, 34(A): 175–185.

Parker, A.H. 1965a. The predatory behaviour and life history of *Pisilus tipuliformis* Fabr. (Hemiptera : Reduviidae). *Ent. Exp. Appl.*, 8: 1–12.

Parker, A.H. 1965b. The maternal behaviour of *Pisilus tipuliformis* Fabr. (Hemiptera : Reduviidae). *Ibid*, 8: 13–19.

Readio, P.A. 1927. Studies on the Biology of the Reduviidae of America north of Mexico. *Ibid.*, 17: 5–29.

Richman, D.P., Hemanway, R.C. and Whitcomb, W.H. 1980. Field cage evaluation of predators of the soybean looper, *Pseudoplusia includens* (Lepidoptera : Noctuidae). *Environ. Ent.*, 9: 315–317.

Richman, R.E. 1951. Recent observations on the cannibalism in *Triatoma* (Hemiptera : Reduviidae). *J. Parasit.*, 37: 433–434.

Rukmani, J. 1992. Multidisciplinary tools in the biosystematics of Reduviidae (Insecta : Heteroptera : Reduviidae). *Ph.D. Thesis*, Madurai Kamaraj University, Madurai.

Sahayaraj, K. 1997. Laboratory rearing of predaceous bugs with special reference to reduviids (Insecta : Hemiptera : Reduviidae). *Zoo's Print*, 13(5): 17–18.

Sahayaraj, K. and Ambrose, D.P. 1997a. Biocontrol potential of *Acanthaspis pedestris* Stal (Insecta : Heteroptera : Reduviidae) to *Helicoverpa armigera* Hubner of bhendi. *Madras Agricultural Journal*, 84(5): 294–295.

Sahayaraj, K. and Ambrose, D.P. 1997c. Field cage evaluation of the predator *Ectomocoris tibials* Distant to control *Dysdercus cingulatus* Fabricius. *Journal of Insect Science*, 10(1) : 65–66.

Sahayaraj, K., Martin, P. and Raju, G. 2003. Thermal stress on the dehydration, water loss and food consumption of the reduviid predator *Rhynocoris marginatus* Fab. *Entomon*. (In press).

Sahayaraj, K. 1991. Bioecology, echophysiology and ethology of chosen predatory heipterans and the potential in biological control (Insecta : Heteroptera : Reduviidae). *Ph.D. Thesis*, Madurai Kamaraj University, Madurai.

Sahayaraj, K. 2004. *Indian Insect Predators in Biological Control*. Daya Publishing House, Delhi, pp. 336.

Schofield, C.J. 1977. Sound production in some triatomine bugs. *Physiol. Ent.*, 2: 43–52.

Schofield, C.J. 1979. The behaviour of Triatomae (Hemiptera : Reduviidae) a review. *Bull. Ent. Res.*, 69: 363–379.

Schuh, R.T. and Slater, U.A. 1995. True bugs of the world (Hemiptera : Heteroptera) classification and natural history. Comstock Publishing Associates, Cornell University Press, Ithac, New York, pp. 336.

Stride, G.O. 1956a. On the mimetic association between certain species of *Phonoctonus* (Hemiptera : Reduviidae) and the Pyrrhocoridae. *J. Ent. Society, South Aft.*, 19: 12–27.

Stride, G.O. 1956b. On the Biology of certain West African species of *Phonoctonus* (Hemiptera : Reduviidae) mimetic predator of the Pyrrhocoridae. *Ibid.*, 19: 52–69.

Subbiah, M.S. and Mahadevan, 1957. The life history and biology of the Reduviid *Acanthaspis siva* a predator of the Indian honey bee *Apis indica*. *Ind. J. Vet. Sci.*, 27: 117–122.

Sundararaju, D. 1984. Cashew pests and their natural enemies in Goa. *J. Plant Crops*, 12: 38–48.

Tallamy, D.W. and T.K. Wood. 1986. Convergence patterns in subsocial insects. *Ann. Rev. Entomol.*, 31: 369–390.

Vennison, S.J. 1988. Bioecology and ethology of assassin bugs (Insecta : Heteroptera : Reduviidae). *Ph.D. Thesis*, Madurai Kamaraj University, Madurai, India.

Vennison, S.J. 1989. Bioecology, ecophysiology and ethology of chosen assassin bugs. *Ph.D. Thesis*, Madurai Kamaraj University, Madurai.

Vennison, S.J. and Ambrose, D.P. 1986. Bioecology of a dimorphic assassin bug *Edocla slateri* Distant (Heteroptera : Reduviidae). *Entomon.*, 11(4): 255–258.

Vennison, S.J. and Ambrose, D.P. 1987. Predatory behaviour of two assassin bugs *Edocla slateri* and *Oncocephalus annulipes*. *Environment and Ecology*, 5: 211–215.

Vennison, S.J. and Ambrose, D.P. 1990e. Biology and behaviour of the assassin bugs *Ectomocoris vishnu* Distant (Hemiptera : Reduviidae : Peiratinae) from South India. *J. Soil Biol. and Ecol.*, 10: 89–102.

Vennison, S.J. and Ambrose, D.P. 1991c. Biology and behaviour of *Ectomocoris xavierei* Vennison and Ambrose, 1990 (Heteroptera : Reduviidae : Peiratinae). *Tropical Zool.*, 4: 251–258.

Villiers, A. 1948. Hemiptera Reduviidaes del Afriquemnir. *Franue Tr.*, 18: 1–49.

Zeldon, R.A., Zuniga, A. and Swartzwadder, 1969. The camouflage of *Triatoma dimidiata* and the epidemiology of changes disease in Costa Rica. *Bio. Chienoprasitol.*, 24: 106–108.

Zeledon, R., Guardia, V.M., Zungia, A. and Swartzwadder, J.C. 1970. Biology and ethology of *Triatoma dimidiata* Latreille I. Life cycle amount of blood ingested resistance to starvation and size of adults. *J. Med. Ent.*, 7(3): 313–319.

Zeledon, R., Guardia, V.M., Zungia, A. and Swartzwadder, J.C. 1973a. The camouflaged phenomenon in several species of Triatominae (Hemiptera : Reduviidae). *J. Med. Ent.*, 10: 209–211.

Zeledon, R., Guardia, V.M., Zungia, A. and Swartzwadder, J.C. 1973b. Biology and ethology of *Triatoma dimidiata* (Laterille) III. Habitat and blood sources. *J. Med. Ent.*, 10: 363–370.

Chapter 3

Interactions of Predaceous Ladybirds: A Review

Omkar* and Geetanjali Mishra
Department of Zoology, University of Lucknow,
Lucknow – 226 007, India

Introduction

The use of natural enemies including insect predators, parasitoids, and pathogens for the biological control of harmful arthropods has been receiving increasing attention over the past several decades, particularly because it reduces the use of pesticides and is environment-friendly. Classical approach to biocontrol has involved introduction of insect predators or parasitoids of a pest from its original, native habitat into its new, domiciled habitat. The importance of such a low-key human interference in pest management programmes came to the fore with the astounding success of introduced ladybird beetle, *Rodolia curdinalis* against scale insect, *Icerya purchasi* in California in the year 1889 (Coppel and Mertins, 1977). This was followed by the *ladybird fantasy period* (Lounsbury, 1940), a period in which enthusiastic but unplanned and haphazard introductions of ladybirds were made the world over. Many of these natural enemy releases failed because of the lack

* Corresponding author: E-mail: omkaar55@hotmail.com

of in-depth factual and practical knowledge of the natural enemies and their prey concerned.

An analysis of these biocontrol programmes reveals that their failures were largely due to our inadequate understanding of the interactions between target arthropods and their natural enemies. Such interactions have been extensively studied in the past several years and the literature on them is widely scattered. There is, therefore, the need to sort and sift through the literature available on this subject, and critically review the salient, pertinent information available on this aspect so as to provide a base for improving the efficacy of future biocontrol programmes. It is in this context that the present article reviews the literature with reference to ladybird beetles, potent biocontrol agents belonging to the Order Coleoptera and Family Coccinellidae.

The scattered literature on ladybirds has been reviewed by Hagen (1962), Hodek (1967), Strand and Obrycki (1996) and Obrycki and Kring (1998). Books by Hodek (1973), Majerus and Kearns (1989), Hodek and Honek (1996), Dixon (2000) dealt with the various aspects of ladybirds. Omkar *et al.* (2002) reviewed the phenomenon of intraguild predation in ladybirds. Omkar and Pervez (2004a) gave a detailed prey-predator catalogue of ladybirds reported from India. Specific reviews on three important and partially successful ladybird biocontrol agents *viz.*, *Coccinella septempunctata* (Omkar and Pervez, 2002a), *Chilocorus nigritus* (Omkar and Pervez, 2003) and *Harmonia axyridis* (Koch, 2003) in recent years have been enlightening.

Except for a few phytophagous coccinellids in the sub-families Epilachinae and Coccinellinae, the members of all seven sub-families of Coccinellidae are predaceous insects. These predatory ladybirds have been reported to feed on aphids (Omkar and Bind, 1993, 1995, 1996; Omkar and Pervez, 1999, 2000, 2002b), coccids (Omkar and Pervez, 1999, 2000), mites (Putman, 1955), adelgids (Delucchi, 1954), ants (Pope and Lawrence, 1990), aleyrodids (Heinz and Zalom, 1996), chrysomelid larvae (Elliot and deLittle, 1980), cicadellids (Ghorpade, 1979), pentatomids (Afroze and Shujauddin, 1998), phylloxera (Pope, 1973) and psyllids (Chazeau *et al.*, 1991, Booth, 1997). They also interact with other co-existing natural enemies. Their interactions may thus be considered and studied under two categories: (1) prey-predator interaction and (2) predator-predator interaction, which are dealt with detail in the following text. The following text aims to provide a comprehensive and critical view of the myriad interactions between predatory ladybirds, their prey and other coexisting predators and wherever necessary their relevance to biocontrol has also been discussed.

Prey-predator Interactions

Prey Suitability and Acceptability

Coccinellids have a wide range of food sources (Hodek and Honek, 1996; Omkar and Pervez, 2004a). Species, which are generalist feeders, are known as *euryphagous* whereas those having high specificity or low range of prey species are called *stenophagous.* The generalist feeders despite feeding on a broad prey range grow, develop and thrive well on only a few limited food types (McClure, 1987; Bind, 1998; Obrycki and Orr, 1990; Lapiz and Gopud, 1995; Omkar and Srivastava, 2003a; Omkar and James, 2004; Omkar and Bind, 2004; Pervez and Omkar, 2004; Figures 3.1 and 3.2). An understanding of prey suitability for potential biocontrol agents provides a basis for evaluating the potential range of predator-prey interactions in the field and interpreting results of field releases (Hagen, 1974).

Hodek (1959, 1962) classified prey on the basis of their suitability and specificity to the ladybeetles. Prey capable of sustaining development and oviposition were considered as *essential* or *suitable* food whereas those that support mere survival but not sustenance of generations were regarded as *alternative* or *acceptable* food. Toxic prey were considered as *rejected* (Hodek and Honek, 1996). Malcolm (1992) used different terms *viz. included, peripheral* and *excluded* as alternates for the terminology suggested above (essential, alternative and rejected, respectively).

Coccidophagous ladybeetles are more selective than others in their prey choice (Jalali and Singh, 1989; Kairo and Murphy, 1995; Strand and Obrycki, 1996), perhaps because of the immobility and protective covering of their prey (Bristow, 1988). Although aphidophagous ladybeetles are largely generalists, they show apparent preferences for certain aphid species (Omkar *et al.*, 1997, 1999; Omkar and Bind, 1998; Omkar and Pervez, 2001; Afroze, 1999), with their larvae growing and developing well on only a selected few species. Of the total prey range of the adults, the immature stages can subsist on only a limited few, which are known as *nursery prey* (Dixon, 2000).

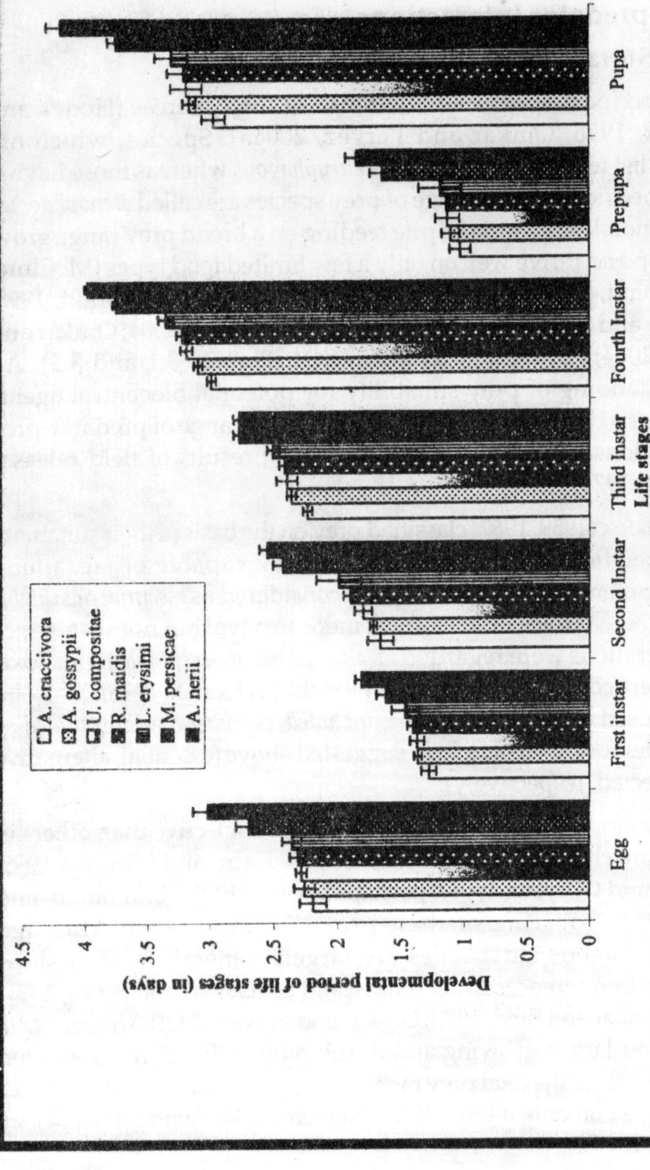

Figure 3.1: Developmental Periods (in days) of Different Life Stages of *P. dissecta* on Seven Aphids (After Mishra, 2004)

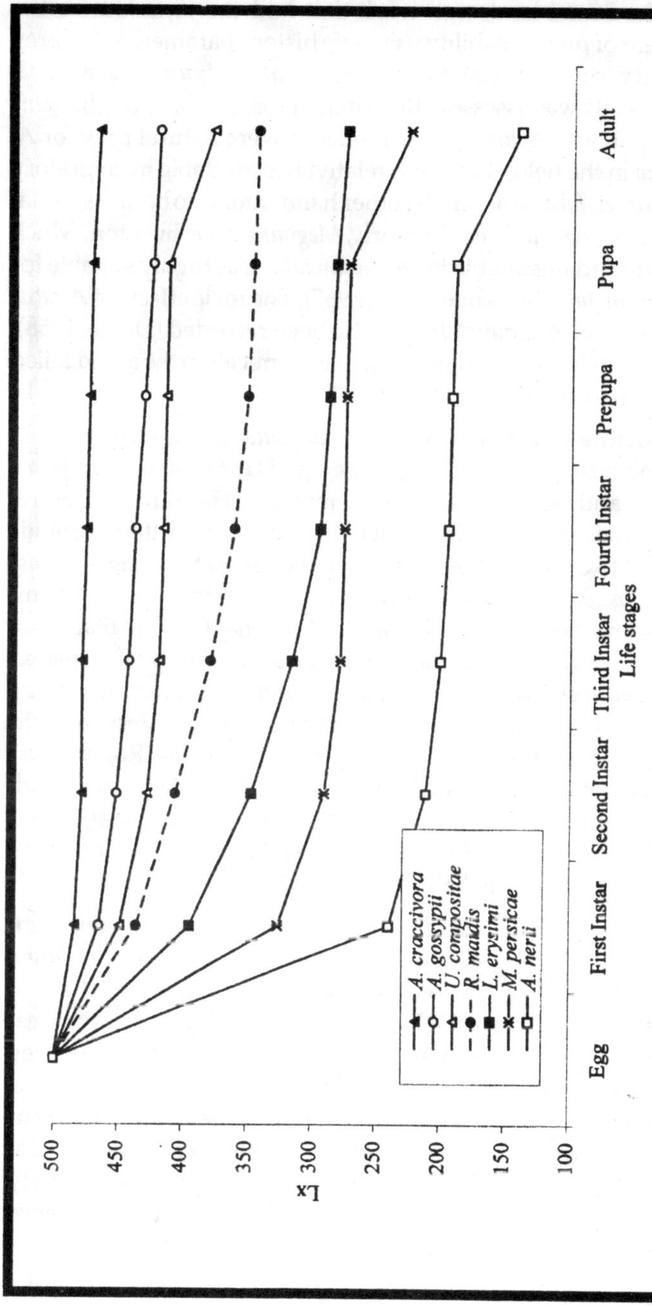

Figure 3.2: Stage Specific Survival of Different Life Stages of *P. dissecta* Starting from Original Cohort Size of 500 Eggs After Feeding on Seven Aphid Species (After Mishra, 2004)

In the initial landmark work that laid the criterion for the assessment of prey suitability using life history parameters, the prey suitability of different aphids to *Adalia bipunctata* and *C. septempunctata* was assessed (Blackman, 1965, 1967). Though, *Aphis fabae* Scopoli and *Aphis sambuci* Linnaeus were natural prey for *A. bipunctata* in the field, they were relatively unsuitable in laboratory conditions. *A. fabae* was, on the other hand, found to be an essential prey for *C. septempunctata*. Similarly, *Megoura viciae* Buckton, which was also found unsuitable for *A. bipunctata*, was highly suitable for *C. septempunctata* (Blackman, 1965, 1967). The toxic effects of *A. fabae* and *M viciae* for *A. bipunctata* had also been recorded (Dixon, 1958), but the search for physiologically active chemicals in the aphid failed to show any result (Dixon *et al.*, 1965).

Immature stages of *Coleomegilla maculata* (De Geer) developed fastest on *Myzus persicae* followed by eggs of *Leptinotarsa decemlineata* (De Geer) and pollen of *Zea mays* Linnaeus (Hazzard and Ferro, 1991). *Hippodamia sinuata* Mulsant when fed on two different aphids was found to complete the development of immature stages fastest on *Rhopalosiphum maidis* than on *Schizaphis graminum* (Rondani) (Michels and Behle, 1991). Instars of *Scymnus frontalis* reared on *Diuraphis noxia* completed development sooner than those reared on *Sitobion avenae* or *Acyrthosiphon pisum* (Gibson *et al.*, 1992). *Propylea quatuordecimpunctata* was found to develop fastest on aphid *S. graminum* in comparison to six other aphid species (Rogers *et al.*, 1994). *Harmonia eucharis* (Mulsant) also developed faster on aphid, *Macrosiphoniella pseudoartemisiae* (Shinji) in comparison to the other two aphids, *Brachycaudus helichrysi* (Kaltenbach) and *Eriosoma lanigerum* (Hausman) (Chakrabarti *et al.*, 1995).

Of ten aphid species, seven new essential aphid prey were identified for *Anatis ocellata* (Linnaeus), while six for *Calvia quatuordecimpunctata* (Linnaeus) on the basis of development of two ladybeetles (Kalushkov and Hodek, 2001). Field studies do not reveal co-occurrence of these two ladybeetles on ten essential prey (Kalushkov and Hodek, 2001). This finding goes against Majerus (1994), who advocates the inclusion of only those prey in the list of essential prey which are used by the ladybeetles as food in the field. He gives the example of *A. pisum*, which while being essential prey for 23 ladybeetle species in the laboratory has a recorded incidence of only seven ladybeetle species in the fields. Kalushkov and Hodek

(2001), however, argue that prey not recorded in the fields, but suitable in the laboratory can be successfully used for maintaining laboratory cultures and mass multiplication, and for use in inundative biocontrol. Kalushkov and Hodek (2004) evaluated thirteen aphid species for their suitability to *C. septempunctata* in terms of larval development and survival and found all of them to be essential prey.

The relative suitability of prey to a predator is determined not only by the prey type but also by the number consumed which may be affected by impeding plant-structures. The stem and leaf dimensions, branch angles, surface complexity and the hierarchical process by which mobile natural enemies locate and capture insect prey influence the routes involved in prey search and thus determine the costs (Ferran and Deconchat, 1992; Frazer and McGregor, 1994; Vohland, 1996; Clark and Messina, 1998; Geitzenauer and Bernays, 1996). By modifying predator mobility, plant structure also affects a predator's residence time or searching efficiency (Kareiva and Perry, 1989). The lower predation rates of the third instar larvae of *C. septempunctata* on the aphid, *A. pisum* on pea plant than on bean plant is attributed to the slippery surface of the pea leaves, which cause the larvae to slip off more frequently (Carter *et al.*, 1984). Plants differing in their architecture also differ in providing suitable sheltered micro environment for harbouring the prey (Clark and Messina, 1998). The differing plant characteristics, thus play major role under field conditions.

The prey contents also play a role in determining prey suitability. Consumption of *Aphis sambuci* by *C. septempunctata* results in almost none (Hodek, 1956) or very few small-sized survivors having low-fitness (Blackman, 1967). *Aphis craccivora* is toxic to *H. axyridis* (Pallas) when it is obtained from the plants, *Robinia pseudoaccacia* and *Vicia faba* (Hukusima and Kameii, 1970). The lethal effect of *A. craccivora* from *R. pseudoaccacia* was a result of amines *canavanine* and *ethanolamine* present in the host plant (Obatake and Suzuki, 1985). The ladybird, *Exochomus quadripustulatus* did not complete its development on aphid *M. viciae* and died within a couple of days of feeding (Radwan and Loveii, 1983). Larvae of *C. septempunctata* also showed abnormal development on consumption of aphid, *Macrosiphum albifrons* from bitter lupins having a high concentration of the alkaloid lupanine and 13-OH- lupanine but showed normal development when fed the same aphid from sweet lupins (Emrich, 1991). The monophagous aphid, *Acrythosiphon nipponicus*, feeding on *Paederia scandens* secreted a potent deterrent, an iridoid glycoside paederoside, accumulated from the plant, which resulted in the immediate release of the prey (Nishida and Fukami, 1989). Ladybirds have been observed salivating and grooming their mouthparts for a long time after an encounter with such a prey.

The suitability of a food type is also influenced by change in seasons and the resulting change in prey, host plant, etc. Furthermore, long-term ecological relationship between prey and predator may also modify the suitability; long term rearing of *A. bipunctata* (six generations) on poor prey *Aphis fabae* habituated and so well established the predator that even a shift to the previously identified better prey, *A. pisum* caused reduction in fitness levels (Rana *et al.*, 2002). The recognition of such flexibility in prey suitability has been considered a major reason for the need for a review of the present prey classification.

Foraging

The foraging of coccinellids is important in considering the application of ladybirds for biocontrol. The nature of navigation, the detection methods used by predators in locating prey patches and the hurdles reducing their searching efficiency are still not well understood, despite the many experiments conducted.

Optimal foraging represents a compromise between benefits achieved (*e.g.* finding of food, mates or shelter) and costs involved (*e.g.* risk of mortality, energy consumption) (Bell, 1990). Theoretically, a forager should try to achieve a maximum of energy gain or reproductive success at a minimum of expense (Charnov, 1976). This however is only true for ladybird larvae in both field and laboratory conditions and for adults in controlled conditions. In field conditions adults tend to fly away to new prey patches in case of food scarcity. This tendency of the adults, especially females, to avoid or fly away from prey patches in the dying phases was explained by the promulgation of the optimal foraging theory (Kindlmann and Dixon, 1993; Hemptinne *et al.*, 1993).

Studies indicate that it is the selection and identification of prey patches by the adults that eventually affects the fitness and survival of the species (Kindlmann and Dixon, 1993). In adults, the males tend to be more interested in the search for a suitable mate than in foraging, a fact proven by the males of *A. bipunctata* who did not show functional response to the differing prey densities (Hemptinne *et al.*, 1996). The females instead tend to search for oviposition sites preferably in proximity to the prey patches. Thus, the foraging behaviour of the females is often equated with the search for oviposition sites. Optimal foraging in ladybirds involves the location of foraging sites through various recognition cues, searching of sites and location of prey and then predation and consumption. The following sections review various steps constituting the foraging behaviour of ladybirds.

Recognition Cues

The ability to recognize prey as suitable or unsuitable and determine the quality and carrying capacity of the particular patch depends on the recognition cues present on the prey or emanating from their clusters. The selection process involves the interplay of recognition cues at different stages: association with habitats, prey patches and individual aphid types (Hassell and Southwood, 1978; Ferran and Dixon, 1993; Hattingh and Samways, 1995).

Hattingh and Samways (1995) identified numerous cues through a detailed experiment on the visual and olfactory location of biotopes, prey patches and individual prey by *C. nigritus.* Vertically oriented parallel lines, horizon with a tree line and simple ovate leaves were identified as important factors for biotope location. Location of prey patches by adults involved olfactory cues while larvae used physical ones. The differences in the ability of larvae and adults to locate prey, stem from the adults being the active locators of biotope and patch, whereas the comparatively immobile larvae depend on their parent's ability for long-range location of prey.

The prolegs of the fourth instars of *Coccinella transversoguttata* play a role in prey recognition (Storch, 1976). The touch of the maxillae or maxillary palps elicits prompt capture and consumption (Kesten, 1969; Nakamuta, 1985; da Silva *et al.*, 1992). *Chilocorus montrouzieri* recognize the prey through physical contact (Heidari and Copland, 1992). *C. septempunctata* orientates towards the prey at a distance of 7–10 mm in day light but does not turn towards the prey in the dark, thus suggesting the role of visual cues in recognition of prey (Stubbs,

1980; Nakamuta, 1984a; Nakamuta and Saito; 1985). Stubbs (1980) suggested the role of olfactory cues along with the visual ones.

Aphid body fluid has been considered to be the final cue in the prey recognition (Nakamuta, 1984b), thus pointing towards chemical detection. Honeydew acts as an arrestant stimulus for the coccinellids and induces intensive and localized search in the area (Carter and Dixon, 1984; Carter *et al.*, 1984). *Coccinella septempunctata* adults are capable of identifying odours of different prey and different predators (Sengonca and Liu, 1994). The aphid alarm pheromone, β-farnesene was identified as a locating factor for coccinellids (Hemptinne *et al.*, 2002).

Coccinellids avoid the eggs of heterospecifics even in the condition of food scarcity, probably as a result of these distinctive alkanes and due to the danger signalled by these unique signatures (Agarwala and Dixon, 1992; Hemptinne *et al.*, 2001). Recognition through physical contact is probably due to the cuticular wax structures present on the prey. Distinctive and species-specific chemicals in the footprints of three coccinellids, *viz. Epilachna vigintioctomaculata, Epilachna vigintioctopunctata* and *C. septempunctata* have been identified. These chemicals consist of hydrocarbons and true waxes and were identical to that in the cuticular waxes (Kosaki and Yamaoka, 1996). They perhaps were responsible for the detection and subsequent avoidance of areas having larval or ladybeetle populations. Though not yet determined, it is possible that similar footprint chemicals secreted from the aphids might guide the ladybeetles to their prey site. The role of volatiles secreted by infested plants in attracting predators to prey patches have been studied in parasitoids and predators other than ladybirds (Hagen, 1986; Strand and Obrycki, 1996).

Comparative sizes of the aphids and the topography and chemical constituents of the host plant too help determine the recognition and acceptance of the prey as a food source. Relatively smaller sized beetle and their larvae may be kicked off the plant or daubed with so much siphuncular wax that the result is eventual death (Dixon, 2000).

The ladybeetles and their prey possess specific signatures, which play a major role in orientating them towards their desired direction. Despite the many risks involved in consumption of certain prey, they are predated and consumed in the face of starvation even as their lethality or unsuitability is signalled. This act is not an adventurous one but an act of desperation to save its life in the face of imminent death either through starvation or consumption of an unsuitable or toxic prey. The consumption of an unsuitable prey leaves the predator with the slightest chance of survival but starvation ensures sure death.

Searching

Ladybird larvae and adults usually walk along leaf edges, interveinal and intraveinal areas to stay in prey vicinity. The search usually takes place in the morning hours though prey venturing into their vicinities at night time are consumed (Nakamuta, 1987).

Prey location involves an extensive search with a switch over to an intensive one in the presence of signals indicating desired and suitable sites. The initial signals help locate the habitat, the following cues the location of prey on a plant and the final cues, a particular prey patch. Each successive cue leads to sublimation in the extensive and random searching mode, with the most extensive search involved in habitat location and the least extensive in locating the prey patch (Carter and Dixon, 1982; Nakamuta, 1985).

Random searching is not observed in field conditions since ladybeetles gravitate towards areas recently or presently occupied by the prey. Prey odour or odours released by infested plants facilitate this process (Hagen, 1986; Ponsonby and Copland, 1995; Sengonca *et al.*, 1995). Having once located specific cues signalling prey presence, adults switch to more intensive search of the surroundings, moving slowly and turning more often. The immediate switch over to intensive search after consumption is however not the norm and is adaptive in nature (Murakami and Tsubaki, 1984). It occurs before the prey consumption (Ferran *et al.*, 1994) and is probably induced by starvation. Sometimes, the ladybeetle does not switch over to an area-restricted search immediately after consumption. Either the onset of area-restricted search is delayed or it simply does not take place. This is probably an insurance against the fluctuating nature of prey patches (Tortorici and Bell, 1988; Ferran *et al.*, 1994). There is evidence that induction of area restricted search is prey specific. Adults of *A. bipunctata* switch to an intensive search only after encountering and consuming the desired prey (Kalushkov, 1998, 1999).

If unsuccessful in locating more prey after the initial switch-over inducing encounter, ladybeetles tend to switch back to an extensive search. Their larvae are known to chase the prey that eludes capture even after an encounter. The pursuit is fast and is attributed to the larval motivation to forage (Carter and Dixon, 1982). Larvae avoid areas bearing their tracks, so as to save precious time and energy (Marks, 1977; Dixon, 2000). The presence of larval tracks is also an indicator to gravid females to desist from foraging and ovipositing (Ruzicka, 1994, 1996, 1997a, b, 2002).

It has been suggested that the switch over from extensive to intensive search occurs in the presence of a number of cues. The cues could be either in the form of honeydew or wax of mealy bugs or in form of physical contact or visual cues or as chemical olfaction. *C. septempunctata* was found to orientate towards crushed *Berberis vulgaris* leaves and towards its methanol extracts inside an olfactometer, indicating the role of olfaction in habitat location (Schaller and Nentwig, 2000). The wax of mealy bugs acts as an arrestant stimulus for *Diomus* sp., *Exochomus* sp. and *Cryptolaemus montrouzieri* (van der Meiracker *et al.*, 1990; Merlin *et al.*, 1996a). Honeydew also acts as an arrestant stimulus (Carter *et al.*, 1984) and induces an intensive search. The ladybeetles can detect and orientate towards the prey prior to physical contact, which has been proved by a number of works on *C. septempunctata* (Stubbs, 1980; Nakamuta, 1984). Change in searching behaviour in the presence of aqueous aphid extract points towards role of pheromones in signalling (Shonouda, 1999). Aphid alarm pheromones released by captured prey attract foraging ladybird instars to the site of prey capture and consumption. This strategy especially helps neonates in prey searching and attaining their first meal (Hemptinne *et al.*, 2000).

Hassell and Southwood (1978) suggested that duration of search was dependent on the search rate of predator and the average sub-patch size. A constant duration of area-restricted search was observed on all aphid patches except those with extremely low densities; this probably maximizes the number of encounters and increases prey consumption. The cues deciding the duration of the area-restricted search are termed the giving up time (GUT) by Nakamuta (1984). Through well thought experiments, it was deduced that the duration is influenced by cues related to the amount and size of prey consumed but the induction is not affected by any such cue. GUT has a linear relationship with the time spent in prey consumption, which in turn has a direct relation to the weight of the prey. GUT was also determined by the size of the latest prey and not the general hunger level (Nakamuta, 1984). Previous experiments contrarily revealed that GUT was inversely proportional to the mean feeding rate (Charnov, 1976). GUT increases with increasing hunger levels (Bond, 1980; Carter and Dixon, 1982).

The efficiency of intensive search is determined by the relative sizes, speeds and densities of predators and their prey (Hassell, 1978) with plant topography playing an important role (Shah, 1982; Geitzenauer and Bernays, 1996). *Scymnus nigritus* spent most of its time in searching prey-infested needles and remained in the upper and outer parts of plant having higher needle density (Vohland, 1996). In this case the searching pattern of both adults and larvae were similar perhaps as a result of unidimensional narrow structure of the needle. This single dimensional aspect of the habitat did not allow induction of area-restricted search due to the lack of space for movement. The encounter time is also governed by plant characteristics, such as hairiness and waxiness (Banks, 1957; Heinz and Parrella, 1994; Carter *et al.*, 1984).

The foraging or searching behaviour when studied in females is discerned through their ovipositional response. Reports indicate that the signal for oviposition may either be the anticipation of the explosion in aphid population or could be the location of prey patches. The females locate and select a prey patch for foraging and oviposition with the aid of visual and pheromonal cues and with the help of factors determining the quality of the patch.

It has been purported that there is a nutritional threshold level below which there is no oviposition. The reduced capturing efficiency in females is probably to this cause, since it reduces the consumption and allows the ladybeetle to determine the quality of the prey patch due to a longer stay (Dixon, 1959). The age of the aphid patch too plays an important role in determining the foraging and subsequent oviposition by the females. Theoretically, the females oviposit only when the quality of the prey patch has been determined. The duration of the aphid colony is almost equal to the developmental period of the ladybeetles. The time of oviposition with respect to the population graph of aphids determines the development and survival of the ladybird progeny.

A window is known to exist in the aphid population graph, which has all parameters favourable for oviposition (Dixon, 1997). The opening of the window is possibly determined by the minimum density of aphid consumption required for oviposition and survival of the first instar larvae (Dixon, 1959). It is the time of oviposition that determines the survival of the ladybeetle progeny. An early but excessive oviposition results in higher levels of mortality at the later stages due to prey scarcity; later oviposition is detrimental to the larval survival for the same reason. It is the early but limited laying of eggs that ensures maximum survival due to limited competition for resources (Hemptinne *et al.*, 1992; Kindlmann and Dixon, 1993; Doumbia *et al.*, 1998). The closing of the window is possibly governed by the response of the adults to the larval tracks, which deter egg laying through pheromones, *viz.* oviposition deterring pheromones present in the larval tracks. These pheromones are alkanes or their complexes with usually methyl-branched (C_{23} to C_{33}) and straight chain alkanes (C_{20}–C_{31}) predominating. These alkanes spread easily on the hydrophobic cuticle of the plants and their effects last for a minimum of at least ten days (Hemptinne *et al.*, 1992; Doumbia *et al.*, 1998; Hemptinne *et al.*, 2001; Ruzicka *et al.*, 1997a; Merlin *et al.*, 1996b). These factors ensure that the instinctive need of the ladybeetles to survive is fulfilled and in addition, the aphid population is also controlled.

There have been reports of the aphid and plant phenology affecting the oviposition. The theory of oviposition in relation to the age of the aphid colony has been based on the above assumption and has been proved in the hoverflies (Kan and Sasakawa, 1986; Kan, 1988a, b; Hemptinne *et al.*, 1993). In a previously conducted study, it was observed that the old aphid colonies had only a weak deterrent effect on the oviposition (Hemptinne *et al.*, 1993) when compared to the larval tracts (Doumbia *et al.*, 1998). However, recent experiments proved that the age structure does not play any important role in oviposition determination (Hemptinne *et al.*, 1999).

Feeding and Predatory Response

Food Intake Patterns

Male ladybeetles consume lesser number of prey than the females (Hodek and Honek, 1996; Dixon, 2000; Omkar and Srivastava, 2001a). The primary goal of males is to mate and a major part of their activity and resources are concentrated on mate search, resulting in reduced prey consumption. Females on the contrary live to fulfill their primary aim of nurturing future generations by laying eggs in a microenvironment that provides the required amount of food. Reproducing adults were found to consume less than the virgin adults owing to distractions in form of sexual activity (Rhamhalinghan, 1987). The newly emerged adults consume minimum of aphids, as they tend to sit quietly while taking considerable time to fold up their moist wings (Rhamhalinghan, 1987). Conditions that might increase the feeding potential are suitable prey, increased prey densities and a higher temperature. The lesser consumption of prey by the predator in the presence of *ad libitum* or higher amounts of prey does not necessarily indicate its maximum feeding capacity as it could be a result of prevailing temperatures (Hazzard and Ferro, 1991). The miniscule prey consumption by *Scymnus marginicollis* is perhaps the reason that it stays on the plant even in the presence of very low aphid population of less than an aphid per plant (Buntin and Tamaki, 1980). Melanics and non-melanics have differential prey consumption (Muggleton, 1978; Rhamhalinghan, 1985, 1987), probably owing to the increased capability of melanics to trap and utilize solar radiation and thus increase their temperatures rapidly leading to a rise in their metabolic rate and thus consumption.

Ladybirds have adapted to different food ranges probably owing to the varying habitats that they occupy. These differing food choices have led to modifications in their primitive and basic biting type of mouthparts. The basic components of the primitive mouthparts are a pair of massive, wide, sickle shaped mandibles having incisor and molar parts on its inner side (Hodek and Honek, 1996). The mouthparts in the phytophagous types are adapted to biting off the plant tissue with a scraping motion and then sucking out the juices from the scraped portions. This leaves a very typical pattern on the leaf. The carnivorous type with both the terminal teeth comparatively smaller and split off near the tip, along with a smaller ventral smaller tooth, is one of the most basic and widely represented mandible types in the family Coccinellidae (Hodek and Honek, 1996). The pollinivorous and micromycetophagous ladybeetle have comb-like mandibles to collect pollen and spores (Ricci, 1982).

The consumption in the aphidophagous ladybeetles involves both biting and chewing the prey or sucking out its fluid through an appendage. The consumption pattern to a large extent depends on the relative size of prey. The smaller sized prey are usually devoured completely; hard and solid body parts and all (Butt, 1951; Hagen, 1962; Kesten, 1969). However, when the prey is too large, its body fluids are sucked out and the skin with the appendages is rejected (Butt, 1951; Harpaz, 1958; Hagen, 1962; Kesten, 1969).

Some ladybird larvae show extra-intestinal digestion/extra-oral digestion (Richards and Golestes, 1991); they periodically regurgitate fluid from the gut into the chewed prey and suck back this pre-digested food. Pre-digestion has been observed in the larvae of *Stethorus* sp. (Hagen, 1962) and *Platynapis luteorubra* (Ricci, 1979). They hold on to the prey via its appendage. The body of aphid is alternately deflated and inflated as the pre-digestion commences (Hodek and Honek, 1996). A detailed description of extra-intestinal digestion in *P. luteorubra* revealed that the labra, mandibles and hypopharynx play an important role in the process (Ricci, 1979). These appendages form a hypopharyngeal bridge, which is indirectly also responsible for the tight clamped grip of the predator on the prey. The mandibular groove, epipharynx and galae also play an important role in the formation of the bridge and of sealing it effectively to the victim. The bridge/channel so formed has the dual purpose of injecting mesentronic juices into the victim's body and of sucking up the digested food. The ingestion and then withdrawal of juices can be easily observed in pale coloured aphids. The time between one injection and the other increases continuously till duration of two minutes when the operation comes to an end. It takes about 50 cycles of injection and reimbition of all the nutrients to be absorbed. *Scymnodes lividigaster* also shows a similar method in the larval stages, which have a sickle shaped unidentate mandibles that pierce and hold the prey still while sucking up the juices; the adults of this ladybeetle have a chewing habit. *Scymnus subvillosus* and *Nephus includens* have a mandibular groove, so it might be assumed that feeding is done by piercing and sucking (Ricci, 1979).

Rhyzobius ventralis larvae feed on females of coccid, *Eriococcus coriaceus* by making a hole in the lower part of the sac and curling around and feeding until the prey is empty instead of attacking through the large and far more vulnerable opening in the sac through which the tip of its abdomen emerges (Richards, 1981). Larvae and adults of a coccidophagous ladybeetle, *Rodatus major* are specialized in eggs of *Monophlebulus pilosior;* the fourth instar feeds almost continuously throughout the day, lying with its mouth close to the female's genital opening so that it can eat the eggs as they are laid (Richards, 1985). *Pseudoscymnus kurohime* attack the nymphs of aphid, *Ceratovacuna langiera* the moment they are born to curtail their induction of soldier rank (Arakaki, 1992).

Food intake patterns also play an important role in the selection of a biocontrol agent. The efficacy of clearing infestation is an important parameter to consider, especially in the coccidophagous ladybeetles, since scale infestation deface the fruit or other consumable parts. *C. nigritus* removes the entire scale, body and covers both, thus keeping the infested parts clean. The characteristic tin opener shape and action of the mandibles enable it to completely remove the whole adult red scale (Samways and Wilson, 1988). These findings suggest that biocontrol projects should take cognizance of the mandible type and the feeding behaviour, where improved short-term cosmetic appearance of the fruit is an added bonus to the long-term prey population control.

Functional Response

The response of predator to varying prey densities in terms of consumption and searching behaviour was first analyzed in a simple manner by Solomon (1949) who termed it as functional response. Holling (1959, 1965) further analyzed this behavioural aspect in a complex manner and proposed three typical responses, which more or less ascribe to all prey density, related feeding behaviour with slight modifications.

The initial response differed in all the three responses, though a saturation threshold was later obtained. The consumption is directly proportional to the prey density in the Type I response resulting in a prominent linear graph. The Type II response is curvilinear in nature with an initial rapid increase in consumption followed by a subsequent slowing down of this increase. The Type III response has a sigmoidal graph, which is obtained as a result of the increasing prey consumption over a limited range of prey densities. The mortality in prey was density independent in Type I, negatively density dependent in Type II and positively density dependent in Type III responses. Simple charting using counted data is not a very reliable method of determining the type of response and logit analysis has been recommended for the discrimination among the functional response and identification of region of density dependent predation (Trexler *et al.*, 1988). In addition to the well-established functional response to prey density, there was observed a functional response to prey distribution (Zebitz, 1988).

Coccinellids have been known to exhibit Holling's Type II response (Sinha *et al.*, 1982; Beg *et al.*, 1987; Babu and Ananthakrishnan, 1993a; Mora *et al.*, 1995: Lapiz and Gopud, 1995; Akhtarruzzaman and Ahmad, 1988; Veeraval and Baskaran, 1997; Omkar and Srivastava, 2001b, 2002; Omkar and James, 2001; Omkar and Pervez, 2004b, 2005; Pervez and Omkar, 2003; Figure 3.3). The various parameters studied in functional response, *viz.* handling time, attack rate, area of discovery and the number of prey consumed are dependent on the hunger levels, which are determined by the contents of the gut: the gut determining hypothesis as suggested by Holling (1959). Carter and Dixon (1982) supported the above hypothesis and stated that the attack rate and handling time were not determined by the average encounter rate or the average rate of food intake but by the amount of food contents in the gut. Cook and Cockrell (1978) were the first to disagree with the gut limiting theory. They found that the feeding time per prey and the amount of food consumed decreases as the interval between the two subsequent captures shortens. Mills (1982) on the basis of a few previous experiments and his own experience postulated a theory, which suggested that the duration of phases within an attack cycle differ with the level of satiation, *i.e.* with the time spent since the consumption of the last prey.

There is a serious drawback in the presently widely used methodologies of investigation. The experimental setup fails to take into consideration the structure of the host plant and the obstacles that it might pose in the search for prey. Significant differences in the consumption were seen in the presence of plant parts (da Silva *et al.*, 1992), since the varying architecture of the plant induces a change in the attack coefficients and the handling time. The host plant influence altered the very type of functional response (Messina and Hanks, 1998). The size of the arena however did not change the functional response in *C. sulphurea* (Hodek *et al.*, 1989). The increase

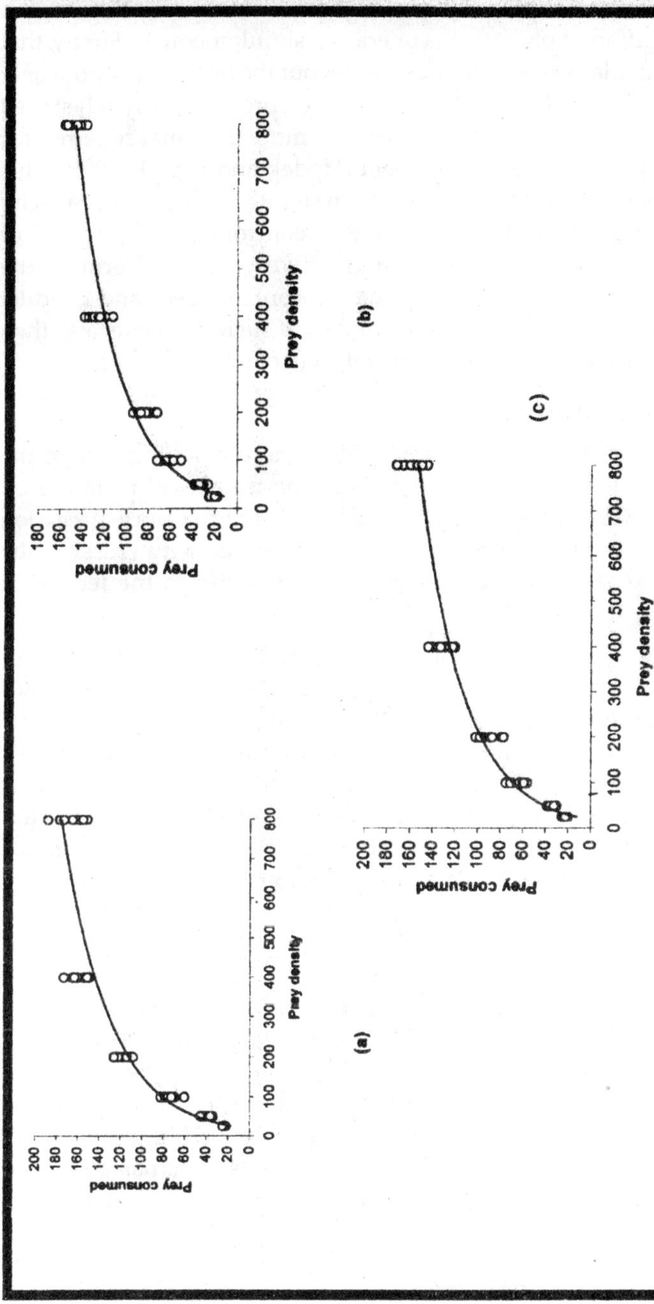

Figure 3.3: Prey Consumed by (a) Fourth Instar; (b) Adult Male, and (c) Adule Female of *P. dissecta* at Different Densities of *A. gossypii* (After Omkar and Pervez, 2004b)

in the number of aphids killed at high prey density may be the result of two different phenomena operating simultaneously. Firstly, that the hungry ladybeetles completely devour the first few prey but only partly consume the subsequently attacked prey. Secondly, when prey is abundant, the larvae may consume more than that required for normal growth and development (Hodek and Honek, 1996). This ability of the predators to stretch even slightly their feeding capacity is essential for it to be an effective biocontrol agent. The study of functional response is thus an effective way to determine the predatory potential of a potential biocontrol agent and conduct comparative studies among a number of such bioagents and then determine the most effective and efficient of them.

Numerical Response

As the density of prey increases, there is an increase in the number of predators and is termed the numerical response by Solomon (1949). The increase could be in the form of an aggregation response of the ladybeetles with the prey density crossing the threshold level or could be an answering rise in the fecundity (Holling, 1959, 1965).

The aggregation response does not lead to a permanent inclusion into the local ladybeetle populations. Even at prey densities above the limiting threshold, there exists a dynamic ladybeetle population with the ladybeetles coming and going almost continuously (Ives, 1981; Frazer and Raworth, 1985). This aggregation or rather the numerical response of the ladybeetles has been extensively studied in a number of ladybeetles: *C. septempunctata* (Pandey *et al.*, 1984; Omkar and Srivastava, 2003b), *C. sexmaculata* (Wagiman, 1996; Agarwala and Bardhanroy, 1997, 1999; Veeraval and Baskaran, 1997; Bind, 1998), *C. transversalis* (Veeraval and Baskaran, 1997; Agarwala and Bardhanroy, 1997, 1999; Omkar and James, 2001), *Hippodamia convergens, Hippodamia tredecimpunctata, Coleomegilla maculata, Theriopsis maculata* (Radcliffe *et al.*, 1976), *Cheilomenes lunata* (Ofuya and Akingbohungbe, 1988), *Scymnus argenticus* (Veiera, 1995), *Propylea dissecta* (Omkar and Pervez, 2004c; Pervez and Omkar, 2003) and *A. bipunctata* (Hemptinne *et al.*, 1992).The numerical response of the ladybeetles has also been studied in terms of their efficacy as a biocontrol agent for their respective prey (Ofuya and Akingbohungbe, 1988; Veeraval and Baskaran, 1997; Agarwala and Bardhanroy, 1997, 1999).

The reproductive numerical response involves the fecundity fluctuations in response to the prey density (Figure 3.4). It was observed that the number of eggs laid increased curvilinearly with an increase in the prey density. The increase took place till an optimum or rather standard fecundity was obtained, though the rate of increase continued to slow down considerably. This is possibly a response to the detected carrying capacity of the patch. A marked reduction was observed in the reproductive numerical response in the presence of conspecific larvae. At the high prey densities, the females were observed to be highly active and thus did not stay on the patch for long. In the presence of conspecific larvae, the female will lay a fewer eggs than when kept alone. The extent of the inhibition of egg laying was negatively correlated with the rate of encounter with larvae. The presence of larvae in the aphid colony signal to the gravid females, the inability of the patch to support development of more immature stages and thus, deter egg laying on that particular patch (Hemptinne *et al.*, 1992).

Numerical response involves the reaction to both fluctuating prey as well as predator densities. It is an instinctive means to ensure good growth and development of its progeny and is dependent on the quality of the prey patch and the number of both con- and hetero-specifics lurking in the vicinity. It may thus be recognized as an extension of the optimal foraging response.

Predator-predator Interactions

Cannibalism

Cannibalism is the act of killing and feeding upon a live conspecific. Cannibalism was earlier regarded to be abnormal and an artifact of laboratory rearing (Dixon, 1959), but proof of its occurrence in the natural ecosystems (Mills, 1988; Elgar and Crespi, 1992) points towards it being a part of the normal foraging behaviour. Cannibalism and interspecific predation were considered one of the most important mortality factors in coccinellid juveniles (Wright and Laing, 1982). It was also identified as a major problem in mass rearing of coccinellids for augmentative biocontrol programmes (Obrycki and Kring, 1998).

The victims of cannibalistic instinct are usually at a vulnerable stage of development, *i.e.* either in the egg or pupal stage or at the ecdysal stage (Majerus, 1994; Agarwala and Dixon, 1992; Stevens,

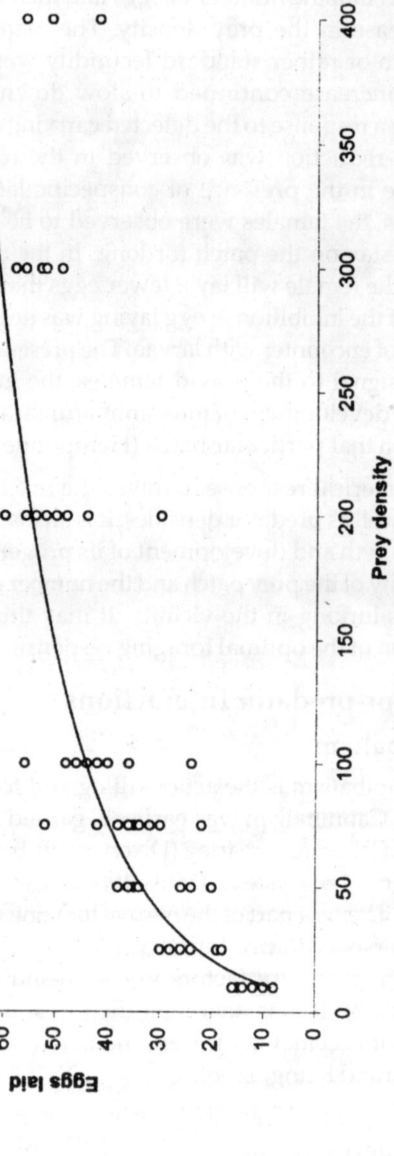

Figure 3.4: Relationship Between Number of Eggs Laid by Female *P. dissecta* at Varying Densities of *A. gossypii* (After Omkar and Pervez, 2004b)

1992). Cannibalism is usually not noticed in the individuals of the same developmental stage. Scarcity of prey or abundance of predators, especially the most voracious stages, leads to the induction of cannibalism in that particular habitat.

The ecology and dynamics of cannibalism have been extensively reviewed (Dong and Polis, 1992; Elgar and Crespi, 1992). Cannibalism is to a large extent influenced by prey scarcity (Takahashi, 1987; Agarwala and Dixon, 1991; Yasuda and Shinya, 1997). In the absence of prey, hungry individuals feed on their fellow con-specific eggs or larvae. A way to reduce cannibalism is to lay larger but lesser eggs, which provide the double benefits of lowered levels of cannibalism due to fewer larvae and shorter development period. Despite these benefits of larger and lesser eggs, ladybeetles opt for laying a large number of eggs; pointing to an innate constraint on egg size (Dixon and Guo, 1993). The resulting levels of cannibalism ensure development of the larvae with higher levels of inclusive fitness.

Cannibalism carries numerous costs with it. Each cannibal faces the disadvantage of being consumed by its intended victim, since the victim and the cannibal are of similar size and predatory instinct. In addition, the predator may be more susceptible to parasitic and viral attacks as a possible result of consumption of infected and vulnerable ladybeetle stages (Dixon, 2000). The consumption of kin also depletes the gene pool, thus placing the fitness of future generations at risk. The risks involved are seemingly big and the costs should be high, yet cannibalism occurs. Why?

The very selection of cannibalism as a part of the natural process indicates the presence of some hidden benefits in it. Interfamilial and interspecific variations in cannibalistic tendencies have been reported (Stevens, 1992; Dixon, 2000). Burgio *et al.* (2002) observed that interspecific interaction between ladybirds *H. axyridis* and *A. bipunctata* and investigated the consumption of interspecific eggs by fourth instar and adult. They found that cannibalism was frequent in *H. axyridis* than *A. bipunctata*. Results showed *H. axyridis* had a high potential for cannibalism. They are reduced in coccidophagous ladybeetles perhaps as a result of their sheltered life cycle and proximity to the prey (Dixon, 2000).

Ladybeetles have been attributed with the ability to detect the carrying capacity of the prey patch. There is, however, very little likelihood of the presence of advanced sensory capabilities in ladybeetles (Dixon, 1958), which allow them to determine the suitability of patch in terms of comparative values of μ/g where, μ is the probability of death per unit time and g the growth rate per unit time (Werner and Gilliam, 1984; Gilliam and Fraser, 1987; Gilliam, 1990). A patch, which has μ/g lesser than the average for all the patches in the habitat ought to be selected for oviposition, but the perceptive abilities are unlikely to be as highly developed as required for such estimation. The probability of consumption of food increases when its benefit to cost ratio (B/C) exceeds that of the prey patch (Dong and Polis, 1992). Cannibalism is prevalent when B/C on consumption of a con-specific is more. The cost is, however, usually high due to the comparative difficulties in the capturing process. Thus, cannibalism is more evident at stages, which are slower or immobile and hence more vulnerable. There is usually disparity in size and vigour of the cannibal and the victim. The likely competitive benefit in terms of reduced competition is the probable cause of cannibalism despite the absence of a high B/C ratio.

Con-specifics are a rich source of nutrition and most ladybeetles are able to complete their lifecycle on them (Koide, 1962; Dimetry. 1974; Agarwala and Dixon, 1992; Hemptinne *et al.*, 2001; Gagne *et al.*, 2002). A proportion of ladybird larvae use some of the total biomass accumulated through early feeding to advance their growth and development. The victims function as live meat for their kin; a phenomenon that is known as the *icebox hypothesis* (Alexander, 1974). When faced with a choice, the female tends to consume the eggs of another con-specific female and avoids her own (Hippa *et al.*, 1982; Agarwala and Dixon, 1993). The con-specific eggs become a source of nourishment but the eggs of heterospecifics are avoided probably due to the presence of surface alkanes and perhaps as a result of the toxicity signalled by con-specific alkanes (Agarwala and Dixon, 1992; Hemptinne *et al.*, 2000; Omkar *et al.*, 2004; Figures 3.5 and 3.6). The cuticular wax and the patterns thereof may play an important role in prey recognition (Omkar *et al.*, 2004).

Eggs in clusters laid by a single female are either siblings or half siblings due to the multiple matings. Osawa (1992) considered the benefits of sibling cannibalism in relation to the loss of fitness. He worked out the increase in the probability of survival from first to second instar of cannibals in *H. axyridis* at two aphid population densities. The probability of survival to the next instar is denoted as

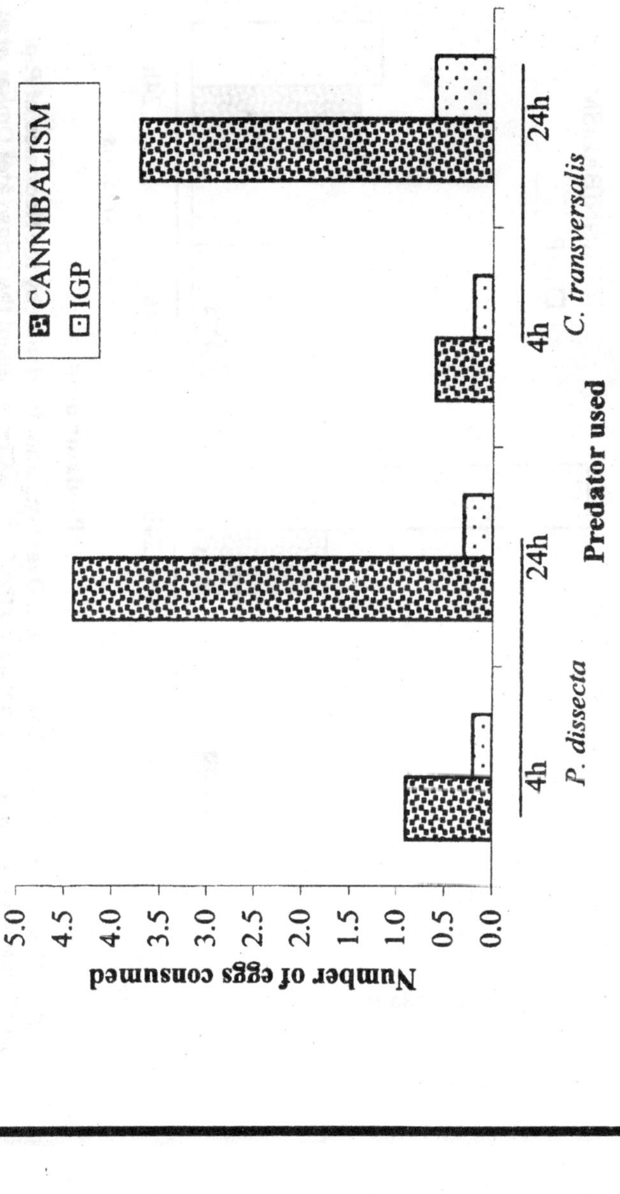

Figure 3.5: Preference for Cannibalism Over Intraguild Predation by the Neonates of *P. dissecta* and *C. transversalis* (Ater Omkar *et al.*, 2004)

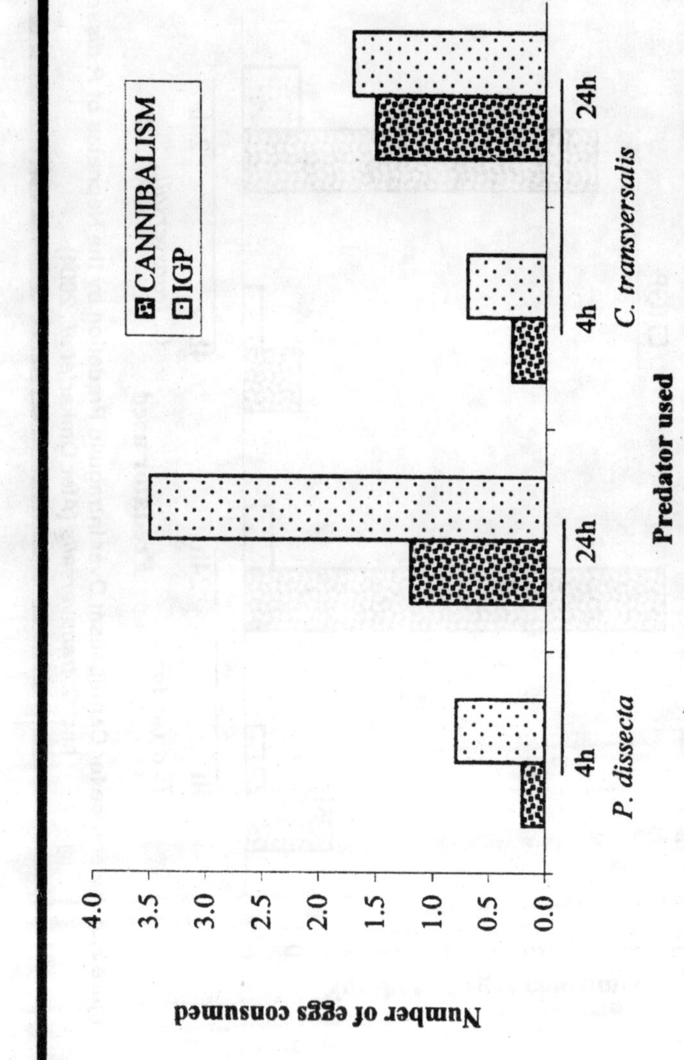

Figure 3.6: Preference for Cannibalism Over Intraguild Predation by the Neonates of *P. dissecta* and *C. transversalis*, after Interchanging the Surface Chemicals of the Eggs (Ater Omkar *et al.*, 2004)

p and the increase in fitness with the consumption of a single aphid as p*, so that the new inclusive fitness level is p+p*. Observations and statistical analysis revealed that p/p* should be less than 2 for siblings and less than 4 for half siblings for cannibalism to be a part of the particular patch (Hamilton, 1964). The results obtained by Osawa (1992) show the adaptiveness of cannibalism when aphids are scarce. Larval cannibalism is advantageous even when aphids are abundant, because it increases the probability of survival markedly.

Cannibalism was identified as one of the maternal investments in which a female controls sibling cannibalism (Osawa, 2003). They found that cluster size significantly affected the number of cannibalized egg per cluster while cluster size was significantly affected by oviposition interval. Clustered eggs were more vulnerable to con-specific attack than to hetero-specific predation, which is probably a result of the aggregation of surface chemicals that act as a deterrent to hetero-specifics and attractant to con-specifics (Omkar *et al.*, 2004).

Cannibalism increases with an increase in predator density at constant prey density and at higher prey densities, there exists the same relation but the increase is comparatively less (Dixon, 2000; Michaud, 2003). On the other hand, there is a decrease in cannibalism with an increase in prey density at constant levels of predators. Cannibalism has, thus, been observed to be density dependent and has the potential to regulate predator numbers.

Intraguild Predation (IGP)

Prey abundance leads to the advent and aggregation of a number of predators belonging to different species, different genera and even different orders. These predators exploit the prey and constitute a guild (Root, 1967). In their joint exploitation of the prey patch, they tend to affect the foraging success of others by competing for food source, through interference and by active predation of guild members. Lucas *et al*. (1998) found that the aggregation of generalist predators in time and space caused overlapping in life stages of predators in a same prey resource. This was because the predators have limited time to use these prey resources.

These interactions or predatory attempts result in immediate energy gain and reduce the potential competitors. Intraguild predation (IGP) is the consumption of guild members belonging to different species or genera. The active aggressor is known as the *intraguild predators*, the vulnerable victim as *intraguild prey* and the primary food source as the *extraguild prey*. Polis *et al.* (1989) and Polis and Holt (1992) extensively reviewed the intricacies and complexities of intraguild predation.

Studies conducted by Honek (1985) give an insight into the habitat specialization of the ladybeetles. The different ladybird populations and their densities are specific to each particular habitat, with one species being dominant. Usually the minor populations consist of smaller sized ladybeetles, which while predating on the prey patches face the risk of consumption by the larger and established ladybeetles. Larger ladybeetles constitute a hazard of major proportions to the smaller and less abundant ladybeetles, thus the smaller ladybeetles invest more in defence and are toxic to the other guild members (Agarwala and Dixon, 1992; Agarwala *et al.*, 1998; Hemptinne *et al.*, 2000). This increased investment in defence is to reduce its window of vulnerability. The smaller ladybeetles tend to consume relatively more of the immature stages of the larger ladybeetles in their fight against the larger ladybeetles for survival (Agarwala and Dixon, 1992; Agarwala *et al.*, 1998). This attempt at self-defence and survival by the smaller ladybeetles is, however, wasted because the consumption by the larger ladybeetles, no matter how little, puts a significant dent in the already less population of the smaller ladybeetles. The consumption of the less abundant predators in the guild despite their displayed defence mechanisms leads to the establishment of the already abundant ladybeetle in the ecosystem.

Simple experiments have proved the presence of intraguild predation, with the larger predator consuming the smaller ones and this is aggravated in the absence of the aphids (Sengonca and Fringe, 1985; Lucas *et al.*, 1997, 1998; Phoofolo and Obrycki, 1998). Intraguild predation, to a large extent, is a direct consequence of the prey scarcity and the need to reduce potential competitors (Kajita *et al.*, 2000). Takahashi (1989) found that when number of aphids declined then larvae of *H. axyridis* preyed on prepupae and pupae of *C. septempunctata*. The experiments conducted by Evans (1991) and Chang (1996) to test the efficacy of ladybird combinations in biocontrol, however, did not reveal the presence of intraguild predation, possibly owing to the relatively similar size of the ladybeetles used in the experiments and also due to the abundance of extraguild prey.

The idea of using a combination of natural enemies has for long had many supporters both in favour and against (Turnbull and Chant, 1961; Ehler, 1990; Benrey and Lamp, 1994). The application of combination biocontrol is possible only if there is confirmed information about the kind of effects likely to be witnessed, *i.e.* synergistic, additive or non-additive. Some workers (Frazer *et al.*, 1981; Murdoch, 1990) have observed additive effects, while reports on non-additive effects are also on records (Rosenheim *et al.*, 1995).

Most of the reports obtained from numerous cage and field studies point towards non-additive results. The efficacy of combined efforts was found to be less than the sum of individual efficiencies. Intraguild predators were found to avoid immature stages of fellow members, though parasitized extra guild prey were readily consumed (Prakasan and Bhat, 1985; Gutierrez *et al.*, 1988; Samways and Wilson, 1988; Colfer and Rosenheim, 1995; Ferguson and Stiling, 1996; Taylor *et al.*, 1998; Takizawa *et al.*, 2000), thus reducing the overall efficacy of biocontrol. The reported non-additive effect may also be a direct result of the reduced oviposition in members of the guild (parasitoids or predators) due to the interference factor. This may be a consequence of the emigration from the patch because of the presence of ladybeetles (Ferguson and Stilling, 1996; Taylor *et al.*, 1998) or as a result of the presence of oviposition deterring allomones (Ruzicka, 1996, 1997a, 1997b; Ruzicka and Havelka, 1998). Studies by Evans (1991) revealed that overlapping oviposition of a ladybird causes more risk of interspecific predation than intraspecific in an unstable aphid colony. Synergistic effect in biocontrol of pest populations has also been reported. This synergistic effect occurs when the foraging activity of the predatory stages of ladybeetle causes alteration in the prey habitats and thus makes it more susceptible to another predatory species that inhabits its new habitat (Losey and Denno, 1998).

The disappearance or reduction in the number of the native ladybeetles on introduction of an exotic ladybird (Subramaniam 1953; Watt, 1986; Calunga and Gage, 1999) is probably a result of intraguild predation and because of the introduced ladybeetle possibly being a *top predator.* Top predators, in this context, are those ladybeetles that have been reported to feed not merely on aphids or coccids but are known to be voracious feeders with a good developmental rate on immature stages of other ladybeetles. *H. axyridis* is a top predator with an excellent development rate on immature stages of *C. septempunctata* (Takahashi, 1989; Yasuda and Sato, 1997; Yasuda *et al.*, 1999, 2001). *H. axyridis* also develop better on *C. maculata* than vice-versa (Cottrell and Yeargan, 1998). This ladybird was also found to survive more number of attacks of the native ladybird, *C. transversoguttata* than vice versa (Snyder *et al.*, 2004).

Ladybirds also interact actively with predators other than the ladybeetles such as chrysopids, erigonid spiders, syrphids, carabids and parasitoids, especially in the aphidophagous guilds. Amongst these interactions, ladybirds are usually the victims when interacting with chrysopids (Dinter, 1998; Lucas *et al.*, 1997, 1998). In the interaction between larvae of *C. sexmaculata* and the lacewing *C. rufilabris*, the former were more susceptible to predation by the latter. The first and second instars of *C. sexmaculata* were more consistently captured than the third and fourth ones. Predation by lacewing was largely unaffected by the dropping attempts of coccinellid since they held their prey in the air or fell with them to the ground where the attack was continued (Lucas *et al.*, 1998). Interspecific predation of *C. carnea* and ladybeetle, *Hippodamia variegata* was investigated in the absence and presence of prey, *A. gossypii*. In both the cases, *i.e.* the presence or absence of prey, the larvae and adults of lacewings emerged superior to those of ladybeetles. The lacewing larvae fed on different life stages of ladybeetles (eggs, larvae, pupae and adults) while the larvae and adults of ladybeetle only fed on the eggs and larvae of lacewing. Presence of prey reduced IGP significantly (Hatami and Zibai, 2000). The syrphids are poor intraguild predators of coccinellids with the latter usually overpowering them (Hindayana *et al.*, 2001; Agarwala and Yasuda, 2001). The wax coating of the ladybirds belonging to *Scymnus* sp. hinders predation by syrphids (Agarwala and Yasuda, 2001). Snyder and Ives (2003) recorded that the generalist predators, *i.e.* carabid beetles, coccinellids and spiders, disrupt biocontrol by a specialist parasitoid of pea aphid, *A. pisum*.

Intraguild predation is the latest concept to take the world of biocontrol scientists by storm. This is a result of its great applicability in the field of biocontrol. The indigenous, local populations of predators in the affected area and their interactions with exotic species are of major consequence to the decision making process regarding the introduction of a predator complex to regulate the prey populations. Brodeur and Rosenheim (2000) observed that intraguild predation was usually unidirectional with the generalist predators feeding upon specialists with the reverse being rarely found. Owing to this behaviour it is possible that ladybeetles may disrupt biocontrol programmes involving parasitoids. Haphazard introductions may also lead to the impeachment of the indigenous species. Therefore, at least a simple additive effect, if not the desired synergistic effect should be obtained from the designed predator complex. Only then will an effective multi-predator biocontrol strategy be formulated.

Conclusion

The above account shows that the interactions prevalent in a complex of prey and its natural enemy, to a large extent, determine the efficacy of biocontrol displayed in a microhabitat. The ladybirds interact with both prey and co-existing predators. These interactions are not simple and the ladybeetles undertake decision-making at many levels, which involve a thorough analysis of all those aspects that will ensure survival and furthering of generations of beneficial species.

Despite a known broad range of prey, the ladybeetles are usually selective with regards to their prey types. Even the generalist feeders limit their preference to a few, which can help sustain a good developmental rate. The prey suitability is determined by the relative ease or difficulty in its capture, by its toxic contents and by the contents and architecture of the plants that they infest. However, this last aspect of plant architecture and contents has, to quite an extent, been ignored.

The orientation towards the habitat involves the instinctive association with a microhabitat other than the chemical cues emanating from the infested plant. The association with the microhabitats is an aspect that needs to be further looked into. The prey patch selection is dependent on its accessibility and the number of encounters within it, since there is almost a constant duration of searching time in the patches of varying densities. The distinctive cues in the form of alkanes, alkaloids and cuticular wax patterns help in the recognition of the prey in the encountered patch. If the predator is not starved, it will consume the prey that meets its specific requirements of suitability. The prey patch is then subsequently exploited through an intensive search, if the prey meets the standards of the predator's biology. The switch over to an intensive search from an existing extensive one need not be a compulsion.

The prey intake patterns of the predators have a wide range of variability and are adapted to the prey and the environments they inhabit. The ability of the predator to increase its prey consumption with an increase in prey as well as predator densities, albeit curvilinearly is an asset to the biocontrol programme. It can be used to calculate the most effective numbers of predators to be released per acreage, keeping in mind, the prevailing prey density.

Other than the prey-predator interactions, the interactions among the con-specifics and hetero-specifics too influence the efficacy of the biocontrol programme. The occurrence of cannibalism in the absence of the prey points towards the need to survive. However, its limited presence in a microenvironment, even when there is an abundance of prey, points towards its evolutionary benefits in terms of survival of only the fittest. Intraguild predation in a habitat is yet another indication of the fight for supremacy and survival. The smaller ladybeetles and their larvae stand the danger of consumption by their larger counterparts, thus they tend to invest more in defence systems, with the intention of deterring their potential predators by poisoning them.

Any attempted introduction of biocontrol agents into an infested microhabitat has to take into consideration not merely the specificity of the predator but also its preferred habitat, its foraging efficiency on the particular infested plant, the prey density, the natural guild and the interactions of the predator with it. It is only after the detailed in-depth study of these aspects that the most effective predator or predator complex can be determined.

Acknowledgements

The authors are thankful to Professor J.J. Obrycki, Department of Entomology, Iowa State University, Iowa, USA for certain valuable suggestions and to Professor K.C. Pandey, Head, Department of Zoology, University of Lucknow, Lucknow–226 007 (India) for encouragements.

References

Afroze, S. 1999. Preferential predation by *Anegleis cardoni* (Weise) (Coccinellidae: Coleoptera) on some aphid species. *J. Aphidol.*, 13: 95-97.

Afroze, S. and Shujauddin. 1998. Bioecology of *Synia melanaria* Mulsant (Coleoptera: Coccinellidae), predating on *Coptosoma ostensum* Distant. *J. Entomol. Res.*, 22: 329–336.

Agarwala, B.K. and Bardhanroy, P. 1997. Oviposition behaviour and reproduction efficiency in ladybeetles (Coccinellidae: Coleoptera. A case study of *Menochilus sexmaculatus* (Fabr.). *J. Aphidol.*, 11: 49–59.

Agarwala, B.K. and Bardhanroy, P. 1999. Numerical response of ladybird beetles (Coccinellidae: Coleoptera) to aphid prey (Homoptera: Aphididae) in a field bean in northeast India. *J. Appl. Entomol.*, 123: 401–405.

Agarwala, B.K., Bhattacharya, S. and Bardhanroy, P. 1998. Who eats whose eggs ? Intraversus inter-specific interactions in starving ladybird beetles predaceous on aphids. *Ethol. Ecol. Evol.*, 10: 361–368.

Agarwala, B.K. and Dixon, A.F.G. 1991. Cannibalism and interspecific predation in ladybirds. In: *Behaviour and Impact of Aphidophaga*, (Eds.) Polgar, L., Chambers, R.J., Dixon, A.F.G. and Hodek, I. SPB Academic Publishing, The Hague, The Netherlands, pp. 95–102.

Agarwala, B.K. and Dixon, A.F.G. 1992. Laboratory study of cannibalism and interspecific predation of ladybeetles. *Ecol. Entomol.*, 17: 303–309.

Agarwala, B.K. and Dixon, A.F.G. 1993. Why do ladybeetles lay eggs in cluster? *Fun. Ecol.*, 7: 541–548.

Agarwala, B.K. and Yasuda, H. 2001. Overlapping oviposition and chemical defense of eggs in co-occurring species of ladybird predators of aphids. *J. Ethol.*, 19: 47–53.

Akhtaruzzaman, M. and Ahmad, M. 1998. Handling time, interference constant, quest constant and area of discovery of coccinellid predators. *Thai J. Agr. Sci.*, 31(3): 342–351.

Alexander, R.D. 1974. The evolution of social behaviour. *Ann. Rev. Ecol. Syst.*, 5: 325–383.

Arakaki, N. 1992. Feeding behaviour of the ladybeetle, *Pseudoscymnus kurohime. J. Ethol.*, 10: 7–13.

Babu, A. and Ananthakrishnan, T.N. 1993. Predatory efficiency, reproductive potential and bioenergetics of *Coccinella transversalis* F. and *Menochilus sexmaculatus* F. (Coleoptera: Coccinellidae) in relation to aphid (Homoptera: Aphididae) prey quality. *Phytophaga*, 5(2): 121–133.

Banks, C.J. 1957. The behaviour of individual coccinellid larvae on plants. *Britain J. Anim. Beh.*, 5: 12–24.

Beg, A.M.A., Hashmi, A.A. and Ashraf, M. 1987. *Coccinella septempunctata* as a component in pest management of *Brevicoryne brassicae. Int. J. Pest Con.*, 29(1): 17–18.

Bell, W.J. 1990. Searching behaviour patterns in insects. *Ann. Rev. Entomol.*, 35: 447–467.

Benrey, B. and Lamp, W.O. 1994. Biological control in the management of planthopper populations. In: *Planthoppers: Their Ecology and Management*, (Eds.) R.F. Denno and T.J. Perfect. Chapman and Hall, New York, pp. 519–550.

Bind, R.B. 1998. Bioecology and behaviour of a ladybird beetle, *Cheilomenes (=Menochilus) sexmaculata (Fabricius) (Coleoptera: Coccinellidae). Ph.D. Thesis*, University of Lucknow, Lucknow, pp. 162.

Blackman, R.L. 1965. Studies on specificity in Coccinellidae. *Annals of Applied Biology*, 56: 336–338.

Blackman, R.L. 1967. The effects of different aphid foods on *Adalia bipunctata* L. and *Coccinella 7-punctata* L. *Ann. Appl. Biol.*, 59: 207–219.

Bond, A.B. 1980. Optimal foraging in a uniform habitat: The search mechanism of the green lacewing. *Anim. Beh.*, 28: 10–15.

Booth, R.G. 1997. A review on the species of *Calvia* (Coleoptera: Coccinellidae) from the Indian sub-continent, with descriptions of two new species. *J. Nat. Hist.*, 31: 917–934.

Bristow, C.M. 1988. What makes a predator specialize? *Trends Ecol. Evol.*, 3: 1–2.

Brodeur, J. and Rosenheim, J.A. 2000. Intraguild interactions in aphid parasitoids. *Ent. Exp. Appl.*, 97: 93–108.

Buntin, L.A. and Tamaki, G. 1980. Bionomics of *Scymnus marginicollis* (Coleoptera: Coccinellidae). *Can. Entomol.*, 112: 675–680.

Burgio, G., Santi, F. and Maini, S. 2002. On intra-guild predation and cannibalism in *Harmonia axyridis* (Pallas) and *Adalia bipunctata* L. (Coleoptera: Coccinellidae). *Biol. Cont.*, 24: 110–116.

Butt, F.H. 1951. Feeding habits and mechanisms of the Mechanism bean beetle. *Cornell Univ. Agr. Exp. Sta., Mem.*, 306: 32.

Calunga, G.M. and Gage, S.H. 1999. Arrival, establishment, and habitat use of the multicoloured Asian ladybeetle (Coleoptera: Coccinellidae) in a Michigan landscape. *Environ. Entomol.*, 27: 1574–1580.

Carter, M.C. and Dixon, A.F.G. 1982. Habitat quality and foraging behaviour of coccinellids. *J. Anim. Ecol.*, 51: 865–878.

Carter, M.C. and Dixon, A.F.G. 1984. Honeydew an arrestant stimulus for coccinellids. *Ecol. Entomol.*, 9: 383–387.

Carter, M.C., Sutherland, D. and Dixon, A.F.G. 1984. Plant structure and the searching efficiency of coccinellid larvae. *Oecologia*, 63: 394–397.

Chakrabarti, S., Debnath, N. and Ghosh, D. 1995. Bioecology of *Harmonia eucharis* (Mulsant) (Coleoptera: Coccinellidae). An aphidophagous predator in Western Himalayas. *Entomon.*, 20: 191–196.

Chang, G.C. 1996. Comparison of single versus multiple species of generalist predators for biological control. *Environ. Entomol.*, 25: 207–212.

Charnov, E.L. 1976. Optimal foraging: the marginal value theorem. *Theor. Pop. Biol.*, 9: 126–136.

Chazeau, J., Bouye, E. and Bonnet De Larbogne, L. 1991. Cycle de développment et table de vie d' *Olla v-nigrum* (Col.: Coccinellidae) ennemi naturel'd *Heteropsylla cubana* (Hom.: Psyllidae) introduit en Nouvelle Calédonie. *Entomophaga*, 36: 275–285.

Clark, T.L. and Messina, F.J. 1998. Plant architecture and foraging success of ladybird beetles attacking the Russian wheat aphid. *Ent. Exp. Appl.*, 86: 153–161.

Colfer, R.G. and Rosenheim, J.A. 1995. Intraguild predation by coccinellid beetles on an aphid parasitoid, *Lysiphelbus testaceipes*. In: *Proc. Beltwide Cot. Conf.*, 2: 1033–1036.

Cook, R.M. and Cockrell, B.J. 1978. Predator ingestion rate and its bearing on feeding time and the theory of optimal diets. *J. Anim. Ecol.*, 47: 529–549.

Coppel, H.C and Mertins, J.W. 1977. *Biological Insect Pest Suppression*. Springer-Verlag, Berlin Heidelberg, New York, pp. 314.

Cottrell, T.E. and Yeargan, K.V. 1998. Intraguild predation between an introduced ladybeetle, *Harmonia axyridis* (Coleoptera: Coccinellidae), and a native ladybeetle, *Coleomegilla maculata* (Coleoptera: Coccinellidae). *J. Kansas Ent. Soc.*, 71(2): 159–163.

da Silva, P.G., Hagen, K.S. and Gutierrez, A.P. 1992. Functional response of *Curinus coerulus* (Col: Coccinellidae) to *Heteropsylla cubana* (Hom: Psyllidae) on artificial and natural substrates *Entomophaga*, 37: 555–564.

Delucchi, V. 1954. *Pullus impexus* (Muls.) (Coleoptera: Coccinellidae), a predator of *Adelges picea* (Ratz.) (Hemiptera: Adelgidae) with notes on its parasites. *Bull. Ent. Res.*, 45: 243–278.

Dimetry, N.Z. 1974. The consequences of egg cannibalism in *Adalia bipunctata* (Coleoptera: Coccinellidae). *Entomophaga*, 19: 445–451.

Dinter, A. 1998. Interactions between spider and lacewing predators (Araneae: Erigonidae) and Neuroptera: Chrysopidae) and their effects on the grain aphid *Sitobion avenae* Fab. (Homoptera: Aphididae). *Bulletin OILB/SROP*, 21(8): 91–101.

Dixon, A.F.G. 1958. The escape responses shown by certain aphids to the presence of the coccinellid *Adalia bipunctata* (L). *Experientia*, 32: 181–182.

Dixon, A.F.G. 1959. An experimental study in the searching behaviour of the predatory coccinellid beetle *Adalia decempunctata* (L). *J. Anim. Ecol.*, 28: 259–281.

Dixon, A.F.G. 1997. Patch quality and fitness in predatory ladybeetles. *Ecol. Stud.*, 130: 205–223.

Dixon, A.F.G. 2000. *Insect Predator-Prey Dynamics: Ladybird Beetles and Biological Control.* Cambridge University Press, pp. 257.

Dixon, A.F.G. and Guo, Y. 1993. Egg and cluster size in ladybird beetles (Coleoptera: Coccinellidae): The direct and indirect effects of aphid abundance. *Eur. J. Entomol.*, 90(4): 457–463.

Dixon, A.F.G., Martin–Smith, M. and Subramanian, G. 1965. Constituents of *Megoura viciae* Buckton. *J. Chem. Soc.*, 1562–1564.

Dong, Q. and Polis, G.A. 1992. The dynamics of cannibalistic populations: a foraging perspective. In: *Cannibalism Ecology and Evolution among Diverse Taxa,* (Eds.) M.A. Elgar and B.J. Crespi, Oxford University Press, Oxford, pp. 13–37.

Doumbia, M., Hemptinne, J.L. and Dixon, A.F.G. 1998. Assessment of patch quality by ladybeetles: role of larval tracks. *Oecologia,* 113: 197–202.

Ehler, L.E. 1990. Introduction strategies in biological control of insects. In: *Critical Issues in Biological Control,* (Eds.) M. Mackauer, L.E. Ehler and J. Roland. Intercept, Andover, pp. 111–134.

Elgar, M.A. and Crespi, B.J. 1992. Ecology and evolution of cannibalism. In: *Cannibalism Ecology and Evolution among Diverse Taxa,* (Eds.) M.A. Elgar and B.J. Crespi. Oxford University Press, Oxford, pp 1–12.

Elliot, H.J. and deLittle, D.W. 1980. Laboratory studies on predation of *Chrysophtharta bimaculata* (Olivier) (Coleoptera: Chrysomelidae) eggs by the coccinellids *Cleobora mellyi* (Mulsant) and *Harmonia conformis* (Boisduval). *Gen. Appl. Entomol.*, 12: 33–36.

Emrich, B.H. 1991. Erworene toxizitat bei der lupinenblattlaus *Macrosiphum albifrons* und ihr einfluss auf die aphidophagen Pradatoren *Coccinella septempunctata, Episyrphus balteatus* und *Chrysoperla carnea. Zeit. Pflanzen. Pflanzen.*, 98: 398–404.

Evans, E.W. 1991. Intra versus interspecific interactions of ladybeetles (Coleoptera: Coccinellidae) attacking aphids. *Oecologia,* 87: 401–408.

Ferguson, K.I. and Stiling, P. 1996. Non-additive effects of multiple natural enemies on aphid populations. *Oecologia*, 108: 375–379.

Ferran, A. and Deconchat, M. 1992. Exploration of wheat leaves by *Coccinella septempunctata* L. (Coleoptera: Coccinellidae) larvae. *J. Insect Beh.*, 5: 147–159.

Ferran, A. and Dixon, A.F.G. 1993. Foraging behaviour of ladybird larvae (Coleoptera: Coccinellidae). *Eur. J. Entomol.*, 90(4): 383–402.

Ferran, A., Ettifouri, M., Clement, P. and Bell, W.J. 1994. Sources of variability in the transition from extensive to intensive search in coccinellid predators (Coleoptera: Coccinellidae). *J. Insect Beh.*, 7:633–647.

Frazer, B.D. and McGregor, R.R. 1994. Searching behaviour of adult female Coccinellidae (Coleoptera) on stem and leaf models. *Can. Entomol.*, 126: 389–399.

Frazer, B.D. and Raworth, D.A. 1985. Sampling of adult coccinellids and their numerical response to strawberry aphids (Coleoptera: Coccinellidae, Homoptera: Aphididae). *Can. Entomol.*, 117: 153–161.

Frazer, B.D., Gilbert, N., Ives, P.M. and Raworth, D.A. 1981. Predator reproduction and overall predator-prey relationship. *Can. Entomol.*, 113: 1015–1024.

Gagne I., Coderre, D. and Mauffette, Y. 2002. Egg cannibalism by *Coleomegilla maculata lengi* neonates: preference even in the presence of essential prey. *Ecol. Entomol.*, 27: 285–291.

Geitzenauer, H.L. and Bernays, E.A. 1996. Plant effects on prey choice by a vespid wasp, *Polistes arizonensis*. *Ecol. Entomol.*, 21: 227–234.

Ghorpade, K.D. 1979. *Ballia eucharis* (Coleoptera: Coccinellidae) breeding on Cicadellidae (Homoptera) at Shillong. *Curr. Res.*, 8: 113.

Gibson, R.L., Elliott, N.C. and Schaefer, P. 1992. Life history and development of *Scymnus frontalis* (Fabricius) (Coleoptera: Coccinellidae) on four species of aphid. *J. Kansas Ent. Soc.*, 65(4): 410–415.

Gilliam, J. 1990. Hunting by the hunted: optimal prey selection by foragers under predation hazard. In: *Behavioural Mechanisms of Food Selection*, (Ed.) R. Hughes. NATO ASI Series, Series A: Life Sciences, Springer-Verlag, New York, pp. 797–820.

Gilliam, J. and Fraser, D. 1987. Habitat selection under predation hazard: test of a model with foraging minnows. *Ecology*, 68: 1856–1862.

Gutierrez, A.P., Neuenschwander, P., Schulthess, F., Herren, H.R., Baumgaertner, J.U., Wermelinger, B., Lohr, B. and Ellis, C.K. 1988. Analysis of biological control of cassava pests in Africa II. Cassava mealybug, *Phenacoccus manihoti*. *J. Appl. Ecol.*, 25: 921–940.

Hagen, K.S. 1962. Biology and ecology of predaceous coccinellidae. *Ann. Rev. Entomol.*, 7: 289–326.

Hagen, K.S. 1986. Ecosystem analysis: plant cultivars (HPR), entomophagous species and food supplements. In: *Interactions of Plant Resistance and Parasitoid and Predators of Insects*, (Eds.) D.J. Boethel, R.D. Eikenbary and E. Horwood, pp. 151–197.

Hagen, K.S. 1974. The significance of predaceous Coccinellidae in biological and integrated control of insects. *Entomophaga*, 7: 25–44.

Harpaz, I. 1958. Bionomics of the 11-spotted ladybeetle, *Coccinella septempunctata* L., in a subtropical climate. *10th Int. Cong. Entomol.* Montreal, 1956, 2: 657–659.

Hassell, M.P. and Southwood, T.R.E. 1978. Foraging strategies of insects. *Ann. Rev. Ecol. Syst.*, 9: 75–98.

Hassell, M.P. 1978. *The dynamics of arthropod predator-prey system. Mono. Pop. Biol.*, 237, Princeton University Press, Princeton, NJ.

Hatami, B. and Zibai, K. 2000. Laboratory study of interspecific predation of green lacewing, *Chrysoperla carnea* (Steph.) and ladybeetle, *Hippodamia variegata* (Goeze). *J. Sci. Tech. Agric. Nat. Res.*, 4(3): 99–110.

Hattingh, V. and Samways, M.J. 1992. Prey choice and substitution in *Chilocorus* spp. (Coleoptera: Coccinellidae). *Bull. Ent. Res.*, 82: 327–334.

Hazzard, R.V. and Ferro, D.N. 1991. Feeding responses of adult *Coleomegilla maculata* (Coleoptera: Coccinellidae) to eggs of Colarado potato beetle (Coleoptera: Chrysomellidae) and green peach aphids (Homoptera: Aphididae. *Environ. Entomol.*, 20(2): 644–651.

Heidari, M. and Copland, M.J.W. 1992. Host finding by *Cryptolaemus montrouzieri* (Coleoptera: Coccinellidae), a predator of mealy bugs (Homoptera: Psuedococcidae). *Entomophaga*, 37: 621–625.

Heinz, K.M. and Parrella, M.P. 1994. Poinsettia (*Euphorbia pulcherrma* Wild.ex Koltz) cultivar-mediated differences in performance of five natural enemies of *Bemisia argentifolii* Bellows and Perring, n.sp. (Homoptera: Alyerodidae. *Biol. Cont.*, 4: 305–308.

Heinz, K.M. and Zalom, F.G. 1996. Performance of the predator *Delphastus pusillus* on Bemisia resistant and susceptible tomato lines. *Ent. Exp. Appl.*, 81: 345–352.

Hemptinne, J.L., Dixon, A.F.G. and Coffin, J. 1992. Attack strategy of ladybird beetles (Coccinellidae): factors shaping their numerical response. *Oecologia*, 90: 238–245.

Hemptinne, J.L., Dixon, A.F.G., Doucet, J.L. and Petersen, J.E. 1993. Optimal foraging by hoverflies (Diptera: Syrphidae) and ladybeetles (Coleoptera: Coccinellidae): Mechanisms. *Eur. J. Ent.*, 90: 451–455.

Hemptinne, J.L., Dixon, A.F.G. and Gauthier, C. 2000b. Nutritive cost of intraguild predation on eggs of *Coccinella septempunctata* and *Adalia bipunctata* (Coleoptera: Coccinellidae). *Eur. J. Entomol.*, 97(4): 559–562.

Hemptinne, J.L., Dixon, A.F.G. and Lognay, G. 1996. Searching behaviour and mate recognition by males of the two-spot ladybird beetle, *Adalia bipunctata*. *Ecol. Entomol.*, 21: 165–170.

Hemptinne, J.L., Doumbia, M. and Dixon, A.F.G. 1999. Assessment of patch quality by ladybeetles: Role of aphid and plant phenology. *J. Insect Beh.*, 13(3): 353–359.

Hemptinne, J.L., Gaudin, M., Dixon, A.F.G. and Lognay, J. 2000a. Social feeding in ladybird beetles: adaptive significance and mechanism. *Chemoecology*, 10: 149–152.

Hemptinne, J.L., Lognay, G., Doumbia, M. and Dixon, A.F.G. 2001. Chemical nature and persistence of the oviposition deterring pheromone in the tracks of the larvae of the two spot ladybird, *Adalia bipunctata* (Coleoptera: Coccinellidae). *Chemoecology*, 11: 43–47.

Hindayana, D., Meyhofer, R., Scholz, D. and Poehling, H.M. 2001. Intraguild predation among the hoverfly *Episyrphus balteatus* deGeer (Diptera: Syrphidae) and other aphidophagous predators. *Biol. Cont.*, 20(3): 236–246.

Hippa, H., Koponen, S. and Roine, R. 1982. Feeding preference of *Coccinella hieroglyphica* (Col.: Coccinellidae) for eggs of three chrysomelid beetles. *Rep. Kevo Subarctic Res. Stn.*, 18: 1–4.

Hodek, I. 1956. The influence of *Aphis sambuci* L. as prey of the ladybird beetle *Coccinella septempunctata* L. *Vest. Cs. Spol. Zool.*, 20: 62–74.

Hodek, I. 1959. Ecology of Aphidophagous Coccinellidae. *Int. Conf. Insect Path. Biol. Cont.*, Praha, 1958, 543–547.

Hodek, I. 1962. Essential and alternative food in insects. *11th Int. Cong. Entomol., Vienna*, 1960, 2: 696–697.

Hodek, I. 1967. Bionomics and ecology of predaceous Coccinellidae. *Ann. Rev. Entomol.*, 12: 79–104.

Hodek, I. 1973. *Biology of Coccinellidae.* Academia Prague, Dr. W. Junk. N.V., The Hague, pp. 259.

Hodek, I., Chakrabarti, S. and Rejmanek, M. 1989. The effect of prey density on food intake by an adult *Cheilomenes sulphurea* (Coleoptera: Coccinellidae). *Entomophaga*, 29(2): 179–184.

Hodek, I. and Honek, A. 1996. *Ecology of Coccinellidae.* Kluwer Academic Publishers, Dordrecht, Boston, London, pp. 464.

Holling, C.S. 1959. Some characteristics of simple types of predation and parasitism. *Can. Ent.*, 91: 385–398.

Holling, C.S. 1965. Functional response of predators to prey density and its role in mimicry and population regulation. *Mem. Entomol. Soc. Canada*, 45: 3–60.

Honek, A. 1985. Activity and predation of *Coccinella septempunctata* adults in the field (Coleoptera: Coccinellidae). *Zeit. Ang. Entomol.*, 100: 399–409.

Hukusima, S. and Kamei, M. 1970. Effects of various species of aphids as food on development, fecundity and longevity of *Harmonia axyridis* Pallas (Coleoptera: Coccinellidae). *Res. Bull. Fac. Agri. Gifu University*, 29: 53–66.

Ives, P.M. 1981. Estimation of coccinellid numbers and movements in the field. *Can. Ent.*, 113: 981–997.

Jalali, S.K. and Singh, S.P. 1989. Biotic potential of three coccinellid predators on various diaspine hosts. *J. Biol. Cont.*, 3: 20–23.

Kairo, M.T.K. and Murphy, S.T. 1995. The life history of *Rodolia iceryae* Janson (Coleoptera: Coccinellidae) and the potential for use in innoculative releases against *Icerya pattersoni* Newstead (Hom., Margarodidae) on coffe e. *J. Appl. Entomol.*, 119: 487–491.

Kajita, Y., Takano, F., Yasuda, H. and Agarwala, B.K. 2000. Effects of indigenous ladybird species (Coleoptera: Coccinellidae) on the survival of an exotic species in relation to prey abundance. *Appl. Entomol. Zool.*, 35(4): 473–479.

Kalushkov, P. 1998. Ten aphid species (Sternorrhyncha: Aphididae) as prey for *Adalia bipunctata* (Coleoptera: Coccinellidae). *Eur. J. Entomol.*, 95(3): 343–349.

Kalushkov, P. 1999. The effect of aphid prey quality on searching behaviour of *Adalia bipunctata* (Coleoptera: Coccinellidae). *Ent. Exp. Appl.*, 92: 277–282.

Kalushkov, P. and Hodek, I. 2001. New essential aphid prey for *Anatis ocellata* and *Calvia quatuordecimguttata* (Coleoptera: Coccinellidae). *Biocontrol Science and Technology*, 11: 35–39.

Kalushkov, P. and Hodek, I. 2004. The effects of thirteen species of aphids on some life history parameters of the ladybird *Coccinella septempunctata*. *Biocontrol*, 49: 21–32.

Kan, E. 1988a. Assessment of aphid colonies by hoverflies, I. Maple aphids and *Episyrphus balteatus* (De Geer) (Diptera: Syrphidae). *J. Ethol.*, 6: 39–48.

Kan, E. 1988b. Assessment of aphid colonies by hoverflies, II. Pea aphids and 3 syrphid species: *Betasyrphus serarius* (Wiedeman), *Metasyrphus frequens* Matsumara and *Syrphus vitripennis* (Meigen). *J. Ethol.*, 6: 135–142.

Kan, E. and Sasakawa, M. 1986. Assessment of the maple aphid colony by the hoverfly, *Episyrphus balteatus* (De Geer) (Diptera: Syrphidae). *J. Ethol.*, 4: 121–127.

Kareiva, P. and Perry, R. 1989. Leaf overlap and the ability of ladybird beetles to search among plants. *Ecol. Entomol.*, 21: 165–170.

Kesten, U. 1969. Zur morphologie und biologie von *Anatis ocellata* (L.) (Coleoptera: Coccinellidae). *Zeit. Ang. Entomol.*, 63: 412–455.

Kindlmann, P. and Dixon, A.F.G. 1993. Optimal foraging in ladybird beetles (Coleoptera: Coccinellidae) and its consequences for their use in biological control. *Eur. J. Entomol.*, 90(4): 443–450.

Koch, R.L. 2003. The multicolored Asian ladybeetle, *Harmonia axyridis:* A review of its biology, uses in biological control, and non-target impacts. *J. Insect Sci.*, 3: 32, pp.16.

Koide, T. 1962. Observations on the feeding habit of the larva of *Coccinella septempunctata brucki* Mulsant. The feeding behaviour and number of prey fed under different temperatures. *Kontyu.*, 30: 236–241.

Kosaki, A. and Yamaoka, R. 1996. Chemical composition of footprints and cuticular lipids of three species of ladybeetles. *Japanese J. Appl. Entomol. Zool.*, 40(1): 47–53.

Lapiz, R.V. and Gopud, V.P. 1995. Life history and functional response of *Acarinus philippinensis* Kapur (Coleoptera: Coccinellidae) on *Aphis gossypii* Glover (Hemiptera: Aphididae). *Asia Life Science (Philippines)* Special Issue (Jan–June), 4(1): 45–62.

Losey, J.E. and Denno, R.F. 1998. Positive predator-predator interactions: enhanced predation rates and synergistic suppression of aphid populations. *Ecology*, 79: 2143–2152.

Lounsbury, C.P. 1940. The pioneer period of economic entomology in South Africa. *J. Entomol. Soc., South Africa*, 3: 9–29.

Lucas, E., Coderre, D. and Brodeur, J. 1997. Instar specific defence of *Coleomegilla maculata lengi* (Col.: Coccinellidae): influence on attack success of the Intraguild predator *Chrysoperla rufilabris* (Neur.: Chrysopidae). *Entomophaga*, 42: 3–12.

Lucas, E., Coderre, D. and Brodeur, J. 1998. Intraguild predation among aphid predators: characterization and influence of extraguild prey density. *Ecology*, 79: 1084–1092.

Majerus, M.E.N. 1994. *Ladybirds.* Harper Collins, London, pp. 367.

Majerus, M. and Kearns, P. 1989. *Ladybeetles.* Richmond Publishing Co. Ltd. England, pp. 103.

Malcolm, S.B. 1992. Prey defence and predator foraging. In: *Natural Enemies: The Population Biology of Predators, Parasites and Diseases* (Ed) M.J. Crawley. Blackwell Scientific Publication, Oxford, pp. 458–475.

Marks, R.J. 1977. Laboratory studies of plant searching by *Coccinella septempunctata* L. larvae. *Bull. Ent. Res.*, 67: 235–241.

McClure, M.S. 1987. Potential of Asian predators, *Harmonia axyridis* Pallas (Coleoptera: Coccinellidae), to control *Matsucoccus resinosea* Bean and Godwin (Homoptera: Margarodidae) in the United States. *Environ. Entomol.*, 16: 224–230.

Merlin, J., Lemaitre, O. and Gregoire, J.C. 1996a. Oviposition in *Cryptolaemus montrouzieri* stimulated by wax filaments of its prey. *Ent. Exp. Appl.* 79: 141–146.

Merlin, J., Lemaitre, O. and Gregoire, J.C. 1996b. Chemical cues produced by con-specific larvae deter oviposition by the coccidophagous ladybird beetle, *Cryptolaemus montrouzieri. Ent. Exp. Appl.*, 79: 147–151.

Messina, F.J. and Hanks, J.B. 1998. Host plant alters the shape of the functional response of an aphid predator (Coleoptera: Coccinellidae). *Environ. Entomol.*, 27: 1196–1202.

Michaud, J.P. 2003. A comparative study of larval cannibalism in three species of ladybird. *Ecological Entomology* 28: 92–101.

Michels, G.J.Jr. and Behle, R.W. 1991. A comparison of *Coccinella septempunctata* and *Hippodamia convergens* larval development on greenbugs at constant temperatures. *Southwestern Entomologist*, 16: 73–80.

Mills, N.J. 1982. Satiation and the functional response: A test of a new method. *Ecol. Entomol.*, 7(3): 305–316.

Mills, N.J. 1988. Voracity, cannibalism and coccinellid predation. *Ann. Appl. Zool.*, 101: 144–148.

Mora, J.G., Gapud, V.P. and Velasco, L.R.I. 1995. Life history and voracity of *Coelophora inaequalis* (Fabricius) (Coleoptera: Coccinellidae) on *Aphis craccivora* Koch. (Hemiptera: Aphididae. *Phil. Entomol.*, 9(5): 523–553.

Muggleton, J. 1978. Selection against the melanic morphs of *Adalia bipunctata* (two spot ladybird): A review of some data. *Heredity*, 40(2): 269–280.

Murakami, Y. and Tsubaki, Y. 1984. Searching efficiency of the ladybeetle *Coccinella septempunctata* larvae in uniform and patchy environments. *J. Ethol.*, 2: 1–6.

Murdoch, W.W. 1990. The relevance of pest-enemy models to biological control. In: *Critical Issues in Biological Control*, (Eds.) M. Mackauer, L.E. Ehler and J. Rowland. Intercept Ltd., Andover, pp. 1–24.

Nakamuta, K. 1984. Visual orientation of a ladybeetle, *Coccinella septempunctata* L. (Coleoptera: Coccinellidae), towards its prey. *Appl. Entomol. Zool.*, 19: 82–86.

Nakamuta, K. 1985. Mechanisms of the switch over from extensive to area concentrated search behaviour of the ladybird beetle, *Coccinella septempunctata brucki. J. Insect Phy.*, 31: 849–856.

Nakamuta, K. 1987. Diel rhythmicity of prey-search activity and its predominance over starvation in the ladybeetle, *Coccinella septempunctata brucki* Mulsant (Coleoptera: Coccinellidae): releasing stimuli and decision of giving-up time. *Japanese J. Appl. Entomol. Zool.*, 29: 55–60.

Nakamuta, K. and Saito, T. 1985. Recognition of aphid prey by the ladybeetle, *Coccinella septempunctata brucki* Mulsant (Coleoptera: Coccinellidae). *Appl. Entomol. Zool.*, 20(4): 479–483.

Nishida, R. and Fukami, H. 1989. Host plant iridoid-based chemical defense of an aphid, *Acyrthosiphon nipponicus*, against ladybeetles. *J. Chem. Ecol.*, 15: 1837–1845.

Obatake, H. and Suzuki, H. 1985. On the isolation and identification of canavanine and ethanolamine contained in the young leaves of black locus, *Robinia pseudoaccacia*, lethal for the ladybeetle, *Harmonia axyridis. Tech. Bull. Fac. Agri. Kagawa Univ.*, 36: 107–115.

Obrycki, J.J. and Orr, C.J. 1990. Suitability of three prey species for Nearctic populations of *Coccinella septempunctata*, *Hippodamia variegata* and *Propylea quatuordecimpunctata* (Coleoptera: Coccinellidae). *J. Econ. Entomol.*, 83: 1292–1297.

Obrycki, J.J. and Kring, T.T. 1998. Predaceous Coccinellidae in biological control. *Ann. Rev. Entomol.*, 43: 295–321.

Ofuya, T.I. and Akingbohungbe, A.E. 1988. Functional and numerical responses of *Cheilomenes lunata* (Fabricius) (Coleoptera: Coccinellidae) feeding on the cowpea aphid, *Aphis craccivora* Koch (Homoptera: Aphididae). *Insect Sci. Applic.*, 9: 543–546.

Omkar and Bind, R.B. 1993. Records of aphids-natural enemies complex of Uttar Pradesh, II. The Coccinellids. *J. Adv. Zool.*, 14(2): 96–99.

Omkar and Bind, R.B. 1995. Records of aphids-natural enemies complex of Uttar Pradesh, IV. The Coccinellids. *J. Adv. Zool.*, 16(2): 67–71.

Omkar and Bind, R.B. 1996. Records of aphids-natural enemies complex of Uttar Pradesh, V. The Coccinellids. *J. Adv. Zool.*, 17(1): 44–48.

Omkar and Bind, R.B. 1998. Prey preference of a ladybeetle, *Cheilomenes (=Menochilus} sexmaculata* (Fabricius) (Coleoptera : Coccinellidae). *J. Aphidol.*, Gorakhpur, 12(1&2): 63–66.

Omkar and Bind, R.B. 2004. Prey quality dependent growth, development and reproduction of a biocontrol agent, *Cheilomenes sexmaculata* (Fabricius) (Coleoptera: Coccinellidae). *Biocontrol Science and Technology*, 14(7): 665–673.

Omkar and James, B.E. 2001. Predation efficacy of a ladybeetle, *Coccinella transversalis* Fabricius in relation to certain aphids. *Biol. Memoirs*, 27(1): 20–26.

Omkar and James, B.E 2004. Influence of prey species on immature survival, development, predation and reproduction of *Coccinella transversalis* Fabricius (Col., Coccinellidae). *J. Appl. Ent.*, 128 (2): 150–157.

Omkar, James, B.E and Srivastava. S. 1999. Prey preference of ladybeetle, *Coccinella transversalis* Fabricius (Coleoptera: Coccinellidae). *Adv. Bios.*, 18(2): 117–122.

Omkar, Mishra, G. and Pervez, A. 2002. Intraguild Predation by Ladybeetles: An Ultimate Survival Strategy or An Aid to Advanced Aphid Biocontrol? *Prof. S.B. Singh Comm. Vol., Zool. Soc. India*, p. 77–90.

Omkar and Pervez, A. 1999. New record of coccinellids from Uttar Pradesh, I. *J. Adv. Zool.*, 20(2): 102–106.

Omkar and Pervez, A. 2000. New record of coccinellids from Uttar Pradesh, II. *J. Adv. Zool.*, 21(1): 43–47.

Omkar and Pervez, A. 2001. Prey preference of a ladybeetle, *Micraspis discolor* (Fabricius). *Entomon*, Trivandrum, 26(2): 195–197.

Omkar and Pervez, A. 2002a. Ecology of Aphidophagous Ladybird Beetle, *Coccinella septempunctata* Linn. (Coleoptera: Coccinellidae): A Review. *Journal of Aphidology*, 16(1&2): 175–201.

Omkar and Pervez, A. 2002b. New record of coccinellids from Uttar Pradesh, III. *J. Adv. Zool.*, 23(1): 63–65.

Omkar and Pervez, A. 2003. Ecology and Biocontrol Potential of a Scale-Predator, *Chilocorus nigritus*. *Biocontrol Science and Technology*, 13(4): 379–390.

Omkar and Pervez, A. 2004a. Predaceous Coccinellids in India: Predator-prey catalogue. *Oriental Insects*, 38: 27–61.

Omkar and Pervez, A. 2004b. Functional and numerical responses of *Propylea dissecta* (Mulsant) (Col., Coccinellidae). *J. Appl. Ent.*, 128(2): 140–146.

Omkar, Pervez, A. and Gupta, A.K. 2004. Role of Surface Chemicals in Egg Cannibalism and Intraguild Predation by Neonates of Two Co-occurring Aphidophagous Ladybirds, *Propylea dissecta* and *Coccinella transversalis*. *J. Appl. Ent.*, 128(9–10): 691–695.

Omkar and Srivastava, S. 2001a. Functional response of a ladybeetle, *Coccinella septempunctata* Linnaeus to three aphid species. *J. Adv. Zool.*, 22(1): 40–45.

Omkar and Srivastava, S. 2001b. Comparative predatory potential of a ladybird beetle, *Coccinella septempunctata* Linn. on six prey species. *Biol. Mem.*, 27(2): 59–63.

Omkar and Srivastava, S. 2002. Functional response of the ladybeetle, *Coccinella septempunctata* Linnaeus on *Uroleucon compositae* (Theobald) and Aphis nerii Boyer de Fonscolombe. *Prof. S.B. Singh Comm. Vol., Zool. Soc.*, 115–128.

Omkar and Srivastava, S. 2003a. Influence of six aphid prey species on development and reproduction of a ladybird beetle, *Coccinella septempunctata*. *Biocontrol*, 48(4): 379–393.

Omkar and Srivastava, S. 2003b. Predation efficacy and searching efficiency of a ladybeetle, *Coccinella septempunctata* Linnaeus. *Indian J. Exp. Biol.*, 41: 82–84.

Omkar, Srivastava, S. and James, B.E. 1997. Prey preference of a ladybeetle, *Coccinella septempunctata* Linnaeus (Coleoptera: Coccinellidae). *J. Adv. Zool.*, 18(2): 96–97.

Osawa, N. 1992. Sibling cannibalism in the ladybeetle *Harmonia axyridis:* consequence for mother and offspring. *Res. Pop. Ecol.*, 34: 45–55.

Osawa, N. 2003. The influence of female oviposition strategy on sibling cannibalism in the ladybird beetle *Harmonia axyridis* (Coleoptera: Coccinellidae). *Eur. J. Entomol.*, 100: 43–48.

Pandey, K.P., Kumar, A., Singh, R., Shanker, S. and Tripathi, C.P.M. 1984. Numerical response and area of discovery of a predator, *Coccinella septempunctata* L. *Zeit. Ang. Entomol.*, 97: 418–423.

Pervez, A. and Omkar 2003. Predation potential and handling time estimates of a generalist aphidophagous ladybird, *Propylea dissecta. Biol. Memoirs*, 29(2): 91–97.

Pervez, A. and Omkar 2004. Prey-Dependent Life Attributes of an Aphidophagous Ladybird Beetle, *Propylea dissecta* (Coleoptera: Coccinellidae). *Biocontrol Science and Technology*, UK, 14(4): 385–396.

Pervez, A. and Omkar 2005. Functional Responses of Coccinellid Predators: An Illustration of a Logistic Approach. *J. Insect Sci.*, USA (In Press).

Phoofolo, M.W. and Obrycki, J.J. 1998. Potential for intraguild predation and competition among predatory Coccinellidae and Chrysopidae. *Ent. Exp. Appl.*, 89: 47–55.

Polis, G.A. and Holt, R.D. 1992. Intraguild predation: the dynamics of complex trophic interactions. *Tre. Ecol. Evol.*, 7: 151–154.

Polis, G.A., Myers, C.A. and Holt, R.D. 1989. The ecology and evolution of intraguild predation: potential competitors that eat each other. *Ann. Rev. Ecol. Syst.*, 20: 297–330.

Ponsonby, D.J. and Copland, M.J.W. 1995. Olfactory responses by the scale insect predator *Chilocorus nigritus* (F.) (Coleoptera: Coccinellidae). *Biocont. Sci. Tech.*, 5: 83–93.

Pope, R.D. 1973. The species of *Scymnus* (SSTR), *Scymnus* (*Pullus*) and *Nephus* (Col., Coccinellidae) occurring in the British Isles. *Entomol. Mon. Mag.*, 109: 3–39.

Pope, R.D. and Lawrence, J.F. 1990. A review of *Scymnodes* Blackburn, with the description of a new Australian species and its larva (Coleoptera: Coccinellidae). *Syst. Entomol.*, 15: 241–252.

Prakasan, C.B. and Bhat, K.P. 1985. Interference of the predator, *Cryptolaemus montrouzieri* with the perfonnance of a newly introduced parasite, *Leptomastix dactylopii*. *J. Cot. Res.*, 15: 29–32.

Putman, W.L. 1955. Bionomics of *Stethorus punctillum* Weise in Ontario. *Can. Ent.*, 87: 9–33.

Radcliffe, E.B., Weires, R.W., Stucker, R.E. and Barnes, D.K. 1976. Influence of cultivers and pesticides on pea aphid, spotted alfalfa aphid, and associated arthropod taxa in a Minnesota alfalfa ecosystem. *Environ. Entomol.*, 94: 169–175.

Radwan, Z. and Löveii, G.L. 1983. Aphids as prey for the coccinellid, *Exochomus quadripustulatus*. *Ent. Exp. Appl.*, 34: 283–286.

Rana, J.S., Dixon, A.F.G. and Jarosik, V. 2002. Costs and benefits of prey specialization in a generalist insect predator. *J. Anim. Ecol.*, 71: 15–22.

Rhamhalinghan, M. 1985. Intraspecific variations in ovariole numbers/ovary in *C. septempunctata* L. (Coleoptera: Coccinellidae). *Indian Zool.*, 9: 91–97.

Rhamhalinghan, M. 1987. Seasonal variation in ovarioler output in *Coccinella septempunctata* L. (Coleoptera: Coccinellidae). In: *Proc. 5th Indian Sym. Invert. Reprod.*, (Ed.) Palanichamy, S. P.G. and Res. Dept. Zool., A.P.A. College of Arts and Culture, Pilani, India, pp. 149–157.

Ricci, C. 1979. L'apparato boccale pungente succhiante della larva di *Platynaspis luteoubra* Geze (Col., Coccinellidae. 10 Contributo alla conoscenza dei (Coleotteri Coccinellidae). *Boll. Lab. Entomol. Agr. "F. Silvestri"* 36: 179–198.

Ricci, C. 1982. Sulla costituzione e funyione delle mandibole delle larvae di *Tytthaspis sedecimpunctata* (L.) e *Tytthaspis trilineata* (Weise). [Constitution and function of the mandibles in

Tytthaspis sedecimpunctata (L.) and *Tytthaspis trilineata* (Weise) larvae.] *Frus. Entomol.*, III: 205–212.

Richards, A.M. 1981. *Rhyzobius ventralis* (Erichson) and *R. forestieri* (Mulsant) (Coleoptera: Coccinellidae), their biology and value for scale insect control. *Bull. Ent. Res.*, 71: 33–46.

Richards, A.M. 1985. Biology and defensive adaptations *Rodatus major* (Coleoptera: Coccinellidae) and its prey, *Monophebulus pilosior* (Hemiptera: Margarodidae). *J. Zool.*, 205: 287–295.

Richards, D.R. and Goletsos, C. 1991. *Feeding behaviour in Australian aphidophagous Coccinellidae. Behaviour and Impact of Aphidophaga* (Eds.) Polgar, L., Chambers, R.J., Dixon, A.F.G. and Hodek, I. SPB Academic Publishers, The Hague, pp. 227–234.

Rogers, C.E., Jackson, H.B. and Eikenbary, R.D. 1994. Responses of an imported coccinellid, *Propylea 14-punctata* to aphids associated with small grains in Oklahoma. *Environmental Entomology*, 1: 198–202.

Root, R.B. 1967. The niche exploitation patterns of the blue-gray gnatcatcher. *Ecol. Mon.*, 37: 317–350.

Rosenheim, J.A., Kaya, H.K., Ehler, L.E., Marois, J.J. and Jaffee, B.A. 1995. Intraguild predation among biological-control agents: Theory and evidence. *Biol. Cont.*, 5: 303–335.

Ruzicka, Z. 1994. Oviposition-deterring pheromone in *Chrysopa oculata* (Neuroptera: Chrysopidae). *Eur. J. Entomol.*, 91: 361–370.

Ruzicka, Z. 1996. Oviposition-deterring pheromone in chrysopids: Intra- and inter-specific effects. *Eur. J. Entomol.*, 93: 161–166.

Ruzicka, Z. 1997a. Persistence of oviposition-deterring pheromone in *Chrysopa oculata* (Neur.: Chrysopidae). *Entomophaga* 42: 109–114.

Ruzicka, Z. 1997b. Recognition of oviposition-deterring allomones by aphidophagous predators (Neuroptera: Chrysopidae, Coleoptera: Coccinellidae). *Eur. J. Entomol.*, 94: 431–434.

Ruzicka, Z. and Havelka, J. 1998. Effects of oviposition-deterring pheromone and allomones on *Aphidoletes aphidomyza* (Diptera: Cecidomyiidae). *Eur. J. Entomol.*, 95: 211–216.

Samways, M.J. and Wilson, S.J. 1988. Aspects of the feeding behaviour of *Chilocorus nigritus* (F.) (Col., Coccinellidae) relative to its effectiveness as a biocontrol agent. *J. Appl. Entomol.*, 106: 177–182.

Schaller, M. and Nentwig, W. 2000. Olfactory orientation of the seven-spot ladybird beetle, *Coccinella septempunctata* (Coleoptera: Coccinellidae): attraction of adults to plants and conspecific females. *Eur. J. Entomol.*, 97(2): 155–159.

Sengonca, C. and Fringe, B. 1985. Interference and competitive behaviour of the aphid predators, *Chrysoperla carnea* and *Coccinella septempunctata* in the laboratory. *Entomophaga*, 30: 245–251.

Sengonca, C., Kotikal, Y.K. and Schade, M. 1995. Olfactory reactions of *Cryptolaemus montrouzieri* Mulsant (Col., Coccinellidae) and *Chrysoperla carnea* Stephens (Neur., Chrysopidae) in relation to period of starvation. *Anz, Schad. PflSchutz.*, 68: 9–12.

Sengonca, C. and Liu, B. 1994. Responses of the different instar predator, *Coccinella septempunctata* L. (Coleoptera: Coccinellidae), to the kairomones produced by the prey and non-prey insects as well as the predator itself. *Zeit. PflKrankh. PflSchutz.*, 101: 173–177.

Shah, M.A. 1982. The influence of plant surfaces on the searching behaviour of coccinellid larvae. *Ent. Exp. Appl.*, 31: 377–380.

Shonouda, M.L. 1999. Aphid aqueous-extract as a source of host searching kairomones for the aphidophagous predator *Coccinella septempunctata* L. (Col., Coccinellidae). *Anze. Schad.*, 72(5): 126–128.

Sinha, T.B., Pandey, R.K., Singh, R., Tripathi, C.P.M. and Kumar, A. 1982. The functional response of *Coccinella septempunctata*, a coccinellid predator of mustard aphid, *Lipaphis erysimi*. *Entomon.*, 7(1): 7–10.

Snyder, W.E. and Ives, A.R. 2003. Interactions between specialist and generalist natural enemies: Parasitoids, predators, and pea aphid biological control. *Ecology*, 84: 91–107.

Snyder, W.E., Clevenger, G.M. and Eigenbrode, S.D. 2004. Intraguild predation and successful invasion by introduced ladybird beetles. *Oecologia*, In press.

Solomon, M.E. 1949. The natural control of animal populations. *J. Anim. Ecol.*, 18: 1–35.

Stevens, L. 1992. Cannibalism in beetles. In: *Cannibalism, Ecology and Evolution among Diverse Taxa*, (Eds.) M.A. Elgar, B.J. Crespi. Oxford University Press, Oxford, pp. 156–175.

Storch, R.H. 1976. Prey detection by fourth stage *Coccinella transversoguttata* larvae (Coleoptera: Coccinellidae). *Anim. Beh.*, 24: 690–693.

Strand, M.R. and Obrycki, J.J. 1996. Host specificity of insect parasitoids and predators. *Bioscience*, 46: 422–429.

Stubbs, M. 1980. Another look at prey detection by coccinellid. *Ecol. Entomol.*, 5: 179–182.

Subramaniam, V.K. 1953. Control of the fluted scale in peninsular India. *Indian J. Ent.*, 17: 103–120.

Takahashi, K. 1987. Differences in oviposition initiation and sites of ladybeetles, *Coccinella septempunctata bruckii* Mulsant and *Harmonia axyridis* (Pallas) (Coleoptera: Coccinellidae) in the field. *Japanese J. Appl. Ent. and Zool.*, 31: 253–254.

Takahashi, K. 1989. Intra- and inter-specific predations of ladybeetles in spring alfalfa fields. *Japanese J. Ent.*, 57: 199–203.

Takizawa, T., Yasuda, H. and Agarwala, B.K. 2000. Effects of parasitized aphids (Homoptera: Aphididae) as food on larval performance of three predatory ladybeetles (Coleoptera: Coccinellidae). *Appl. Entomol. Zool.*, 35(4): 467–472.

Taylor, A.J., Muller, C.B. and Godfray, H.C.J. 1998. Effect of aphid predators on oviposition behaviour of aphid parasitoids. *J. Insect Behav.*, 11: 297–302.

Tortorici, C. and Bell, W. 1988. Search orientation adults *Drosophila melanogaster.* Response of rovers and sitters to resource dispersion in a food patch *J. Insect Behav.*, 1: 209–233.

Trexler, J.C., McCulloch, C.E. and Travis, J. 1988. How can the functional response best be determined? *Oecologia*, 76: 206–214.

Turnbull, A.L. and Chant, D.A. 1961. The practice and theory of biological control of insects in Canada. *Can. J. Zool.*, 36: 697–753.

van der Meiracker, R.A.F., Hammond, W.N.O. and van Alphen, J.J.M. 1990. The role of kairomones in prey finding by *Diomus* and *Exochomus* sp., two coccinellid predators of the cassava melybug, *Phenococcus manihotii. Ent. Exp. Appl.*, 56: 209–217.

Veeraval, R. and Bhaskaran, P. 1997. Functional and numerical responses of *Coccinella transversalis* and *Cheilomenes sexmaculata* Fabr. Feeding on the melon aphid, *Aphis gossypii* Glov. *Insect Sci. Appl.*, 17(3/4): 335–339.

Veiera, G.F. 1995. Functional and Numerical responses of *Scymnus* (*Pullus*) *argentinicus* (Weise, 1906) (Coccinellidae: Coleoptera) at different densities of green bug, *Schizaphis graminum* (Rondani, 1852) (Homoptera: Aphididae. Lavras MG (Brazil), pp. 58.

Vohland, K. 1996. The influence of plant structure on searching behaviour in ladybird *Scymnus nigrinus* (Coleoptera: Coccinellidae). *Eur. J. Ent.*, 93: 151–160.

Wagiman, F.X. 1996. Functional response of *M. sexmaculatus* Fabricius against *A. gossypii*. *J. Per. Tanaman Indonesia*, 2(2): 38–43.

Watt, J.C. 1986. Beetles (Coleoptera) of the offshore islands of northern New Zealand. In: *The Offshore Islands of Northern New Zealand*, (Eds.) A.E. Wright and R.E. Beever. Department of Lands and Survey, Wellington, pp. 221–228.

Werner, E. and Gilliam, J. 1984. The ontogenic niche and species interactions in size-structured populations. *Ann. Rev. Ecol. Syst.*, 26: 619–633.

Wright, E.J. and Laing, J.E. 1982. Stage specific mortality of *Coleomegilla maculata lengi* Timberlake on corn in Southern Ontario. *Environ. Entomol.*, 11: 32–37.

Yasuda, H., Kikuchi, T., Kindlmann, P. and Sato, S. 2001. Relationships between attack and escape rates, cannibalism, and intraguild predation in larvae of two predatory ladybirds. *J. Insect Behav.*, 14: 373–384.

Yasuda, H. and Sato, K. 1997. Cannibalism and inter-specific predation in two predatory ladybirds in relation to prey abundance in the field. *Entomophaga*, 42(1/2): 153–163.

Yasuda, H. and Shinya, K. 1997. Cannibalism and inter-specific predation in two predatory ladybeetles in relation to prey abundance in the field. *Entomophaga*, 42: 155–165.

Zebitz, C.P.W. 1988. The dependence of the efficiency of coccinellid larvae on the distribution of their prey. *Mit. Deut. Ges. All. Ang. Entomol.*, 6: 454–457.

Chapter 4

A Review on the Honey: A By-product of Bees (Insecta: Hymenoptera) Nutritional and Medicinal Aspects

S.K. Gahlawat,[1] Bhawana Chhabra,[2] K.L. Jain[1], R.K. Gupta[1], Mukesh Dhillon[3] and R.C. Sihag[1]*

[1] Department of Zoology and Aquaculture, [2] Department of Foods and Nutrition, [3] Department of Entomology, CCS Haryana Agriculture University, Hisar – 125 004

ABSTRACT

Honey is a substance produced by honeybees after collecting and processing the nectar from different flowers. It is the most wonderful gift of God and is of high esteem, next only to "Amrit". It is the mini power house packed with carbohydrates providing 3000 calories/kg. It's composition constitutes 70–80 sugar, 1–2 per cent proteins and amino acid which are highest in Indian honey, 1 per cent mineral salt, 16–17 water and enzymes and other chemicals are in trace amount. There is no specific time and age for taking honey and but its doses vary according to age group as 0–15 g/day (child), 30–35 g (youth), 30–50 g (healthyman), 20–30 g (old people) and patients are given as per the recommendation of the physician.

* Corresponding author.

Keeping in view the nutritional composition of the honey, its consumption advised for atheletes, sportsman, mountaineers, deep sea divers, patients in hospitals and workers in factories for quick replenishment of energy lost in their activities. Honey is valued not only as a food but also for a number of its therapeutic attributes. The use of honey as a medicine is recorded in the most ancient human writings, *Vedas* and *Quran*. The honey has Antibacterial, Antiviral and Antifungal properties and is used in the treatment of wounds and ulcers, hay fever and pollen asthma fournier's gangrene.

Keywords: *Honey, Composition, Properties, Nutritional value, Medicinal value.*

Introduction

Honey is a natural sweet substance produced by the honey bees from nectar of blossoms or from the secretions of other parts of the plants, which they collect, converts into honey and store in the wax comb to ripen. It is the most wonderful gift of God and mother nature to mankind. It has played an important role in the spiritual, social and economic life of man since ancient times. It has an honoured place in several religions of the world. Hindus use honey in worship and in sacrifices. Honey is given as the first feed to the new born child. Jews and Catholics prepare special cakes and mead from honey for religious festivals. Ayurvedic and unani system of medicine use honey in several formulations. However, the most important use of honey is as a natural and unrefined food that has many advantages--as a laxative, blood purifier, and preventive against cold, cough, and fever, curative for sores, eyes, ulcers, sore throat and burns. Greeks believed that liberal use of honey delayed the onset of old age and abolished fatigue. Ancient physicians long before bacteria also used it in surgery and burn therapy. An important mineral *i.e.*, chromium present in honey plays an important role in our body. Chromium value alone is a proof that honey is a preferable source of simple carbohydrates to refined cane sugar (Sucrose). Chromium is most essential mineral in our diet. It assist in the absorption and regulation of sugar in the blood. It regulates cholesterol, assist in the body's own manufacture of insulin. It also assists in the processing of alcohol. Lack of chromium is believed to contribute towards adult onset diabetes and heart diseases (Floyd, 2000).

All over the world, except India and other developing countries, honey is used as a food or in the food products. But in India honey is consumed as a medicine only. The rate of its consumption per person per year is only 25 g whereas in countries like Germany and Switzerland the average human consumption of honey is 1.5 kg. In England, Japan and Italy the average consumption is 250 kg of honey per year. There is a great scope to increase our per capita consumption to get rid of malnutrition and other health problems.

The consumers in India need to be educated about the qualities of honey. Most of the people have the conception that only the translucent honey is the best but as a matter of fact it is the granulated honey, which is the best. There are various different types of honey for which we need to develop a taste to sunflower, wild honey and litchi honey, etc. One should not forget the fact that consumption of pure honey makes an individual healthy, strong and sturdy. Honey has also been documented as the best food for the old, infirm and senile persons and to get rid of various body ailments.

Definition

Honey is "the unfermented, sweet substance produced by honeybees from the nectar of blossoms or from secretion of or living parts of plants, which they collect, transform and combine with specific substance and store (ripen and mature) in honey combs". Honey shall not have any objectionable flavour, aroma or tint absorbed from foreign matter during its processing and storage and shall not contain natural plant toxins in an amount, which may constitute a hazard to health.

Honey is also defined as "a substance produced by bees and some other special insects from nectar or honeydew that they collect from living plants which they transfer by evaporating water by action of enzymes they themselves secrete". As a rule honeybees seal the finished honey in cells of their comb (White *et al.*, 1994).

Categories of Honey

The honey can be categorized on the basis of extraction methods these are:

Extracted Honey (Table Honey)

It is extracted by centrifugal machine from bee frames with optimal moisture content and is superior, transparent and free from foreign materials.

Squeezed Honey

It is obtained by squeezing or pressing the honey combs of wild bees with high moisture content liable to fermentation (Ombir and Kaushik, 2001).

Grading on Honey

The honey will be graded on the basis of moisture content:

1. Special grade (less than 20 per cent)
2. A grade (less than 22 per cent)
3. Standard grade (less than 25 per cent).

Table 4.1: Specification for Different Grades of Honey in India

Characteristics	*Agmark 1986*	*ISI-4941, 1974*			*PFA Act Rule A07.03*
		A	*B*	*C*	
Moisture (max. %)	20–25.0	20	22	25	25
Specific gravity at 21°C (min.)	1.35	1.41	1.39	1.37	ND
Ash (max. per cent)	0.5	0.50	0.50	0.50	0.5
Acidity (max. per cent)	0.2	0.20	0.20	0.20	ND
Reducing sugar (min. %)	65	70.0	65.0	65.0	65.0
Sucrose (max. per cent)	ND	5.0	5.0	5.0	5.0
Fructose/glucose ratio (min.)	0.95	1.0	1.0	1.0	0.9
Fieche's Test	Negative	Negative	Negative	Negative	Negative
Aniline chloride test	Negative	Negative	Negative	Negative	Negative

A: Special grade; B: A grade; C: Standard.

ND: Not detected.

Source: Shamala and Jyothi, 1999.

Though honey with more than 25 per cent moisture content is rejected for consumption purpose. On the basis of moisture content different agencies gave the specification for the grades specification for the different grades of honey by Shamala and Jyothi (1999).

Honey is healthier compared to refined sugar since honey has traces of minerals and vitamins to its credit. It is the mini power house packed with carbohydrates providing 3000 calories per kg.

One tablespoon of honey contains about 100 calories. It contains fewer calories and carbohydrates than sugar since it contains water. But when substituting honey for sugar in recipes, due to high density of honey compared to sugar, substituting honey for sugar by volume rather that by weight would give you more calories. Honey contains sugar in the form of glucose and fructose that gives the sweetness.

Chaudhary and Wakhle (1993) recommended the dosage for child, youth, healthy man and old people as 10–15, 30–35, 30–50 and 20–30 g/day, respectively.

Table 4.2: Calorific Value of Honey vis-à-vis Common Food Stuffs

Product	*Calories/kg*
Carrot	330
Apple	481
Milk	670
Potatoes	970
Egg	1272
Honey	3000
Peas	3150
Cheese	3480

Source: Chaudhary and Wakhle, 1993.

Table 4.3: Food Value of Honey vis-à-vis Different Syrups

Product	*Measure for 100 Calories*	*Distribution*		
		Fat	*Protein*	*Carbohydrates*
Honey	1 tbsp.	1.1	1	99
Corn syrup	1¾ tbsp.	1.5	1	100
Maple syrup	1½ tbsp.	1.2	ND	100
Maple sugar	4 tbsp.	1.1	ND	100
Molasses	1½ tbsp.	1.2	3	97
Sugar (white)	2 tbsp.	0.9	ND	100
Sugar (Brown)	2 tbsp.	0.9	ND	100

Source: Chaudhary and Wakhle, 1993.

ND: Not detected.

Composition and Chemical Properties

The composition of honey is given by different authors. Nutritional value of some natural sweeteners is discussed by the aspect–concentration of sweet tasting carbohydrates, average composition of essential nutrients like mineral, vitamins and occurrence of other substances like organic acids etc. The nutritional value of natural sweeteners is not significantly different from "sugar" (saccharose). Natural sweetness does not contribute to fulfill the need for minerals, trace elements or vitamins in an important manner. The sensory properties are different and the intensity of sweet taste is usually higher in sucrose than in syrups. Description of such products as "naturally healthy without sugar" are not justified.

Nutritive Value

Nutritive value of honey per 100 g contains some of the nutrients (Honey.htm.@bawarchi.com, 2002).

Table 4.4: Nutritive Value of Honey

Nutrient	*Value/100 g of Honey*
Protein	0.3
Fat	–
Mineral	0.2
Carbohydrates	79.5
Energy	319 kcal
Calcium	5 mg
Phosphorus	16 mg
Iron	0.696 mg

Source: (Honey.htm.@bawarchi.com, 2002).

Nutrients and Chemicals Properties

Carbohydrates

A report related to all the three types of sugar *i.e.*, monosaccharide, disaccharide and polysaccharide are present in honey. Simple sugar (dextrose and levulose) diasaccharide (maltose, lactose) and other higher sugars are present. Fifteen types of sugar are present in honey, many of them are identified easily and many do not occur in nectar but are formed during ripening and storage by effect of enzyme action and honey acids.

Table 4.5: Average Composition of Honey

Characteristics	*Average*
Moisture (per cent)	17.20
Fructose (per cent)	38.19
Glucose (per cent)	31.28
Sucrose (per cent)	1.31
Maltose (per cent)	7.31
Higher sugar (per cent)	1.50
pH	3.91
Free acid (meq/kg)	22.03
Ash (per cent)	0.17
Nitrogen (per cent)	0.04
Lipids (mg/kg)	361

Source: Sihag and Gahlawat, 1999.

The rapid absorption of honey, monosaccharides and the slower metabolism of the fructose content appear to be the basis of its popularity as a source of quick energy for athletes, scuba divers and mountains climbers. The honey supplementation in infants and diabetic patients is as under:

1. Infant and geriatric diet
2. Honey and diabetes.

Infant and Geriatric Diet

The facile absorption of the monosaccharides, improved weight gain, relief from constipation, decrease of diarrhoea, and good tolerance by the infants at special risk is cited. Improved calcium retention and utility in feeding prematures are also cited. Most of the articles concluded that honey should have a wider use in infant feeding.

At the other end of the human experience, honey appears to have some particular utility in geriatric feeding. Utilization of glucose is markedly decreased with aging, while that of fructose is only

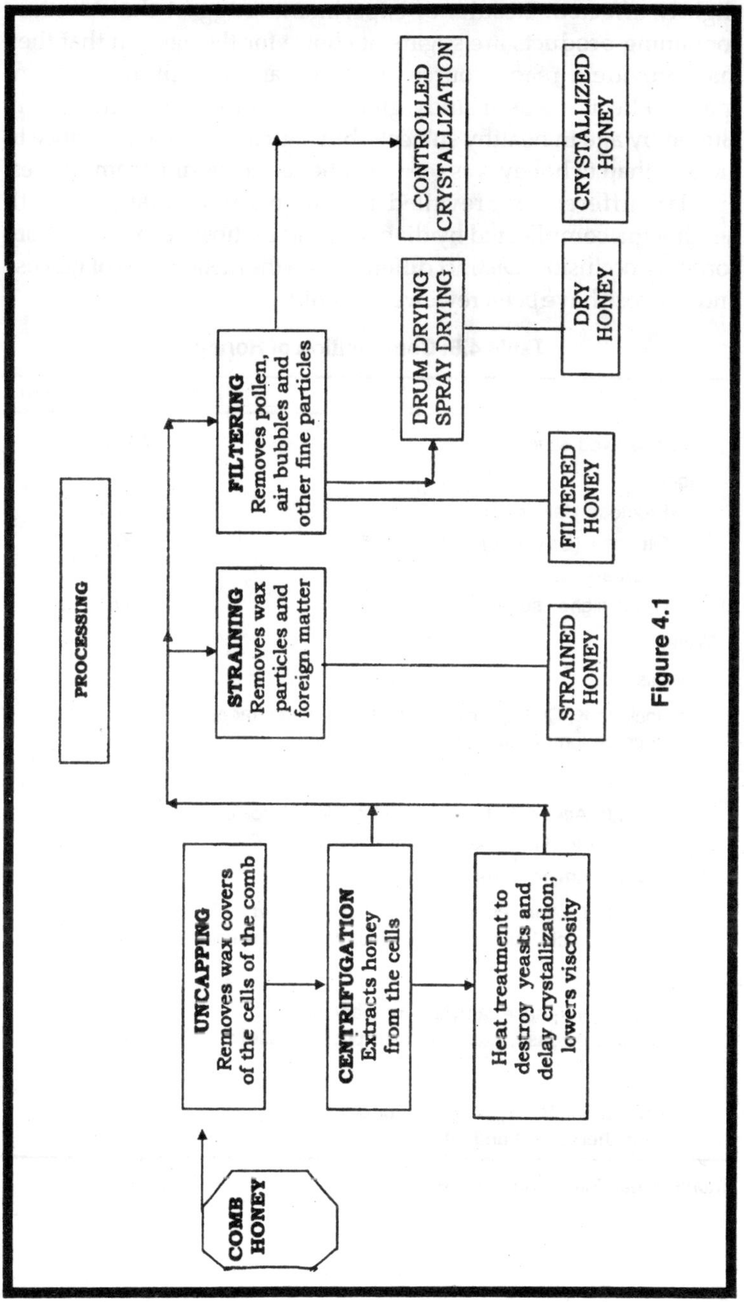
COMB HONEY
UNCAPPING Removes wax covers of the cells of the comb
CENTRIFUGATION Extracts honey from the cells
Heat treatment to destroy yeasts and delay crystallization; lowers viscosity
PROCESSING
STRAINING Removes wax particles and foreign matter
FILTERING Removes pollen, air bubbles and other fine particles
DRUM DRYING SPRAY DRYING
CONTROLLED CRYSTALLIZATION
STRAINED HONEY
FILTERED HONEY
DRY HONEY
CRYSTALLIZED HONEY

Figure 4.1

slightly affected. Results of experiments suggest that levulose containing products are sugars of choice for the aged in that they may provide a ready source of energy and an optimal protein-sparing effect. In a later study, glucose tolerance is not significantly altered by age in healthy subjects, but a greater loss of tolerance to glucose than to honey was seen in patients recovering from strokes. Similar differences prevailed in patients with diabetes, with hemiplegia complicated by diabetes, and in those recovering from coronary occlusion. Distinct differences in the metabolism of glucose and fructose have been reviewed recently.

Table 4.6: Composition of Honey

Constituents	*Percentage (per cent)*
Total dissolved solids	70–80
Sugars	
Fructose (levulose)	38
Glucose (Dextrose)	37
Sucrose	2
Other higher sugar	0.5
Water	20
Minerals	
(include K, Cl, Mg, Fe, Cu, Mn, P, S, Cl and traces of chromium, nickle, tin, silver, gold etc.)	0.5
Acids	
(include Acetic, Buteric, Citric, Formic, Gluconic, Maleic, Lactic, Succinic, Tartaric etc.)	0.2
Proteins and Amino acids	
(include Proline, Phenylalanine, Aspartic acid, Glutamic acid, Asperigin, Leucine, Valine, Isoleucine)	0.25
Enzymes	
(include invertase, amylase or diastase, glucose oxidase, catalase, acid phosphotase)	Traces
Vitamin	
(include Vit. B_1, B_2, Niacin, B_6 or pyridoxal, pantothenic acid and Vit. C)	Traces

Source: Chaudhary and Wakhle, 1993.

Honey and Diabetes

Beekeeping and lay publications at times contain suggestions that diabetics can use honey without incident. This is non-sense, since honey contains a considerable proportion of glucose. Because on the weight basis, honey (at 80 per cent solids) is about as sweet as sugar and provides an average of 31 per cent as glucose, compared with sugar's 52 per cent, some advantage in "sweetening power" is theoretically available to the stabilized diabetic. Selection of tupelo honey, averaging about 25 per cent glucose, would approximately double the "sweetness" intake without increasing glucose. It must be emphasized, however, that any such substitution be undertaken only upon the advice of the physician.

Aspects considered include: classification of honey extraction and processing method, storage of honey, specific standards for honeys that apply in India, nutritional values for various honeys: factors influencing sensory properties of honey and conclusions. Honey and its characteristics etc. are reviewed recently. Besides the above aromatic compounds, colloids, pollen etc, contribute to the colour, taste and flavor of honey.

Sugar

Sweetness to honey depends upon sugars, which are its major components (70–80 per cent). Sugars are source of metabolic energy that is essential for life processes. The sugars include mainly glucose and fructose. They are easily digestible and give instant energy. They are directly absorbed in to the blood that provides rapid sources of energy.

Surveys of floral honey composition have established that the three major components are fructose, glucose and water averaging, 38.2, 31.3 and 17.2 per cent respectively. Glucose and fructose are the only monosaccharide in honey and these sugars combined in various form that comprise the di- and trisaccharide fractions of floral honey.

Genuine honey sample of unifloral, multifloral and commercial varieties were obtained from different sources and the glucose was determined by hypoiodite and enzymatic method. The data revealed that the glucose values obtained by the hypoiodite method were consistently higher than those by enzymatic (true-glucose oxidase) method due to the interference of maltose and other sugar present in honey, this discrepancy in the determination of glucose affected in turn, the fructose value as well as fructose: glucose ratio which are reflected in quality standards for honey.

The use of new instrumental analytical method for separation and identification revealed several sugars in honey. Total dissolved solids in honey are as high as 85 per cent and almost all are carbohydrates and sugars. Presence of over 19 minor sugars in honey has been confirmed by HPLC (High Performance Liquid Chromatography). Honey chiefly contains two reducing sugars *i.e.*, Dextrose and Levulose (fructose). The two sugars account for 70 per cent of the total solids. Sucrose is most common diasaccharide and varies between 0–5 per cent. A reducing sugar maltose is also present in higher amount that is 3–16 per cent but it may not be present initially, the quantity increases with storage and this probably happens by the conversion of other sugars by the action of enzymes and acids.

Proteins and Amino Acids

They are mainly of plant origin and their contents varies in different honey. Protein and all essential amino acids required by the human body for growth are present in honey. Honey from Karvi (*Carvia collosa*) contains 1.2 per cent proteins perhaps the highest among Indian honeys. About 18 free amino acids are known to occur in honey. The most abundant among the amino acids is Proline.

Honey is known to be acidic due to presence of tartaric, citric, maliec, succinic, formic, formic, acetic, butyric acids, and gluconic acid which is most common, pH (3.91) of Indian honey contributes to taste but is not evident due to great sweetness of honey. Honey gives sour or acidic taste due to increased acidity on long storage or on fermentation. Proteins are found in small quantities in different honeys. Proteins originate mostly from bees and some from the forage plants. Indian honey contains as high as 0.2 per cent proteins (Suryanayana *et al.*, 1981).

Vitamins

Honey contains several vitamins of little significance to man. The vitamins present in honey are ascorbic acid, niacin, pantothenic acid, riboflavin, thiamin and vitamin-B complex, a group of various vitamins which improve metabolism and appetite. It also stabilizes the nerves of the human body.

Pigments

Carotenes, chlorophyll, xanthophyll are mainly present in honey.

Enzymes

Enzymes are most important part of honey as these are added during the conversion of nectar by bees. The most important enzymes reported in Indian honey are:

Invertase

For conversion of sucrose to simple sugars (Glucose and fructose).

Diastase

Converts starch to dextrins when added by the bees during ripening. Diastase is very heat sensitive hence used as a criteria to estimate heating of honey (Wakhle and Desai, 1983).

Catalase

Decomposes Hydrogen peroxide in honey.

Phosphatase

Decomposes Glycerophosphate.

Glucose Oxidase

It is added by the bees during ripening, oxidises small amount of glucose to Glucone, lactone and to gluconic acid. The acidity thus formed contributes to the stability of the ripening nectar against fermentation. In this reaction one molecule of Hydrogen peroxide is produced for each molecule of glucose oxidized and this promotes or shows the antibacterial activity in honey.

Flavour and Aroma Substances

Hydroxymethyl furfural (HMF) seems to be responsible for the characteristic honey aroma. In honey, fructose in the presence of acids produces HMF. Fresh honey from the hive contains no or very low HMF and 30 years old honey contains 200 mg per kg (Wakhle and Phadke, 1985).

Characteristics of Honey

The honey has three following main characteristics:

1. Hygroscopic Nature

2. Granulation or Crystallization
3. Fermentation.

Hygroscopic Nature

This allows it to absorb atmospheric moisture until it reaches equilibrium based primarily on the moisture content of honeys. BIS (Bureau of Indian Standard) have grouped them in the 3 categories.

1. Special grade (> 20 per cent moisture)
2. 'A' grade (> 22–22 per cent moisture)
3. Standard grade (> 22–25 per cent moisture)

According to BIS any honey with more than 25 per cent moisture is rejected. The honey with more moisture is thinner in consistency especially from coastal region and rock bee honey where moisture contents may be 28 per cent or even more. On the contrary, North and Central Indian honey are thicker with moisture content ranging from 17–20 per cent.

Granulation or Crystallization

Granulation is a natural phenomenon and all honey granulates sooner or later. A number of factors influence granulation of which the most important is perhaps the fact that honey is a super saturated solution of naturally granulating sugars. Honey from different plant sources differs greatly as to their granulating tendency. Many attempts have been made to delay the granulation of honey. The use of ultrasonic waves (9 kilocycles/sec) for 5–30 minutes which destroy all yeast cells and delay granulation for 15 months. They also described the use of two additives, reporting that 0.3 per cent of Isobutyric acid or of Sorbic acid delayed granulation in honey or more over that feeding either of these materials to the bees during the nectar flow (so that it can mixed with honey) also delayed granulation.

Irrespective of heating, clover honey did not granulate during winter (where glucose, separate out in crystalline form) and heating it at even lower level (50°C for 5 min, increased its clarity, translucence and flowability. Heating Indian rape seed honey at 50–70°C for 5 min in a water bath did not solve the problem of its granulation. However, heating at 70°C for 30 min or 77°C for 20 min, completely may deteriorate honey quality. There is thus, need to educate the consumers in India to shed their belief that granulated honey is an adulterated one and accept granulated honey for their domestic use.

Storage at very low temperature greatly retards the granulation but do not eliminate. Moisture is removed by use of vaccum drying where honey is heated to 40–44°C and subjected to vaccum drying.

Fermentation

Fermentation of honey is caused by the multiplication of sugar tolerant yeast cells which are present in all the honey resulting in the formation of alcohol which is degraded in to Acetic Acid and water giving sour taste to honey. During this process, CO_2 gas is liberated in the form of bubbles. Presence of yeast cells, high moisture content, storage conditions and the presence of granulation lead to fermentation. Honey with less than 17.1 per cent moisture will not ferment at all in a year; those between 17.1 and 18 per cent with a yeast count of 1000 g or less are safe for consumption up to one year. For the honey between 18.1 and 19 per cent moisture and yeast count up to 10/gm for its safety for one year and for honeys above the moisture level even 1 yeast cell/gm honey poses a danger of fermentation. Thus in India, most of the honey is liable to be adversely affected after extraction. Store at 50° F will prevent fermentation. Heating at 145°F for 30 min will destroy honey yeast and thus prevent fermentation.

Table 4.7: Per Capita Consumption of Honey in Various Countries

Country	*Consumption (g/person/year)*
India	8.4
China	120.0
USA	550.0
Germany	1800.0
Switzerland	1200.0
Other European countries	540.0–1200.0
USSR	6400
Japan	340.0

Source: Wakhle, 1998.

Medicinal Value of Honey

The earliest recorded medicinal prescripition of honey were dated back to about 2000 B.C. Honey is valued not only as a food but also for some of its therapeutic attributes. 700 descriptions with honey for both internal and external uses have been recorded in "*Ebers papyrus*" about 1500 B.C. (Crane, 1996). Honey is used most commonly as a vehicle or one of the ingredients of medicine in Ayurveda, which enhances the effects of medicine. Today honey is used as one of the component in many cough syrups. Honey enhances the resistance against diseases by improving the biological process of organs and systems. Its medical properties neutralize fatigue, compensatory hypotonia when used in the preparation of beverages (Mihailesu *et al.*, 1974). It (honey) has also some antibiotic properties. The medicinal uses of honey are:

Antibacterial

Pure honey is bactericidal. Several scientists reported effects of honey. Dold *et al.* (1937) for the first time reported that antibacterial effects of honey were due to a material, which they called Inhibine. The Inhibine effect was due to Hydrogen peroxide produced by glucose oxidase to form gluconalactone from honey glucose (White *et al.*, 1963). The inhibitory action of honey was more on gram positive bacteria. Honey has bactericidal effect on many pathogenic bacteria including enteropathogens that is *Salmonella* species, *Shigella* species, *E. coli* and other gram-negative organisms. (Ibrahim, 1981; Jeddler *et al.*, 1985). Honey given with ORS solution shortens to the patients of gastroenteritis the duration of bacterial diarrhoea and may safely be used as a substitute for glucose (Haffejee and Moosa, 1985).

Antifungal and Antiviral Activity

Honey is a good and ideal topical dressing material for healing wounds, burns and ulcers. Honey cleared 60 per cent septic wounds caused by *Staphylococci, proteus, coliform, enterococci, E. coli, haemophilus, pseudomonas, klebsiella* as compared to Savlon which cleared only 36 per cent wounds in the same period (Sugana *et al.*, 1992). Sugar syrups with similar physical properties as honey did not inhibit any of the bacteria and fungi but unprocessed honey inhibited most of the fungi and bacteria responsible for wounds and post surgical infections except *Pseudomonas aeruginasa* and *Clostridium oedematiens* (Efem *et al.*, 1992). Efem (1988) tested unprocessed honey on 59 patients suffering from wounds and ulcers, 80 per cent of which had failed to heal with conventional treatment were treated. Fifty-eight cases showed remarkable improvement following topical application of honey. One case, later diagnosed as Buruli ulcer, however, failed to respond.

Honey's antibiotic, viscous and hygroscopic properties make it an ideal dressing because honey absorbs water from surroundings oedematous tissues, cleans the wound and protect it against further infections. Honey debrided wounds rapidly, replacing slough with granulation tissue. Chemical debridement with honey apart from wound healing spared many patients the necessity for surgical debridement under general anesthesia.

Wound treatment with honey was not painful, reduced the risk of offensive smelly wounds and the incidence of plastic skin graft areas (McInerney, 1990; Postmes, 1995).

Histological observations showed fibroblastic and angioblastic proliferation of thin wavy collagenous fibres on the 15th day in honey treated wounds in buffaloes (Kumar *et al.*, 1993). However, De carvolho *et al.* (1991) found that topical dressing with Calendula's cream is even more effective than honey.

In those patients who had failed to respond to conventional treatment, wound become sterile within one week of topical treatment with honey, necrotic tissues were gradually replaced by healthy tissues and rapid epithelialization and absorption of oedema occurred. In a clinical study foul smelling wounds were rendered odourless and early burn wound healed quickly. In this clinical study, while only one patient, suffering from Buruli ulcer, caused by *Mycobacterium ulcerans* failed to respond, all the other cases showed remarkable improvement after this treatment.

Fournier's Gangrene Disease

Efem (1993) divided cases of Fourniers gangrene in to two groups. One group received the orthodox treatment in which patients underwent wound debridement, wound excision, secondary suturing and in some cases scrotal plastic reconstruction in addition to receiving a mixture of systemic antibiotics dictated by culture and sensitivity results. The other group received systematic antibiotics and topical application of unprocessed honey. Honey treatment showed distinct advantages over the orthodox method. In honey treated group, no death was reported. The need for anesthesia and expensive surgical operation were also obviated in this case and response to treatment and alleviation of morbidity were faster. With these results, it seems that in future honey may revolutionize the treatment for this dreadful post injury aberration by reducing morbidity and mortality.

Hay Fever and Pollen Asthma

Regular consumption of locally produced natural honey alleviates the effect of hay fever and pollen asthma, because the pollen grains in the honey cause the body to develop immunity. Hay fever patients were advised to eat 10–20 g. honey each day for a period lasting from autumn up to and through the following hay fever season (Riches, 1987).

Dental Health

Honey is one of the important dietetic factors in the cause of dental caries that occur as a result of demineralization of dental surface due to interaction of four factors: sugar, plaque, the host and time (Gomez *et al.*, 1992). On the other hand, Grobler *et al.* (1994) observed through electron microscopic observations that honey does not have erosive effects on tooth enamel over a period of 30 minutes.

Pharmacological Activities

Various methods of honey treatment were used for certain non-specific acute and chronic diseases of respiratory organs (Mladenov and Mladenova, 1986) and gastrointestinal disorders (Salem, 1986).

Many medical problems can be reduced or eliminated by the eating of honey (Berthold, 1997).

Table 4.8: Pharmacological Activities

Sl.No.	*Problem*	*Etiology*
1.	Bed wetting	Reduce fluid intake 6 hours before bedtime, one tsp of honey at bed time.
2.	Sleeplessness	Two tsp of honey at bed time either with warm milk or as such.
3.	Cough remedy	Boil one lemon juice for 10 min + 2 tsp of glycerin + honey dose of 1 tsp as needed.
4.	Muscle cramp	Two tsp of honey + 2 bone meal tablets with each meal at bedtime. Take about 1 week to work.
5.	Atheletics	Honey is a great source of energy for athletics when eaten with meals and between contests.

Source: Berthold, 1997.

Therapeutic Value of Honey

As honey contains high content of sugar it is an energy giving foods, par excellence, but not suitable for diabetics. The minor constituents of honey confer undeniable dietetic and even medicinal properties when:

1. Administered orally
2. Used externally
3. Intravenous injections

Administered Orally

Honey cures or provides relief in intestinal troubles, stomach ulcer, insomnia and throat infection. It increases the haemoglobin content in the blood as well as muscular strength. Honey assists in the retention of calcium: It activates bone formation and cutting of teeth. It also has a slightly laxative effect. Some people are allergic to honey and, in turn, it may cause urticaria.

Used Externally

It hastens the healing of burns, sores and rhino-pharyngeal infections. The essential ingredients in this antibiotic activity of honey, an enzyme, glucose oxidase, cause a release of hydrogen peroxide.

Intravenous Injections

Honey that has been specially treated for this purpose is effective against jaundice, urinary troubles and itching. It regularizes the heart rate.

Honey–Risks and Adverse Effect

The use of honey as a potential remedy for a number of diseases makes it necessary to also ask about its potential adverse effects. From our point of view, it is clear that every substance or mixture may show certain adverse effects under certain conditions, in certain people and if falsely applied. Fortunately, those adverse effects are rare.

Nevertheless, to us some knowledge about such adverse effects concerning normal consumption of honey or medical treatment with honey seems important in order to advise consumers and patients adequately.

To meet the needs of a colony, bees collect honey, honey dew, pollen and water from an environment exposed to various bacterial and chemical contaminants, which might be incorporated in products for human consumption.

Poisonous Honey

One of the earliest reports of honey poisoning was the mass intoxication that resulted from the ingestion of honey produced from the nectar of *Rhododendron ponticum*, which was described by Xenophan in the Anabars during the expectation of Cyrus in 401 B.C. near Trebizond (Lampe, 1988).

"The number of bee hives was extraordinary and all of the soldiers that ate the honey combs lost their senses, vomited and were affected with purgins and none of them was able to stand upright, such as had only eaten a little were like man greatly intoxicated and such as had eaten much more were mad men and some persons at the point of death. They lay upon the ground, in consequence, in great numbers as if they had been a defeat and there was general dejection. The next day one of them was found dead and they recovered their senses about the same hour they had lost them on the preceding day."

Infant Botulism

Infant botulism, with constipation being almost always the first symptoms was first described in 1976. Early studies (Arnon *et al.*, 1979) have found associations between botulism and honey feeding to infants in 27–29 per cent. Recently Spike *et al.* (1989) found this association to be true in about 15 per cent of the cases. Microbiological surveys of honey have found that 40–25 per cent of the samples contain *Clostridium botulinium* spores. These spores are heat resistant and may only be destroyed by temperatures greater than 110°C for more than 30 min.

One hypothesis regarding the etiology maintains that in the infant's intestine qualitative and quantitative differences in the composition of the intestinal bacterial flora of formula fed and breast fed infants exist. These differences may influence the germination, multiplication and outgrowth of ingested *Clostridium botulinium* spores. Honey should not be fed to infants younger than one year old and especially not to those younger than 6 months old for safety reasons (Karsten and Lang, 1998).

1. Infants botulism is rare but dangerous.
2. *Preventive measure*: Infants must not be given honey (Med Monatsschr Pharm, 1999).
3. *Infant botulism with down syndrome*: A four months old boy with Down syndrome and Infant botulism is reported. He was admitted to hospital with strider and developed respiratory failure and generalized paralysis. Laboratory examination revealed botulinal toxin type A in his blood and feaces. *Clostridium botulinum* organism was found in the faeces and in the honey which he had fed. Antibotulinal toxin was administered on the 10th day of illness. Artificial ventilation was given for 87 days and he survived. This is the 11th known case of infant botulism in Japan. It is possible that other cases remain diagnosed.

Allergy to Honey

Allergy to honey makes up about 2.3 per cent of all food allergies (Hofer and Wuthrich, 1985). Honey allergic patients are most often sensitized to either pollen related or bee-related allergens or both. Sensitization may be due to honey itself. Other honey allergic patients may develop reactions due to primary sensitization to either compositae pollen or honey bee venom, mold spores, yeast, algae etc, may be responsible for sensitization in some honey allergic patients.

Food allergy due to honey has only been reported occasionally, since honey may be present in hidden forms in various foods (Chocolates bars, candies, gingerbread). Honey allergy should be considered in any patient with suspected but unresolved food allergy (Helbing *et al.*, 1992). Honey allergy is a very rare condition which shows a clinical picture ranging from cough to Anaphylaxis after ingestion of honey. Here five cases are reported (Bawer *et al.*, 1996).

To characterize the allergenic components of honey, 23 patients allergic to honey were investigated. All displayed allergic symptoms after ingestion of honey or honey containing products ranging from itching in the oral mucosa to severe systemic symptoms to Anaphylactic shock. Immunoblast analysis of these patients revealed IgE binding to proteins at a molecular mass of 54 kdr, 60 kdr and 70 kdr or to a 30–33 kdr double bond or to both in sunflower honey extracts. It is concluded that both proteins derived from secretions of pharyngeal and salivary glands of honeybee heads and pollen proteins contained in the honey cause allergic reactions.

Honey–Testing Guides

Now a day when most of the things were found to be adulterated, there is a need for testing guide which proves useful to the consumers. They are in case of appearance, smell, aroma and flavor. They are:

Appearance

1. *Cleanliness*: How pure and free from wax or insect fragments it is?
2. *Clearness*: How clear it is when in a stable physical state (liquid or crystallized)?
3. *Fluidity*: How does it "flow"? (Above 19–20 per cent water content, honey is too fluid and will not conserve well).
4. *Colour*: How light/dark it is? (Varies with floral source and age of comb).
5. *Homogeneity*: A recently, well-harvested honey is generally homogeneous (either liquid or crystallized); layers can form if honeys are mixed.
6. *Crystallization*: Differs among honeys, and does not necessarily reflect quality; can be partial or complete, wholly or fragmented, fine or thick.

Taste and Aroma

1. Taste a small sample with a clean spoon.
2. Aroma is the olfactory sensation, through the retronasal pathway, that is associated with taste.
3. *Acid Savor*: a slight pinch sensed on the tongue's fringes. Certain honeys are characterized by natural acidity; fermentation is also perceived as such.
4. *Bitter savor*: sensed by the back of the tongue. A bitter savor is characteristic of honeys that contain small quantities of polyphenols.
5. *Salty savor*: sensed by the back fringes of the tongue. Not a common perception of honey.

6. *Sweet savor*: sensed by the tip of the tongue. The ratio among the different types of sugars affects the sweetness.
7. *Texture*: Mechanoreceptors in the lips, gums, and palate perceive tactile sensations: fine, coarse crystals, paste-like, etc.

Smell

1. A rounded glass allows best smell emission, with a fine stem for holding.
2. Temperature must be above 20°C to facilitate diffusion of fragrance.
3. Breathe in the honey fragrance through the nose slowly, for several seconds; after the intense primary smell, a secondary one may be perceived.
4. Smell is highly subjective, and difficult to describe. Seek for analogies with well-known references, perfumes of plants, fruits etc.

Conclusion

1. It is delicious.
2. Honey a good appetizer.
3. Cures burns and ulcers.
4. Strengthen heart and circulatory system.
5. Beneficial to several diseases.
6. A good source of energy.

References

Arnon, S.S., Midura, T.F., Damusk, Thompson, B., Wood, R.M. and Chin, J. 1979. Honey and other environmental factors for infant botulism. *J. Rediatr.*, 94: 331–336.

Bauer, L., Kohlich, A., Hirschwehr, R., Siemann, U., Ebnertt, Scheiner, D., Kraft, D., Ebner, C. 1996. Food allergy to honey: Pollen or bee products? Characterization of allergenic proteins in honey by means of immunoblotting. *J. Allergy Clin. Immunol.*, 97(1): 65-73.

Berthold, R.J.R. 1997. Medicinal uses of honey in fold medicine, *Am. Bee J.*, 872–874.

Chaudhary, O.P. and Wakhle. 1993. Honey: The most nutritios food. *Khadi Gramadyog*, 34(11–12): 742–746.

Crane, E. 1996. The past and present importance of bee products to man. In: *Bee Products: Properties, Application and Apitherapy*, (Eds.) Mizrahi and Lensky. Plenum Press, New York, pp. 1–13.

De-carvalho, P.S., Perri, De DGT and Roberto, L.T. 1991. Cutaneous cicatrization after topic application of calendula's cream and comfrey, propolis and honey association's in infected wound of skin: clinic and histologic study in rats. *Rev. Clien. C. Biomed.*, 12: 39–50.

Dold, H., DU, DH, Dziqo, S.T. 1937. Nachweis antibaketerieller hitze and licht emfindlicher hemmungs stoffe (inhibine) in Nature honey. 2 *Hyg Infektionskar*, 120: 155–167.

Efem, S.E.E. 1988. Clinical observation on the wound healing properties of honey. *British J. Surgery*, 75: 679–381.

Efem, S.E.E. 1993. Recent advances in the management of Fournier's gangrene: Preliminary observations. *Surgery* (St. Louis), 173(2): 200–204.

Efem, S.E.E., Udoh, K.T. and Iwara, C.T. 1992. *The Antimicrobial Spectrum of Honey and its Clinical Significance Infection*, 20(4): 227–229.

Floyd, B. 2000. *The Honey/Chromium Value*. National Beekeepers' Association of New Zealand Incorporated, pp. 9.

Gomez, S., Marante, G.R.A. and Sierra, A. 1992. Diet and dental caries. *Alimentaria*, 29(230): 69–72.

Grobler, S.R., DuToit, I.J. and Basson, N.J. 1994. The effect of honey on human tooth enamel *in vitro* observed by electron microscopy and microhardness measurement. *Arch. Oral. Biol.*, 39(2): 147–153.

Haffejee, I.E. and Moosa, A. 1985. Honey in the treatment of infantile gastroenteritis. *Brit. Med. J.*, 290: 1866–1867.

Helbling, A., Peter, Ch., Berthold, E., Bogdanov, S. and Muller, U. 1992. Allergy to honey: relation to pollen and honey bee allergy. *Allergy*, 47: 41–49.

Hofer, T., Wuthrich, B., Nahrungsmittelallergien, II. 1985. Haufigkeit der organ-manifestationen under allergie-auslosenden Nahrungsmittel, *Schwiz Med Wochenschr*, 115: 1437–1442.

Honeyhtm, Bawarchi. 2002. health and nutrition-honey. *The Nature's Gift*, p. 1–3.

Ibrahim, A.S. 1981. Antibacterial action of honey. *Bull. Islamic Med.*, 1(2): 363–365.

Jeddar, A., Kharsany, A., Ramsaroop, U.G., Bhamjee, A., Haffeejee, I.E. and Moosa, A. 1985. The antibacterial action of honey: An *in vitro* Study. *S. African Med. J.*, 67: 257–258.

Karsten Munstedt, M.D. and Lang, U. 1998. Honey: Risks and adverse effects. *Am. Bee J.*, p. 355–356.

Kumar, A.S., Singh, V.K., Prakash, H.P. and Singh, S.P. 1993. Efficacy of some indigenous drugs in tissue repair in buffaloes. *Indian Vet. J.*, 70(1): 42–44.

Lampe, K.F. 1988. Rhododendrons, Mountain laurel and mad honey. *Jama*, 259: 2009.

Mclnerney, R.J.F. 1990. Honey: A remedy rediscovered. *J. Royal Soc. Med.*, 83: 127.

Med Monotsschr Pharm. 1999. 22(5): 152–153.

Mihailescu, N.N., Palos, E., Volenski, T. and Gorgas, C. 1974. Medicinal beverages of honey and natural substances. In: *The Hive products: Food, Health and Beauty*, (Eds.) V. Haranaj, *et al.* Proc Apltherpy; Maric, Spain, Apimondia, pp. 71–74.

Mladenov, S. and Mladenova, D. 1986. New apitherapeutic methods for the treatment of certain non-specific acute and chronic diseases of the respiratory organs. *Proc 30th Int. Cong. Apic. Nagoya Apimendra*, p. 453–454.

Ombir and Kaushik, H.D. 2001. Nutritional and Medicinal value of honey. In: *Compendium of Refresher Course on Advancer in Food Science and Technology*, Held at GJU, Hisar, p. 143–149.

Postmes, T. 1995. Honig und Brandwenden, *Die Biene*, 131(146): 45–48.

Riches, H.R.C. 1987. Honeyand Hay fever. *Bee Craft*, 69(8): 12.

Salem, S.N. 1986. Honey regimen in gastro-intestinal disorders. *Bull. Islamic Med.*, 1: 358–362.

Shamala, T.R. and Jyothi, Y.S. 1999. Honey: It is more than just sweet. *Indian Food Industry*, 18(6): 349–357.

Sihag, R.C. and Gahlawat, S.K. 1999. Beekeeping for producing honey: a food supplement. In: *Recent Advances in Bioscience for Sustainable Food Security*, p. 71–73.

Spika, J.S., Shaffer, N., Hargrett-Bean, N., Callins, Macdonald, K.L. and Blake, P.A. 1989. Risk factors for infant botulism in the United States. *AJDC*, 143: 828–832.

Sugana, L., Chandrakasan, H. and Koithara, T.J. 1992. Influence of honey on collagen metabolism during wound healing in rats. *J. Clin. Biochem. Nutr.*, 13(1): 7–12.

Suryanarayana, M.C., Mohana Rao, G. and Lakshmi, K. 1981. The sandle tree a nectar source for *Apis cerana* and *Apis mellifera. Bee Wld.*, 74: 108–110.

Wakhle, D.M. 1998. Value addition to honey and bee products. *FAO Workshop on Sustainable Bee keeping Development and All India Honey Festival* (APIEXI-98), 1–5 August.

Wakhle, D.M. and Desai, D.B. 1983. Enzymes in Indian honeys. 2nd *International Conference on Apiculture in Tropical Climates*, New Delhi, p. 703–710.

Wakhle, D.M. and Phadke, R.P. 1985. Need to abolish data of expiry marking on AGMARK honey as HMF content in honey-not harmful. *Indian Bee J.*

White, J.W. and Subers, M.H. 1963. Studies on honey inhibine 2: A chemical assay. *J. Apic. Res.*, 2: 93–100.

White, J.W., Rithof, M.L., Subers, M.A. and Kaushik, I. 1994. Composition of American honeys. *US Dept. of Agric. Tech. Bull.*, 1261(4): 1–25.

Chapter 5

Study on Biology of Sheep Body Louse *Damalina ovis* Linn. (Mallophoga) in District Etah (U.P.)

R.K. Dubey* and Anil Paliwal**

**Department of Zoology, Shri Chitra Gupta (P.G.) College, Mainpuri (U.P.)*

***Department of Zoology, Ganjdundwara P.G. College, Ganjdundwara, Distt. Etah (U.P.)*

ABSTRACT

The *Damalina ovis* Linn. is a permanent wingless ectoparasite which is commonly called as wool louse of sheep. It caused the infestation on the sides of neck behind the ear of sheep. The present study reports on various observations of different stages of life cycle of *Damalina ovis* Linn. It was observed that various stages of life cycle of the louse was influenced by seasonal changes. The life cycle of this louse completed more rapidly in the month of March–April and also in the month of July–August. The longevity of the lice (maximum 80 days) recorded in spring months while the longevity increased in laboratory condition at 37.5°C temperature and 65–90 per cent relative humidity. So the 37.5°C temperature and 65–90 per cent R.H. revealed good result in the development of the lice to reach adult stage and louse required three months to reach adult stage.

Introduction

The *Damalina ovis* Linn. is a common wool louse which is found throughout in India on the skin of sheep as a permanent wingless ectoparasite. This parasite feeds mainly epithelial debris of the skin, wool fibres, blood serum, dried serum and own egg shells with the help of biting mouth parts. This ectoparasite also eats wool and effected the growth of sheep, decrease in milk yield and growth per day and therefore, causes great loss to our countries' economy.

Very few reports are available on the biology of *Damalina ovis* Linn. Lamson (1917) and Marthysse (1944–46) made some observations on the life history of *Bovicola bovis*. Martin (1934) described the life history of pigeon louse, Croufurd–Benson (1941), Scott (1952) and Murray (1957) gave few observations on the life cycle of *Damalina ovis* Linn. The existing knowledge of various developing stages of *Damalina ovis* Linn. is quite scanty in literature. So this communication reports in details the study of the biology of *Damalina ovis* Linn. from Distt. Etah (U.P.).

Materials and Methods

The specimens were collected from the sheep of urban area of Distt. Etah by employing a water suction pump through collecting bottle.

Damalina ovis an ectoparasite on sheep was kept in the laboratory by method of Wilson (1936). The influence of temperature and humidity was studied in the laboratory on the number of eggs laid, on the hatching of the eggs, the nymph and adult stages. The lice in presence of suitable food were kept in small celluloid with flat basis, which rested on petridishes supported on glass stands within glass watch boxes. These were covered with vaseline shield lids and humidity was maintained by standard methods using concentration of sulphuric acid (Wilson, 1936).

For the study of the biology sheep louse (mature lice) which were spread over on the neck behind the ear were collected. Various stages of life cycle were studied by confining the lice by means of aluminium cells to tuft of hairs in circumscribed portion of the hair of the animal, the number of eggs laid by female, number of maturity days with the interval between stages of nymph were also studied.

Results

Copulation period of *Damalina ovis* Linn. was 5–7 minutes. After Copulation female laid eggs which were ovoidal in shape, white in colour and translucent at first but turned pale-brown at the end of incubation. The free end of the egg was provided with an operculum which consists of small air cells or cellulae which are 15–20 in number. Each cellulae is elongated tube like in structure and attached with other in longitudinal manner.

The biology of *Damalina ovis* Linn. required three moults (Figure 5.1 A, B, C and D) to reach the adult stage (Figure 5.2A and B). After hatching the Ist instar immediately moves away from the egg shell. The Ist nymphal instar is light in colour, in which only the mandibles are slightly chitinized and the segmentation of the abdomen is very indistinct with seven dorsal transverse row of setae. A pair of dark eye spot and occular sclerite are also quite visible. In IInd nymphal instar the lateral margin of abdominal segments are partly chitinized and appear yellowish in colour. Clypeal lobe band, auricularian process of both mandibles and occular sclerite becomes dark. The abdomen increase with eight dorsal transverse row of setae. The sclerotization of various parts become increased in IIIrd nymphal instar (Table 5.1). The abdomen also become much elongated with nine dorsal transverse row of setae. So the body of adult is heavily sclerotized with nine dorsal transverse row of setae.

Table 5.1: Different Instar Stages of *Damalina ovis* Linn.

Stages	*On Sheep (Days)*		*In Laboratory 37.5°C Temperature 65–90% R.H.*
	Dec.–Feb.	*Mar.–Aug.*	
Incubation	9–11	7–8	7–9
Ist Instar	7–8	6–7	7–8
IInd Instar	6–8	5–6	5–6
IIIrd Instar	10–11	6–8	7–8

The measurement of the whole body, head, thorax and abdomen in various stages of development of *Damalina ovis* Linn. brings out the fact that the size of nymphal instar gradually increased at each moult (Table 5.2). The ratio reveals that the value of each *i.e.*, length of whole body, head/abdomen and head/total length of the body is maximum in Ist instar and minimum in IIIrd instar (Table 5.3).

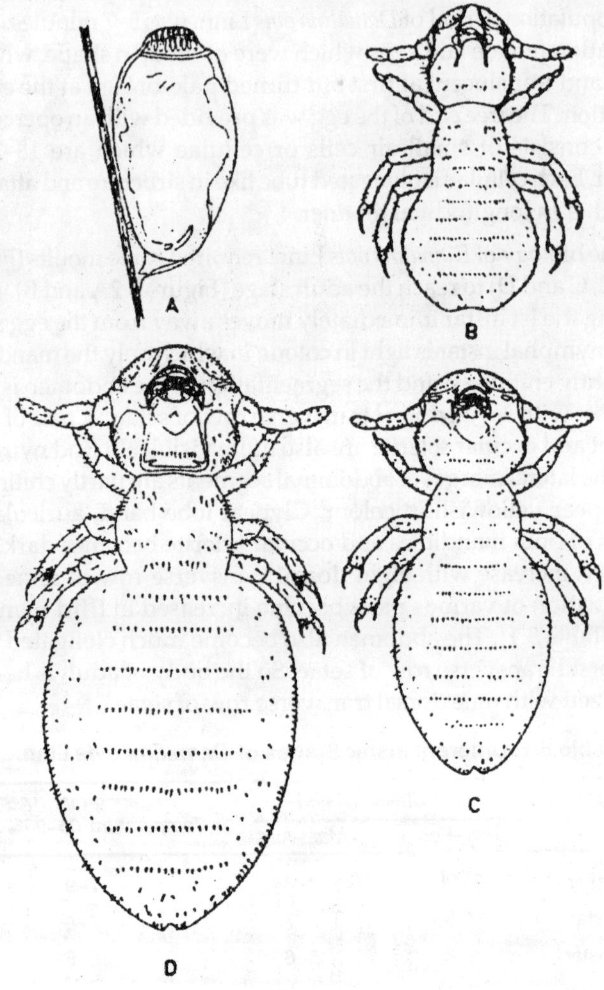

Figure 5.1: Stage of Life Cycle of *D. ovis* (Linn.)

(A) The egg; (B) The Ist Nymphal instar; (C) The IInd Nymphal instar; (D) The IIIrd Nymphal instar.

The time required to the hatching of the egg, each moult of the nymph and to reach adult stage on sheep at natural as well as in laboratory conditions were also recorded. The time required for the completion of life cycle of *Damalina ovis* Linn. is minimum in March–

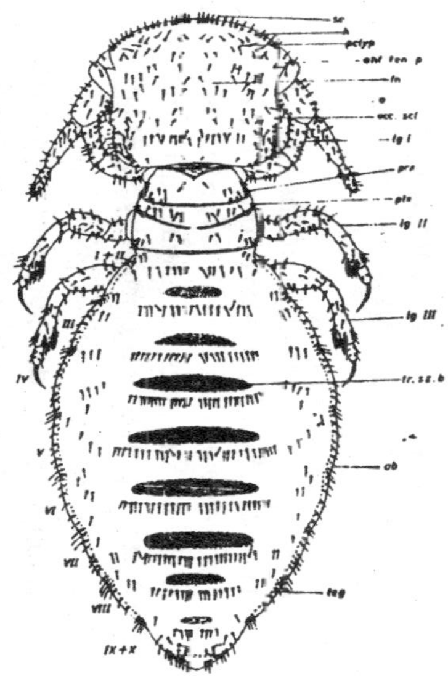

Figure 5.2(A): Dorsal View of Male *D. ovis* (Linn.)

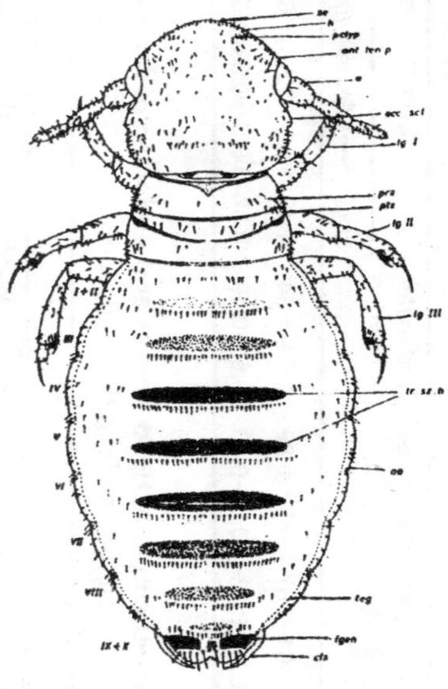

Figure 5.2(B): Dorsal View of Female *D. ovis* (Linn.)

a: Antenna; ant. tem.p.: Anterior teporial pit; ab: Abdomen; cls: Claspers; Fgen: Female genitalia; fn: Frons; lg: Leg; occ. scl: Occualr Sclerite; pclyp: Post clypeus; prs: Prothorax; ptx: Pterothorax; se: Setae, tr.; sz. b: Transverse sclerotized band; teg: Tergite

August but maximum in December–February. The time required in laboratory condition is minimum at 37.5°C temperature with 65–90 per cent relative humidity.

Table 5.2: Measurement of Various Stages of *Damalina ovis* Linn.

Measure-ment	*Ist Instar (mm)*	*IInd Instar (mm)*	*IIIrd Instar (mm)*	*Adult*	
				Male (mm)	*Female (mm)*
Length	0–61–0.67	0.87–0.89	1.29–1.34	1.49–1.55	1.56–1.62
Width	0.31–0.33	0.37–0.40	0.55–0.59	0.57–0.60	0.63–0.66

Table 5.3: Measurement of Head, Thorax and Abdomen of Various Stages of *Damalina ovis* Linn

Stages	*Head*		*Thorax*		*Abdomen*	
	Length (mm)	*Width (mm)*	*Length (mm)*	*Width (mm)*	*Length (mm)*	*Width (mm)*
Ist Instar	0.21-0.23	0.20-0.22	0.11-0.13	0.20-0.23	0.27-0.30	0.26-0.30
IInd Instar	0.22-0.025	0.21-0.24	0.12-0.14	0.22-0.20	0.50-0.54	0.32-0.36
IIIrd Instar	0.23-0.26	0.22-0.27	0.21-0.25	0.24-0.28	0.81-0.87	0.51-0.57
Adult	0.35-0.37	0.34-0.36	0.24-0.27	0.35-0.36	0.95-0.98	0.61-0.65

Discussion

The present communication reports on a detailed structure and measurements of each nymphal instar and adult with head, thorax and abdomen of *Damalina ovis* Linn.

The eggs of *Damalina ovis* Linn. are glued to the base of the hair by means of a drop of cement. Similar observations were produced by Florance (1921) in *Haemotopinus suis* and Croufurd-Benson (1941) in *Haematopinus eurystarnus*. The eggs are always arranged single along the hair. Similar statement was reported by Scott (1952) and Murry (1957) in *Damalina ovis* Linn. from elsewhere.

Lamson (1917) reported that the development from egg to adult in *Bovicola bovis* required 29–31 days while Matthysse (1946) reported only 26 days in cattle louse. Scott (1952) observed 31 days in sheep and 32 days in laboratory condition at 36.5°C temperature with 70–90 per cent R.H.

Present investigation revealed that the development of *Damalina ovis* Linn. from egg upto adult stage required 31 days in the month of March–August, 37–41 days in the month of December–February and 33 days in laboratory conditions at 37.5°C temperature with 65–90 per cent releative humidity.

Acknowledgement

The authors are thankful to Dr. Zafar Khan, Head, Department of Zoology and Dr. R.S. Saxena, Principal, Ganjdundwara PG College, Ganjdundwara Distt. Etah for providing the laboratory facilities during this investigation.

References

Croufurd-Benson and Hixon, H. 1941. The Cattle lice of Great Britain part 1, 2 biology with special reference to *Haematpinus eurysterus*. *Parasitology*, 33: 331–342.

Florence, L. 1921. The hog louse *Haematpinus suis* (Linn.) its biology, anatomy and histology, mom. *Cornell Agri. Exp. Sta.*, 51: 635–743.

Lamson, G.N. 1917. The life history of Cattle lice. *J. Eco. Ento.*, 10: 446

Mathysse, John. G. 1946. Cattle lice their biology and Control. *Cornell Univ. Agri. Exp. Sta.*, Ithaca, New York, p. 1–67.

Murray, M.D. 1957. The distribution of the eggs of *D. ovis* (Linn) on the sheep. *Aust. J. Zool.*, Melbourne, 5(2): 173–182.

Scott, M.T. 1952. Observations on the bionomics of the sheep body louse *D. Ovis* (Linn.). *Aust. J. Agri.*, 3: 60–67.

Chapter 6

Latest Scenario of Fish Diseases in India

S.K. Gahlawat, R.K. Gupta, R.C. Sihag and N.K. Yadava

Department of Zoology and Aquaculture,
CCS Haryana Agriculture University, Hisar – 125 004

Introduction

The intensification and diversification of freshwater aquaculture practices in years have led to increased incidence of fish diseases in the country. A similar picture is prevelent with regard to brackish water and mariculture system. Fisheries suffer from a wide variety of diseases of infectious and non-infectious etiology in their natural environment. Abundant literature is available with regard to fish mortalities due to infection by viruses, bacteria, parasites, fungi, nutritional deficiencies, metabolic disorders and faulty environmental conditions like oxygen deficiency, high or very low water temperature, chemical pollutants, over crowding, malnutrition, algal blooms, etc. However, fish mortalities are not the only criterion to evaluate the effect of fish diseases. Even the morbidity which leads to weight losses and poor growth in surviving fish contribute substantial losses to the farmers. The need for maintaining healthy stock as a prerequisite to enhance fish production levels has been fully appreciated by planners and farmers alike in the recent years.

DNA-based methods have been used in diagnosis and for detection of many economically important viral pathogens of cultured finfish and penaeid shrimp. For finfish, tests have been developed for pathogens such as channel catfish virus (CCV), infectious hematopoietic necrosis virus (IHNV), infectious pancreatic necrosis virus (IPNV), spring viremia of carp virus (SVCV), viral hemorrhagic septicemia virus (VHSV), viral nervous necrosis virus (VNNV) and *Renibacterium salmoninarum* (Muroga, 1997; Plumb, 1997). DNA-based methods for detection of penaeid shrimp viruses are now used routinely in a number of laboratories around the world. DNA-based techniques will have an important role to play in efforts to develop sustainable shrimp culture in Asia and elsewhere. Production facilities in Thailand are currently using PCR techniques to screen shrimp post-larvae for white spot syndrome virus (WSSV). Culturing such larvae in closed (biosecure) or semi-closed culture systems can prevent or minimize viral infections, leading to a viable shrimp industry.

Viral Diseases

Infectious Pancreatic Necrosis Virus (IPNV)

IPNV is a highly contagious viral disease of young fish of the salmonid species. It is widely reported in Europe, Asia, America and Canada. Significant mortalities can occur from IPN outbreaks in fry. Disease signs include: Pale liver and spleen, stomach and intestine devoid of food, darker skin and trailing faecal casts. IPNV can cause an asymptomatic carrier state in salmonid fish. These carriers can shed the virus and transmit the virus to their progeny and other susceptible species. It is therefore important to detect asymptomatic carriers in order to limit the spread of infection.

Two new methods *i.e.*, non-destructive (Munro *et al.*, 2004; Gahlawat *et al.*, 2004a) and destructive (Gahlawat *et al.*, 2004b) were developed for the detection of IPNV carriers in Atlantic salmon (*Salmon salar* L.) and Atlantic halibut (*Hippoglossus hippoglossus* L.). These methods are easy, inexpensive and very sensitive as compared to OIE (Office International des Epizooties, 2000) standard tissue culture and RT-PCR. No information was available on the longer-term persistence of IPNV in halibut and the fish were detected carriers by these methods. The destructive methods of viral isolation involve the need to kill the fish to obtain tissue (kidney, liver, gills). As halibut are repeat spawners and for broodstock testing there was a great need to develop a sensitive non-destructive method for isolation of IPNV. Macrophages are phagocytic cells capable of killing wide range of pathogens. These play a vital role in immune response of fish and hence utilised in developing these methods.

Spring Viremia of Carp (SVC)

SVC is a viral disease of fish, primarily common carp (*Cyprinus carpio*) caused by ssRNA Vesiculovirus (Rhabdoviridae), known as spring viraemia of carp virus (SVCV) or *Rhabdovirus carpio* (RVC) (Fijan, 1999). It was first described in 1971 (Fijan, 1972). Before the disease was recognised, it was variously called infectious dropsy, infectious ascites, hemorrhagic septicemia, or rubella. In Europe, the disease has had substantial impact on the production of carp, with estimated losses of 10–15 per cent of 1-year-old carp or about 4,000 tons annually. In some cases, mortality rates of young carp can reach 70 per cent. This impact of the disease has led to its listing by the Office International des Epizooties (OIE) as notifiable. In Europe, it was quite some time before SVCV was adopted as a viral disease; many people thought it was caused by bacteria especially *A. hydrophilla*.

The disease has been reported to occur in most European countries, the Middle East, Australia and Asia. Recently the disease has been reported from Brazil and the United States of America (Dixon, P., 2003 EU workshop, CEFAS). The virus has also been isolated in United Kingdom from fish exported from China, although the disease has not been reported from China (Dixon, P. 2003 EU workshop, CEFAS) and India (Asia-Pacific Quarterly Aquatic, Animal Disease Reporting System 1991–2000).

First signs of the disease may be a change in behavioral patterns of the fish. Fish may congregate in slow-flowing water, near pond banks, or lie on the bottom. Over time, the rate of respiration will decrease, as will reaction to stimulation and swimming speed. As the disease progresses, the fish become sluggish and may swim and lie on their sides.

Externally, the fish can exhibit a number of non-specific physical signs including darkening of the skin, swollen abdomen, exophthalmia (pop-eye), haemorrhages in the skin, gills and anterior eye chamber, anaemia and pale gills, and a protruding vent.

Internally, the signs are dominated by building up of fluid (oedema) in all organs and in the body cavity, haemorrhages in the swim bladder, and inflammation of the intestines (Fijan, 1999).

Koi Herpes Virus

In 1988 a disease emerged that caused mass mortalities in populations of cultured common carp (*Cyprinus carpio carpio*) and koi carp (*Cyprinus carpio koi*). Devastating losses occurred on farms culturing carp in Israel and Germany and outbreaks of disease with similar etiology occurred in koi carp in USA, England, Germany and Netherlands. A herpes virus was isolated from carp during many of these outbreaks that was later shown, in transmission studies, to be the cause of the disease. Outbreaks of the disease tend to be seasonal and occur at water temperature between 18 and 22°C.

The disease is highly infectious and the most prominent clinical sign is discolouration of the gills. Other signs include lethargy combined with short bouts of hyperactivity, endothalmia (sunken eyes), pale patches on the skin and an increase in mucus production on the skin and gills. Microscopic examination reveals severe hyperplasia and necrosis of the gill epithelium, and nuclear hypertrophy and margination of chromatin can be observed in some epithelial cells.

Bacterial Disease

Innumerable diseases are caused in fishes due to bacterial pathogens and several of them are reported in Indian literature. They occur in nursery, rearing and grow out ponds causing serious concern to fish farms. Some of the important bacterial pathogens are *Aeromonas hydrophila, A. salmonicida, Pseudomonas fluorescens, Pseudomonas* sp., *Flexibacter columnaris, Edwarsiella tarda, Vibrio alginolyticus* and *V. parahaemolyticus* which have been identified as the most commonly occurring bacterial agents of fish disease. Symptomatology, clinical pathology and anatomic pathology have been carefully correlated arrive at the diagnosis of the diseases caused by these organisms.

Fin and tall rot in young fish are reported to be due to a mixed infection of *A. hydrophila* and *P. fluorescens*. Bath treatment in 1 : 2000 copper sulphate for 1–2 minutes or painting of concentrated copper sulphate solution in the affected fish act as a good control measure.

However, now-a-days, iodine solutions and antibiotics are used to contain fin and tail rot.

Dropsy is another important fish disease in India where rohu, catla and mrigal are affected mostly in the late winter and have been reported in composite fish culture ponds. The epidermis and body cavities get filled with fluid and scales protrude out from their pockets leading to severe anaemic condition. Terramycin in the feed for about 2 weeks is reported to have good effect. A mixed infection of *A. hydrophila* and *Myxosporidian* parasite is often reported in cases of *Catla catla.* Some workers in the country have attributed cases of dropsy to *A. hydrophila* in conjunction with malnutrition.

Another bacterial disease in the epidemic form has been the pop-eye-disease the causative organism of which is identified as *A. liquifaciens* and *E. tarda* the cornea of the eyes get vascularized leading to opacity and complete necrosis resulting into death of the fish Corneal opacity in silver carp (*Hypophthaimichrhys molitrix* Val) and *Channa narulius* due to gram-positive bacterium, *Staphylococus aureus* has also been reported. Chloromycetin bath @ 5-10 mg/litre of water has been found effective in controlling the disease at an early stage by some workers. Disinfecting the environment with potassium permanganate ($KMnO_4$) at a dose of 0.1 ppm followed by liming @ 300 ppm check the disease considerably.

Myxobacterial organisms have been attributed to cause a condition known as gill hyperplasia syndrome in common carp (*Cyprinus carpio*). The infectious nature of the disease has been proved by transmitting the disease in Asiatic major carp species through contact. Ulcerative disease in catla has been reported by many workers where bilateral ulcerations of the opercular bones and cranium and *A. hydrophila* isolated from the lesions. Another disease of similar nature causing external lesions over the surface of the body has been described by several workers. The causative organism has been identified as *F. columnaris.* Topical application of $KMnO_4$ or short bath in 500 ppm of $KMnO_4$ has been found to be very effective in completely curing the disease.

A septicaemic disease called Edwardsiellos has been reported affecting brood fish population. *E. tarda* has been isolated from the disease fish showing anaemia, cutaneous lesions and gas filled abscesses in the muscle. With considerable morbidity even large scale mortality of spawn of Indian major carps has been reported in organized hatcheries. The young ones showed anatomical deformities and greyish white discolouration of the body within 48 hours post hatching and died within 72 hours. Although treatment with lodophor has been found to be effective, water quality improvement in the hatchery is the most essential component for keeping this disease away.

Epizootic Ulcerative Syndrome (EUS)

EUS has become a matter of tremendous concern not only among the fisherman and fish farmers, but also among general public, entrepreneurs, administrators and planners. Initially, the epizootics were mainly reported from the North-Eastern states of the country in the year 1988. Gradually, it spread throughout the country and has engulfed the entire length and breadth of the country barring only few states like the Punjab, Haryana, Jammu and Kashmir, Himachal Pradesh and Gujarat. The disease has been reported not only in the freshwater fishes, but also brackish water fishes in Orissa, Kerala, West Bengal, Tamil Nadu, Karnataka and Goa. One common feature of the disease is that it initially affects the bottom dwelling species like murrels followed by catfishes and weed fishes. Subsequently other fishes including Indian major carps are affected.

A number of bacteria *viz., A. hydrophila, A. punctata, Flavobacterium* sp., *Pseudomonas* sp., *E. tarda, V. parahaenrolyticus* and *Streptococcus* sp. have been isolated from the affected specimens. Histopathological studies revealed complete loss of epidermis in the ulcerative area of the skin where the dermis and hypodermis showed characteristic granulomatous changes. Gill tissue showed hyperphlasia of the secondary lamella epithelium, fusion of lamella and clubbing of gill tips. Haemopoietic tissue of the kidney showed proliferation of macrophages. Besides bacteria, virus and fungus parasites have also been reported to be associated with epizootic ulcerative syndrome.

Several methods have been tried or are being tried to control the disease, many antibiotics, sulfonamides, chemicals, herbal preparations, etc. have been advocated to preventive and curative measures. Lure was accepted widely among the fish farmers of the country until the formulation of a medicine named CIFAX was developed by the scientist of CIFA. Marked improvement of the ulcerative condition is noticed within. 7 days of application of medicine and the ulcers are healed up within 10–14 days. Records reveal that there is no recurrence of the disease for at least eleven months in the ponds where the medicine is applied.

Protozoan Disease

Myxosporidia are one of the most important group of pathogens capable of producing disease in fish causing heavy loss in the juveniles. Several species of this pathogen have been reported from various parts of the country causing heavy losses in intensive fish culture and fish seed rearing due to emaciation and asphyxiation. Myxobolus cysts of varying sizes have been reported on the gills and kidney of *Catla catla.* Diagnosis of the disease can be made on the basis of gross appearance of the pin head sized greyish cysts and large number of myxosporidian spores under the microscope.

The other group of protozoa infecting fish belong to cillophora. The carp fry and fingerlings sometimes carry heavy infection of the gills and skin with *Tricodina indica* when there is excessive mucus secretion on the gills which hampers respiration and the fish start surfacing. Some species of *Tricodina* occur inside the urinary bladder. Fish affected with tricodiniasis eventually turn sluggish. A short bath in 2–3 per cent salt solution for 5 minutes can control the infection effectively

Whirling disease in *Cirrhirus mrigala* (Ham) has been attributed to a protozoan parasite, *Myxosoma cerebralis.* The gross changes are characterized by deformed spinal column as report in European and American salmonids suffering from whining disease. Ichthyophthiriasis or "white spot" disease has been reported causing large scale mortality in rearing ponds in different parts of the country. Presence of numerous pin head shaped white spots on the skin, fins and gills are common diagnostic features.

Helminthes Parasites

Fishes are infected on their gills and skin by many species of monogenetic and digenetic trematodes and cestodes. The important monogenetic species are *Gyrodactylus* and *Dactylogyrus.* They remain attached to the skin or gills by the help of anchors which cause wounds. Their presence cause hypersecretion of mucous and haemorrhage from blood capillaries. These wounds later get infected with pathogenic bacteria and fungi. Alternate bath in 1 : 2000 acetic acid solution followed by bath in 2 per cent sodium chloride solution has been found effective.

The digenetic trematodes generally infect the alimentary canal but are rarely harmful. However, at the metacercarial stage, they are harmful as they get encysted in any part of the body, *i.e.*, skin, muscle, abdominal cavity, liver, eyes and brain. In these cases, fish serve as intermediate hosts while the final host is generally a fish-eating bird. In many carp fingerlings, the black spot disease is due to metacercaria of diplostornatidae. Catla and silver carp are the most susceptible species.

Cestodes are also internal parasites in alimentary canal, with higher infestations observed in catfishes than in carp. The infected fish show softening of the muscle tissue probably due to accumulation of water. In cases of heavy infestations, the abdomen bursts resulting in death of the fish. Acanthocephala have been recorded from major carps and catfishes in India, but the life history or biology of the parasites has not been studied in detail. The parasite causes lesions in the intestinal wall by its proboscis.

Crustacean Parasites

Heavy mortality due to Argulosis or fish lice infestation has been reported in Indian major carps. Several species of *Argulus* have been reported to infect cultured fishes. These parasites inhabit the skin, fins and gills of host and cause extensive pathological lesions in the skin showing circular haemorrhagic patches which become ulcerated. Mucous cells proliferate and copious mucous is produced. These lesions promote secondary infections like fungi and bacteria. Rohu has been observed to be the most susceptible species among the carps.

Weekly applications of malathion @ 0.25 ppm for three consecutive doses are advised for the control. Bath in potassium permanganate @ 500 ppm for one minute once a week the combination of malathion treatment is more helpful in the eradication of the parasite.

Fungal Disease

Saprolegania parasitica in the fry and fingerlings of major carps is characterized by a white to brown cotton like growth consisting of colonies of mycelium and filaments which appear as small to large patches on various parts of the body like fins, gills, mouth, eyes or muscle. The infection starts due to netting injury and over-crowding or lesions caused by other diseases. *Saprolegania* often infects the fertilized eggs. They attack the dead eggs first and then spread to the viable living eggs resulting in their spoilage. Infection due to *Aphanomyces pisci* is another fungal disease where the scales become black and later fall off. The hyphae of this species are not visible like those of *Saprolegnia.*

Branchiomyces infection causes "gill rot" in fish. These pathogens invade the blood vessels of gills causing necrosis of the surrounding tissues also. Apart from these, *Aspergilus flavas, A. ochraceous* and *Fusiformis momiliforme* have been reported from some of the IRUS affected specimens.

Several different treatment measures in fungal infection in fish have been formulated, which include dip treatment in 3 per cent common salt solution, 1 : 2000 copper sulphate solution or 1 : 1000 potassium permagnate solution for 5–10 minutes. A flush treatment for 1 hour in malachite green (0.1 ppm) is found to be quit effective in curbing fungal infection.

References

ADB/NACA. 1991. Fish Health Management in Asia-Pacific. Report on a regional study and workshop on fish disease and fish health management. ADB Agriculture Department Report Series No. 1. Network of Aquaculture Centres in the Asia-Pacific, Bangkok, Thailand.

Dixon, P.F. 2003. *EU Workshop on Diagnostic Techniques with Special Emphasis on Carp Diseases*, CEFAS Weymouth, UK from June 2–4.

Fijan, N. 1972. Infectious dropsy in carp: A disease complex. In: *Diseases of Fish ,Symposia of the Zoological Society of London, London,* (Ed.) Mawdesley Thomas LE, p. 39–51.

Fijan, N. 1999. Spring viremia of carp and other viral diseases of warm water fish. In: *Fish Diseases and Disorders, Vol. 3,* (Eds.) Woo, P.T.K. and Bruno, D.W. CAB International, Oxon, p. 177–244.

Gahlawat, S.K., Munro, E.S., Ellis, A.E. 2004a. A non-destructive test for detection of IPNV-carriers in Atlantic halibut. *J. Fish Dis.*, 27: 233–239.

Gahlawat, S.K., Munro, E.S., Ellis, A.E. 2004b. A simple method for isolation of IPNV from kidney macrophages of Atlantic salmon which greatly improves the detection of asymptomatic carriers. *Dis. Aquat. Org.* (Communicated).

Munro, E.S., Gahlawat, S.K., Ellis, A.E. 2004. A sensitive non-destructive method for detecting IPNV carrier Atlantic salmon by culture of virus from plastic adherent blood leucocytes. *J. Fish Dis.*, 27: 129–134.

Muroga, K. 1997. Recent advances in infectious diseases of marine fish with particular reference to the case in Japan. In: *Diseases in Asian Aquaculture III: Fish Health Section*, (Eds.) T.W. Flegel and I.H. MacRae. Asian Fish. Soc., Manila, p. 21–31.

OIE 2000. *Diagnostic Manual for Aquatic Animal Diseases*. Office International des Epizooties, Third edition. Paris, France.

Plumb, J.A. 1997. Trends in freshwater fish disease research. In: *Diseases in Asian Aquaculture III: Fish Health Section*, (Eds.) T.W. Flegel and I.H. MacRae. Asian Fish. Soc., Manila, p. 35-47.

Chapter 7

Industrial Pollutant Effects on Fisheries: A Review

K.L. Jain, R.K. Gupta and Mukta Sharma

Department of Zoology and Aquaculture,
CCS Haryana Agricultural University, Hisar

Introduction

Water is one of the most important natural resources. But the quality of water has deteriorated due to contamination by industrial effluents and agricultural runoff. These effluents often contain various metal compounds and pesticides which constitute a major group of aquatic pollutants due to their non-biodegradable and bioaccumulative nature.

Heavy metals like arsenic, mercury, cobalt and nickel, and pesticides such as endosulfan, DDT, HCH, malathion etc. can cause abnormalities in fish physiology, morphology and behaviour.

Various parameters have been studied under natural conditions with regards to fish toxicology by several workers within India and abroad *viz.*, Rambhaskar and Rao (1987) on ten species of marine fish; Kaviraj and Ghosal (1997), Ghosh and Vass (2000) on *Macropthalmus erato.*

Industrial Effluents and Pollutants

Central Inland Fisheries Research Institute and National Environment Engineering Research Institute in 1981 reported that

Hoogly estuary received 28.5ppm of chromium from tannery and 9.6 ppm of chromium in the sediments, 120.4 ppm of zinc, 127.8 ppm copper and 20 ppm chromium in fish and 112.2 ppm zinc, 10 ppm copper and 0.3 ppm chromium in Shrimps was reported by Kaviraj (1989) after he surveyed Hooghly estuary during 1977–1981.

Ghosh *et al.* (2000) reported DDT, HCH and endosulfan residues in the lower stretch of river Ganga. The average value (µg/l) of DDT, HCH and endosulfan in the lower stretch of river Ganga was found to be 0.006, 0.003 and 0.0016, respectively, while the same in sediments (µg/kg) measured as 0.047, 0.300 and 1.900, respectively. A bit higher values (µg/kg) of these residues as a result of biomagnification were recorded in planktons (0.33, 0.027 and 0.014), benthos (0.628, 0.496 and 0.072) and the fish (0.436,0.408 and 0.074).

Renu *et al.* (1991) studied the water quality and metal speciation of Yamuna river from Dakpatpar to Agra. They observed that pollution increased along side it down the river and cadmium, copper, zinc, lead were available in high percentage.

Ghosh *et al.* (1990) assessed the impact of Darshana sugar mill effluents on water quality and biota of the Mathabhanga-Churni river system, West Bengal and also suggested suitable measures for effluent treatment.

Pardo *et al.* (1990) found that the behaviour of lead is bound to reducible phases and interfractions in the Piswerga river. The residual metal fractions can be taken as a guide for pollution.

Saikia *et al.* (1988) studied the water and wet sediments of river Ganga from a stretch of 480 kms extending from Badrinath to Narora and reported that Mn, Fe, Co, Cu and Zn existed below toxic limits, indicating the water to be unpolluted. As this stretch of river is free from anthropogenic activities which contribute to heavy metals pollution, thus they attribute the occurrence of metals in small amounts to geo-chemical source.

Bioaccumulation

Bhagwant (2003) determined the concentrations of copper and iron in tissues and organs of four edible fish species from the bay of Pourdre D' by flame atomic absorption spectrophotometry. The extent of metal accumulation differed among 4 species, possibly due to differences in behaviour and feeding habits. Copper levels among the species considerably varied ranging from 0.038 to 42.7 mg/kg wet weight. High copper content recorded from the liver of *Mugil cephalus* far exceeded the prescribed permissible limit and thus, requires close monitoring. Copper levels in the muscle tissue (edible part) were below the permissible limit. The distribution trend of Fe in various tissue/organs of species namely, *Mullioidicthys flavolineatus, Lutjanus notatus* and *Lutjanus fulvus* was liver > gut > gill > muscle. Conversely, the bioconcentration order of Fe in *Mugil cephalus* was gill > gut > muscle > liver. Bioaccumulation of Fe showed the following species differences in decreasing order: *Lutjanus notatus* > *Lutjanus fulvus* > *Mulloidicthys flavolineatus* > *Mugil cephalus*, thus suggesting a possible use of *Lutjanus notatus* fish species as biological indicator for Fe monitoring. Cu concentrations in the gills, gut, liver and muscle tissues of the four species considerably varied and were in the following decreasing order: *Lutjanus fulvus* > *Mulloidicthys flavolineatus; Lutjanus notatus* > *Lutjanus fulvus* > *Mulloidicthys flavolineatus. Mugil cephalus* > *Lutjanus notatus* > *Mulloidicthys flavolineatus/Lutjanus fulvus; Lutjanus fulvus/Lutjanus notatus* > *Mulloidicthys flavolineatus* respectively.

Zhou *et al.* (2003) demonstrated that accumulation of copper in fish tissues was related to metallic species (zinc, lead or cadmium) and concentration and type of tissue. The interaction became more significant with the increase in the concentration of metals. The accumulation of copper decreased in gills in the mixture-metal interaction, but increased in the liver and brain. The metal mixture could not affect the accumulation of copper in the muscle. The other metals did not change the order of copper accumulation in tissues. The order of copper accumulation in the fish tissues was liver > gill > brain > muscle.

Faraq *et al.* (2000) studied the distribution of metals during digestion by cut throat trout. Lead was lowest of all metals in the fluid portion of intestinal contents. However, more than 80 per cent of all metals in the hind gut were associated with the particulate fraction where they may still be available for uptake. The metals studied were copper, arsenic, cadmium, lead and zinc.

Lin *et al.* (2000) studied the cadmium tolerance in larval *Oreochromis mossambicus* and revealed that exposing female tilapia to cadmium did not affect reproduction and egg quality nor the survival of the offspring.

Tiwari *et al.* (2000) studied the growth and accumulation of heavy metals in green mussel, *Mytilus viridis*. Specific growth rate of Mussel was 13.25 (L), 31.61 (total weight) and 3.38 per cent per month (mean weight) during the study period and it was revealed during studies that the metal load increased with increasing size.

Kargin (1998) reported accumulation of iron, zinc, copper, lead and cadmium in fresh waster fish *Capoeta banoisi*. Accumulation of cobalt in marine fishes was studied by Udea and Nakashera (1983). Embryos and larvae of seven fresh water fish species exposed to low concentration of cadmium in soft water showed mortality as retardation in growth at concentrations ranging from 4 to 12 μg Cd/l (Eaton *et al.*, 1978).

Sultana and Rao (1998) studied the bioaccumulation pattern of zinc, copper, lead and cadmium in grey mullet, *Mugil cephalus* from harbour waters of Vishakhapatnam, India and reported that alimentary canal and gills can be considered as the interface of the organism and its ambience. The former is the site directly exposed to the ambient condition and is also known for its excretory functions even for some metals like zinc.

Vijayraman *et al.* (1998) determined the uptake and loss of Cd and Zn in various tissues of the fresh water prawn *M. malcolmsonii*. A dose and time dependent uptake of Cd was noted in all the tissues. However, prawns were able to regulate the tissue concentrations of Zn up to a threshold level of dissolved metal exposure. The rate and degree of depuration varied for the two metals. Since the Cd concentrations found in the tissues reflect the external concentrations, these prawns are ideal tools for monitoring environmental Cd level.

Deb and Santra (1997) analyzed the bioaccumulation of copper, lead, zinc and chromium in the liver, intestine, muscle and brain in four fishes, *Channa punctatus, Tilapia mossambicus, Catla catla* and *Labeo rohita* in a sewage fed jheel ecosystem.

Rema and Philip (1997) studied the pattern and extent of bioaccumulation of zinc and mercury in various tissues monitored by analyzing the metal content of liver, gill, caudal muscle, scale and skin of the fish exposed to the respective metals for different periods. It is seen that the levels of the essential metal, zinc, is under metabolic control, whereas that of the non-essential metal, mercury, is the least regulated.

Kureishy and Silva, (1992) conducted experiment on *Tilapia mosambica, Perna viridus* and *Villorita cyprinoides* on the uptake and loss of Hg, Cd and Pb in various tissues of these organisms. Hg was found to be highly toxic to the clams as compared to the fish and mussels. Pb was found to be least toxic. Hg was found to accumulate in greater amounts as compared to Cd and Pb. Preferential accumulation and subsequent retention was observed for each metal to a particular tissue, thus Hg showed affinity towards muscles, Cd towards liver and Pb towards gills. Similar trends of distribution were observed in tissues of fishes collected from the Indian coast except that Pb showed affinity towards gills as well as kidney. The rate and degree of depuration varied for different metals. Since, the metals concentrations found in tissues of these organisms reflect the external concentrations, therefore, make them ideal tools for monitoring environmental levels.

Mohan and Chaudhary (1991) using Atomic Absorption Spectrophotometer studied uptake of zinc in few tissues of fish, *Puntius sophore,* exposed to 7.5 mg/l of $ZnCl_2$ for 48 and 96 hours. Intestine accumulated the highest amount of Zn on dry weight basis followed by gills, liver and kidney.

The accumulation of cadmium in different tissues of *Heteropneustes fossilis* was studied by Das and Kaviraj (1990) after exposure to different concentrations of cadmium (2.5, 6.4 and 55.5 mg/l). Liver accumulated highest amount in longest exposure, followed by gills and kidney.

Schultz and Hermanutz (1990) studied the transfer of toxic concentrations of selenium from parent to progeny in the Fathead Minnow (*Pimepohales promelas*).

Carpene *et al.* (1989) studied the uptake of cadmium, zinc and copper in liver and bile of Gold fish, *Crassius auratus,* after exposing to three metals individually. They reported higher accumulation of cadmium in liver, where as zinc accumulation was less evident and copper was absent.

Radhakrishanaiah (1988) studied the degree of accumulation of copper in the gills, liver, brain and muscle of *Labeo rohita.*

Olassen *et al.* (1988) observed that when Rainbow trout was exposed to cadmium, kidney accumulated the maximum amount as compared to other tissues.

Mirenda (1986) studied the Acute Toxicity and accumulation of zinc in the crayfish, *Orconectes virilis* (Hagen). He maintained temperature at 18°C, pH 7.1, hardness at 26 mg $CaCO_3$/l, alkalinity 15 mg $CaCO_3$/l, dissolved oxygen 8.0 mg/l and found that the gills are the main site of zinc uptake from the water. Appreciable levels were found in the gills at all exposure concentrations (130 mg, 63.3 mg, 26.8 mg, 12.2 mg and 5.2 mg Zn/l) and increased as the concentration in the water increased. The abdominal muscles were the least variable tissue with regard to the changes in the zinc concentration over the range of test concentrations used. Under normal conditions, the crayfish appears capable of regulating zinc content through mechanism for absorption across the body surface and loss via faecal production.

Ney *et al.* (1985) concluded that treating Blue gill sun fish, *Lepolnmis macrohirus* with medium concentrations of zinc, cadmium and lead, freezing of sun fish prior to sample preparation can depress metal concentration in both tissues and whole fish.

Oladimeji *et al.* (1984) studied the long term effects of arsenic accumulation in rainbow trout, *Salmo gairdneri* and found that the arsenic accumulation in rainbow trout tissues increased with increasing levels of arsenic in the diet. The liver, contained the highest levels of arsenic. A high level of arsenic in the liver have been associated with functional change and it may cause histological changes.

Kaviraj and Konar (1983) determined the LC_{50} and LC_{95} values of Hg-G-Cd mixture for the fish *Tilapia mossambicus* which was in the range of 90.0 to 190.5 ppm. Sehgal and Saxena (1986) examined the acute toxicity of cadmium and zinc to *Puntius ticto* and reported the LC_{50} for 24, 48, 72 and 96 h as 35, 30, 28 and 26 ppm for cadmium and 75, 70, 66 and 60 ppm for zinc respectively.

Stromberg *et al.* (1983) studied pathology of lethal and sub-lethal exposure of Fathead minnow to cadmium. The radiotracer studies using cadmium 115 showed a rapid uptake and two-phase elimination of cadmium by the fish.

Greig *et al.* (1983) studied the concentration of silver, cadmium, copper and lead in the liver of Windopane flounder and reported values which were typical for fin fish.

Bull *et al.* (1981) studied growth rates and mercury content of Roach, Perch and Pike living in sewage effluent and recorded slightly elevated mercury content in Roach and Perch. Methyl mercury represented a high proportion of mercury found in all species.

Bonnell *et al.* (1960) observed that cadmium accumulates preferentially in the kidney of the fish and on prolonged exposure, cadmium is deposited initially both in the liver and kidney in comparable concentration. In this study it did not accumulate as much arsenic as the liver. The gills did not seem to accumulate any substantial amount of arsenic during the continued exposure. This observation might be due to the fast elimination of arsenic via the gills.

Kaviraj and Ghosal (1997) studied the bioassays lasting 180 days with common carp, *Cyprinus carpio* exposed to eight different treatments of cadmium (2.5 mg/l) alone or with compost manure. Cadmium alone did not show any adverse effect on the growth of fish but compost manure, irrespective of its dose and combination with cadmium, markedly increased the growth of fish and primary productivity of water.

Gills always are the main site of water movement (Evan's, 1969; Motais *et al.*, 1969). When the fish is exposed to heavy metals in the ambient medium, gill tissue rapidly accumulates the metal (Kumada *et al.*, 1972).

Rita rita was tried as test species for toxicity bio-assay of some heavy metals. As per the added doses, Cu (LC_{50}–2.5 mg/l) was most toxic followed by Zn (LC_{50}–11.22 mg/l), Cr (LC_{50}–213.8 mg/l), while the incipient concentration of these metals indicated different toxicity patterns like Cu > Zn > Pb > Cd > Cr (Ghosh and Mukhopadhyay, 2000).

Effects of Heavy Metals

Bhatia and Sandhu (2004) determined that *Heteropneustes fossilis* exposed to different graded concentration of the pesticides for 30 days toxicity induced copious mucous secretion, loss of pigmentation, exophthalmia and haemorrhagic condition. Anomalies in body proportion were also observed. Eye diameter showed an increase and maximum body width revealed a decrease.

Abbas *et al.* (2003) studied that copper and lead are among the most widely common pollutants distributed in the aquatic environment. The 96-hr LC_{50} bioassays were carried out to the determine the toxicity of these two heavy metals (Cu and Pb) on the blue tilapia; *O. aureus*. The results showed that the 96-hr LC_{50} of copper and lead were 8.17 and 25.86 mg/l, respectively. Groups of adult and healthy *O. aureus* were subjected to the acute (lethal) concentration (96-LC_{50}) of each metal and examined after 24 and 96 hours. Other groups were subjected to the chronic (sub-lethal) concentration (1/10 LC_{50}) of each metal and examined after 6 and 12 weeks. The lethal and sub-lethal concentrations of copper and lead altered the behavioural activities of the treated fish. The gills, liver, kidney and spleen tissues were irritated and injured causing dysfunction in these organs. Haemoglobin contents were increased in fish toxicated by copper and decreased in those toxicated by lead. Serum glucose was increased in all groups treated with copper or lead. The serum AST and ALT recorded the same trend of increase. Moreover, the creatinine and the uric acid levels were elevated.

Anita, Susan and Tilak (2003) evaluated the toxic sub-lethal concentration of fenvalerate (technical grade) in the gill, liver and kidney, of one of the Indian major carps *Cirrhinus mrigala*. There were marked pathological changes like necrosis, progressive degeneration in the gill tissue, liver tissue revealed atrophy, appearance of blood streaks among hepatocytes. Changes in haemopoietic tissue, severe necrosis, vacuoles around renal tubules and hemorrhage etc. were evident in the renal tissue. All these pathological changes make fish less fit for survival. Separation of epithelial layer from the central sinus of filaments was noticed. The degree of dilation of primary gill lamella was more when compared to secondary gill lamella.

Dinodia *et al.* (2003) studied the effects of cadmium toxicity on proteolytic enzyme activity in some fresh water carps and it was found that proteolytic enzyme activity declined maximally in *C. mrigala* from 26 per cent at 2.5 ppm to 55.6 per cent at 5.0 ppm. In *L. rohita* it showed maximum decline at 5.0 ppm (32.6 per cent) as compared to the control fish.

Palaniappan *et al.* (2003) have reported that histopathological effect of the two different sub-lethal concentrations 1.08 ppm and 3.61 ppm of nickel on *Cirrhinus mrigala* fingerlings produced severe damage and changes in cellular levels in gills leading to the death of the fish. The changes in the histology of gills led to the disturbance in basement membrane, degeneration of gill lamellae, swelling of base, inter-lamellar space, pycnotic and necrotic chloride cells, hyperplasia and fusion of secondary lamellae of fish to higher concentration causing failure of respiratory mechanism.

Mikryakov *et al.* (2002) studied that the levels of lysozyme in tissues of spleen, liver and heart of Siberian sturgeon fingerlings, (*A. baeri*), kept in sub-lethal concentrations of heavy metal salts consisting of $HgCl_2$ (3 µg/litre), $CUSO_4$ (15 µg/litre) and $CdNO_3$ (30 µg/litre) were estimated at 15–17°C for 60 days. The highest reaction to toxicants based on the average level of lysozyme was recorded in spleen. In liver, increase in the level of lysozyme through 7 and 14 days exposure was also marked, but it was significant only in the presence of Hg ions.

Sindhe *et al.* (2002) studied *N. notopterus* caught from Simoor Nala near Gulbarga city, India, subjected to toxicity tests by exposing them to different concentrations of heavy metals for 24, 48, 72 and 96 h *i.e.*, mercuric chloride (0.02), cadmium chloride (4.5) and a mixture of both (0.03 ppm) for 2 months, respectively, with replacement of the water and heavy metal concentrations every 48 h. Results showed that the GSI was highest in the mercuric chloride and lowest in the cadmium chloride exposed fish while HIS was lowest in the mercuric chloride exposed fish. Protein, lipid and cholesterol contents in both the ovary and liver also decreased.

Vincent *et al.* (2002) conducted a study to determine the effect of cadmium, 2.5, 3.0, 3.5 and 4.0 mg/litre for 120 days, on the bioenergetics of freshwater fish, *Catla catla*. An inverse relationship ($P < 0.05$) between cadmium concentration and food utilization parameters like food consumption, assimilation metabolism, assimilation efficiencies and production efficiencies was observed. Heavy metal intoxication was also found to exhibit reduction in biomass.

Ahilan and Jeyaseelan (2001) studied the effect of vitamin B_{12} and cobalt chloride on the growth and maturation of goldfish. Cobalt chloride incorporated at 7.5 ppm gave positive results both for gonadal development and growth. Among the doses tested for vitamin B_{12}, the 25 ppm concentration had shown highest response for both maturation and growth.

Jia (2001) determined the effect of four heavy metals (copper, mercury, zinc and lead) on the respiration intensity of juvenile *Misgurnus anguillicaudatus* and results showed that the fish were sensitive to heavy metal pollution. The respiration intensities of different groups of fishes treated with heavy metals were higher than those of control in which Cu showed maximum and Pb the minimum effects with the respiration.

Kannupandi *et al.* (2001) studied effect of endosulfan, HCH, copper and zinc on the phototactic and swimming rate of larvae of the mangrove crab *Macropthlamus erato* (De Man). The different contaminants tested, altered the positive phototactic response and swimming rate of *M. erato* larvae. Positive phototactic response was observed only in the early period of study, ranging from 6–12 h under endosulfan, HCH, copper and zinc stress. A negative photoactic response, followed by no response was observed at later stage of toxicity tests. The severity in the impairment of the response was more pronounced in the highest test concentration. Swimming rate decreased during exposure to endosulfan, HCH, copper and zinc in different Zoe larval stages of development. Decreased swimming rate was directly related to test concentrations.

Patil and Dhande (2001) studied that the effect of mercury and copper at sub-lethal concentrations on ovarian architecture and cholesterol content in the fresh water fish, *C. punctatus* was evaluated. 25 healthy *C. punctatus* were exposed to sub-lethal concentrations of $HgCl_2$ and $CuCl_2$ separately for 30 days. A significant reduction in gonosomatic index (GSI) after 10 ($P < 0.05$) and 30 ($P < 0.01$) days of exposure to $CuCl_2$ and $HgCl_2$ was observed. Extensive vacuolation in the oocortex, necrosis of the oolemma and hypertrophy of follicular cells were observed. Degeneration of vitellogenic and large pre-vitellogenic oocytes were observed after 30 days of exposure to the 2 heavy metals. A significant ($P < 0.01$) decrease in cholesterol level was observed in the ovary exposed to $CuCl_2$. Reduction in cholesterol level was observed in the ovary exposed to $HgCl_2$. It is suggested that heavy metals possibly block vitellogenesis. Moreover, damage in ovarian architecture and depleted ovary cholesterol can be used as one of the index of heavy metal pollution.

Espina *et al.* (2000) evaluated the effect of 96 h exposure to sub-lethal cadmium concentrations on oxygen consumption rate (VO_2 mg^{h-1}) and the oxygen extraction efficiency (OEE' %) was determined in juvenile *Ctenopharyngodon idella* acclimated to 26, 29 and 32°C for at least two weeks. The relationships among the physiological rates (VO_2 and OEE′), body weight and temperature, were fitted by second order polynomial models; the response surfaces generated allowed the evaluation of stress induced by the interaction of those variables. Both VO_2 and OEE′ invariably increased as temperature increased. In small animals kept in 0.5 mg Cd L^{-1}, the VO_2 increase was compensated by a proportional increase in OEE′, whereas in larger fishes (> 1 g) no compensatory response was observed. Both small and large fishes were unable to compensate for a higher VO_2 when exposed to higher cadmium concentration. It is concluded that OEE′ in young fish is an early response of stress due to sub-lethal cadmium exposure.

Golovanova *et al.* (1999) studied *in vitro* effects of cadmium (0.5–50 mg/l) and DDVP (dischlorvous, at 0.2–100 mg/l) on the total amylolytic, sucrose and proteinase activities of intestinal mucosa for the first time in eleven fresh water toleosts. Total amylolytic activity in bur bot, *Lota lota*, crucian carp, *C. auratus* and common carp, *C. carpio*, sucrose activity in blue bream and total proteolytic activity in bur bot and pipe were significantly decreased by cadmium at 50 mg/l, DDVP (at 0.2 mg/l) caused a significant decrease in total proteolytic activity in pipe, but had no effect on either protease or carbohydrase activities in other fish species.

Usha Rani (1999) studied the effect of lethal (50 ppm) and sub-lethal (5 ppm) concentrations of cadmium chloride ($CdCl_2$) on the gills of fresh water fish *Oreochromis mossambicus* by routine histological techniques. Many pathological changes were observed in gill lamellae, such as congestion of primary rachis, primary and secondary gill lamellae, bulging of primary and secondary gill lamellae, hypertrophy and hyperplasia of interlamellar cells, separation of respiratory epithelial layer, atrophy of several of the secondary gill lamellae and necrosis leading to severe haemorrhage of gill lamellae.

Hidalgo *et al.* (1999) conducted a comparative study on the digestive, proteolytic and amylolytic activities in six species of fish with different nutritional habits. Trout and carp showed the highest digestive proteolytic activity. Eels showed the lowest digestive proteolytic potential among the species studied. The ratio of total amylase to the total proteolytic activity was higher in omnivorous fish species.

Increase in total protein in the fry of *C. carpio* was reported by Obula Reddy *et al.* (1999) in thirty days old fries, exposed to 2.32 ppm of liquor ammonia for 7, 14 and 21 days.

Rangaswamy and Naidu (1999) determined that the oxygen equilibrium curves of the blood of the control *O. mossambicus* was sigmoid and P_{50} and 'n' values were 12.5 mm Hg and 3.57 respectively. Upon exposure to sub-lethal and lethal concentration of endosulfan the oxygen equilibrium curves of the blood moved to the right. Thus, the P_{50} value in the treated fish tended to increase signifying the reduction in the affinity of the pigment for oxygen. A decrease in blood pH due to increase in CO_2 content was noticed. This caused increase in P_{50} and consequent shifting of oxygen equilibrium curves to the right in the endosulfan treated fish and showed the existence of normal or negative Bohr effect. Thus endosulfan exposure causes not only the direct harmful effects on the fish, but also the indirect effect on gas exchange property of blood.

Sahoo and Mukherjee (1999) with rohu (*L. rohita*) fingerlings, weighing between 29–38 g made a two months study to establish normal ranges for several haematological parameters and serum protein levels. The morphology of blood cells was also described and it showed wide variations in heamatocrit, mean corpuscular volume and mean corpuscular values of individual healthy fish were also observed.

Gupta and Soni (1998) studied dietary requirement of zinc for the fingerlings of *C. mrigala*. Zinc levels of 10 and 20 mg/kg of diet had significant effect on growth and conversion in *C. mrigala* fingerlings. Low dietary Zn also increased whole body protein and carbohydrate but moderately the fat values. Zinc deficient diet in fish showed certain deficiency symptoms.

Thiruvalluvam *et al.* (1997) stated that heavy metal, cadmium and organophosphate methylparathion when tested on *C. carpio* var. *communis* showed that it was more sensitive to latter than to former. Both are genotoxic, inducing the formation of micronuclei in the erythrocytes.

Ahmad *et al.* (1996) reported that comparison of cadmium, nickel, lead, copper, manganese and zinc in muscle, liver and skin of *Epinepliehus tauvina* and *Lutjanus fulviflamma* indicated that, in general, muscle tissue had markedly lower concentration of metal as compared to liver but higher than in skin.

Hota (1996) mentioned significant fall in the levels of proteins, cholesterol and ascorbic acid in the liver, brain and intestine of *Channa punctatus* on account of arsenic toxicity at 5 ppm for 75 days. The loss of protein has been attributed to proteolysis and utilization of amino acids in various catabolic reactions. Another possibility might be blocking of protein synthesis.

Mukhopadhyay *et al.* (1994) studied the toxicity of industrial heavy metals, Zn, Cu, Cd and Cr to estuarine fishes: *Liza parsia, Appocryptis hato, Therapon jarbua* and *Mystus gulio,* Prawn: *Macrobrachium rude* and fish food organisms *Cyclops* sp., *Daphnia carinata, Chrionomus* sp. and *Lymnae leuteola* were evaluated. Amongest estuarine organisms, crustaceans were least resistant to heavy metal toxicity. Fishes showed wide fluctuations in tolerance. Toxicity of heavy metals was inversely related to salinity and in combination Zn, Cu and Cr were more toxic to fish.

Brafield and Koodie (1994) studied the effect of high dietary zinc on trypsin activity in carp, *C. carpio.* It was observed that carp fed with pellets containing a range of zinc concentration showed a significant increase in trypsin activity in the intestine with increasing zinc in diet.

Midha and Singhal (1994) determined that the cobalt chloride by acting as a constituent of vitamin B_{12}, increases the production of zooplanktons, which subsequently increases the growth of *Labeo rohita* and improves their quality.

Visvikio and Rachlin (1994) observed that metals may act synergistically where the toxic response is greater than the sum of individual toxicities. James *et al.* (1992) studied the effects of sub-lethal doses (Cu, Cd, Zn) mixtures on the activities of key respiratory enzymes (succinate dehydrogenase, SDH and glyceraldehyde dehydrogenase, GDH) and their recovery following withdrawal of treatments in the freshwater fish *O. mossambicus.* On the basis of 96 hr LC_{50}, Cu was highly toxic followed by Zn and Cd, and the trimetal combination (Cu + Zn + Cd) was extremely toxic than any other combinations; Zn + Cd was least toxic. A significant gradual decrease in SDH with a concomitant increase in GDH activity observed in liver, brain, muscle and gill of animals exposed to metal suggest a metabolic shift from aerobiosis to anaerobiosis due to metal action. Exposed individuals when transferred to metal impoverished water showed an improvement in SDH activity and decline in GDH activity suggesting slow reversal to aerobic metabolism in *O. mossambicus* needs more time for complete recovery.

Gupta and Chakarbarty (1993) studied the toxicity of zinc to fresh water teleost, *Notopterus notopterus* (Pallas) and *Puntius javanicus* (Blkr). The comparative account of behavioural response against zinc toxicity exhibited that *Puntius javanicus* was more sensitive than *Notopterus notopterus*.

Kirunbagaran and Jay (1992) exposed *Clarius batrachus* to selected sub-lethal concentrations of mercuric chloride (0.05 mg/l), methyl mercuric chloride (0.04 mg/l) and emisan (organic mercurial fungicide; 0.5 mg/l) for intervals of 45, 90 and 180 days. The gonadosomatic index showed a significant decrease after 90 and 180 days exposure to mercurials. Histologically, the seminiferous tubules were smaller in size and contained mostly spermatids in comparison to the control fish in which they were greatly distended and full of spermatozoa. Observations suggested that impairment of testicular lipid metabolism by mercury is one of the possible factors that lead to the inhibition of steriodogenesis and spermatogenesis.

Varghese *et al.* (1992) studied *Barytelphusa cunicularis* (Westwood), a freshwater crab was exposed to mercuric chloride, copper sulphate and zinc sulphate from 0 to 12 hr and the oxygen consumption of the animal was measured in order to study the stress caused by these heavy metals. Normal oxygen consumption was affected by the three heavy metals. Similarly, crabs exposed to sub-lethal concentrations of the same pollutants for 24, 48, 72 and 96 hr showed an elevation in the blood sugar level with a maximum increase at 48 hr. The results indicate a switch towards glycolysis in order to overcome the anaerobic stress caused by the heavy metals.

James *et al.* (1991) studied the individual and combined effects of zinc, copper and cadmium on growth of *Oreochromis mossambicus*. The combined effect of metals on food utilization in *O. mossambicus* was more severe than when used individually. The reduction in growth rate in animals exposed to Zn, Cd and Cu was about 32, 42 and 47 per cent respectively and in those exposed to metal combinations of Zn + Cd, Cu + Zn and Cu + Cd, it was 65, 73 and 86 per cent, respectively as compared to the control. Animals exposed to metals showed significant decline in food intake. Metabolic rate was increased by 34 and 70 per cent in the animals exposed to Cu alone and Cu + Cd. The combination of Zn + Cd was less toxic than the combinations of Cu + Zn and Cu + Cd on food utilization parameters as observed by James *et al.* (1991). The combinations of Cu + Zn and Cu + Cd were highly toxic because of their synergistic effect (Metelev *et al.*, 1971 and James *et al.*, 1991).

Khanee *et al.* (1991) determined the sub-lethal effects of mercury on growth and biochemical composition and their recovery in *Oreochromis mossambicus*. The medium lethal concentration of mercury for 96 hr was 0.06 ppm and the test animals were exposed to sub-lethal concentration of mercury (0.01 and 0.02 ppm) for four weeks and they were then transferred to metal impoverished water for another four weeks (recovery period). Protein, lipid and carbohydrate content were estimated in tissues of muscle, liver and gill. Significant reduction of growth and biochemical components was observed during exposure period and the loss was regional up to certain extent during recovery period.

Pande and Saxena (1990) examined various histopathological changes in gills and liver of the fingerlings of an Indian major carp, *Labeo rohita* exposed to sub-lethal concentration of copper. The results showed an increased hyperglycemia accompanied by complete fusion of adjacent secondary gill lamellae, detachment of epithelium from underlying connective tissues, increased number of mucous cells. The vacoulation in the hepatocytes and degeneration of cell membrane were observed in the liver.

Benedetti *et al.* (1989) observed the histomorphological changes in some organs of brown bullhead, *Ictalurus nebulosus* following short and long term exposure to copper. Organs studied for histology were skin, liver, gills, gut and kidney and some enzymatic tests were also performed on the liver which confirmed a mild onset and slow evaluation of liver distress.

Dimitrova and Tishinova (1989) studied common carp, *C. carpio*, approximately one year old, kept in water containing cadmium sulphate in concentration of 0, 2.5, 5.0 and 7.5 mg/l for 24 hours. Lipid peroxidation was found reduced in liver tissues in control fish. It increased with increasing cadmium concentration in water. It was concluded that cadmium participates in oxidation reactions in the liver cells to produce lipid peroxides.

Paulose (1989) observed fusion of adjacent lamellae due to hyperplasia, vaculation and desquamation in lamellae and necrosis were the histological changes in relation to accumulation of inorganic and methyl mercury in the gills of *Labeo rohita*.

Sarkar (1989) evaluated the effects of $HgCl_2$ and $CrCl_2$ on the oxygen consumption of a fresh water fish, *Tilapia mossambica*. Results indicated that oxygen consumption of fish exposed to $CrCl_2$ is higher than that of $HgCl_2$. At 0.5 ppm of $HgCl_2$ and 0.7 ppm $CrCl_2$ onwards oxygen consumption of fish progressively decreased.

Gill *et al.* (1988) described that pathogenesis of branchial and renal lesions in common fresh water fish, *Puntius conchonius*, exposed chronically to sub-lethal levels of methoxy ethylmercuric chloride (MEMC). The secondary gill lamellae in control fish were found to be slender, erect with an uninterrupted epithelium covering the pillar cells and contained interspersed sinuses. MEMC chronic intoxication caused dropping of secondary lamellae as well as capillary congestion. In seven days exposure of 3.6 ppm MEMC showed aneurysm and lysed RBC's in lamellar sinuses and epithelial necrosis, hypertrophied chloride cells and cellular debris was revealed after studying the slide on seven days exposure to 6.0 ppm MEMC. Secondary lamellae was showing fusion, lifting of epithelium and wilting of pillar cell system, depleted and atrophied mucocytes on 56 days exposure to 6.0 ppm MEMC. Renal histopathology results did show necrosis of tubular epithelium and pycnotic and karyohetic nuclei on 7 days exposure to 3.6 ppm MEMC; renal tubules with hyperchromatic deposition in epithelium on 7 days exposure to 6.0 ppm MEMC. Part of kidney were showing edematous spaces and several shrunken glomeruli on less than 56 days exposure to 3.6 ppm MEMC.

Dogfish subjected to sub-lethal cadmium exposure (25 mg/l) for 24 and 96 h. by Tort and Torres (1988) showed decreased haematocrit, leucocrit and mean corpuscular volume and increased haemoglobin concentration, red blood cell count and mean corpuscular haemoglobin concentration after 24 h exposure. Some parameters which were unchanged at 24 h showed a significant variation at 96 h (plasma glucose and lactate levels).

Beena *et al.* (1987) studied the effects of Cd and Hg on the fish *C. carpio*. Bhattacharya *et al.* (1987) concluded that blood glucose and hepatic glycogen profiles in *C. punctatus* exposed to non-lethal concentrations of single and mixture of pollutants responded diversely. In short term test hyperglycemia in hepatic glycogen was recorded while in long term exposure, hyperglycemia with persistent depletion of hepatic glycogen in treatment with Cd, Hg and mixture of pollutants was observed.

Chaudhry and Nath (1985) estimated the effect of nickel sulphate at sub-lethal dose of 64 ppm on the blood glucose levels of fresh water fish *Colisa fasciatus* from 3 to 96 hr. The blood glucose level exhibited a steady increase due to nickel toxicity. A maximum increase of 85.08 per cent was observed at 96 hr. It was suggested that the hyperglycemia in *Colisa fasciatus* caused by exposure to nickel sulphate, was possibly a reflection of stress induced hormone mediated response. It appears that blood glucose level is a reliable indicator of nickel toxicity to fish.

Chhattarjee and Bhattacharya (1984) demonstrated that in teleosts, pollutants influence thyroxine release mediated via the peroxidase system of iodination. Mixtures of toxicants either prepared in the laboratory or collected from a factory discharge, are equally detrimental towards the thyroxine homeostasis. Inhibition of head kidney peroxidase was usually associated with a decrease in iodide peroxidase activity and blood thyroxine level.

Sastry and Sunita (1983) reported alterations in the intestinal absorption of xylose in *C. punctatus* after exposure to cadmium, lead, mercury, nickel and malate dehydrogenase in liver and muscles of *C. punctatus* in both acute and chronic exposure to cadmium (1.12 mg/l) for 120 days. Glutamate oxaloacetate transaminase and glutamate pyruvate transaminase activities increased in blood, liver and muscles.

Sastry and Subhadra (1982) reported elevations in the activities of lactate dehydrogenase and succinate dehydrogenase in liver and sub-lethal concentration of cadmium for 15 and 30 days. Further Sastry and Subhadra (1985) reported decreased activities of acid phosphatase, alkaline phosphatase, hexokinase, xanthine oxidase, glutamate dehydrogenase in liver of *H. fossilis* exposed to cadmium (0.26 ppm) for 15, 30, 60 days.

Gill and Pant (1981) reported that juveniles and adults of *Puntius conchonius* were exposed to mercury in soft and hard waters, Juveniles were intolerant to high mercury concentrations and toxicity varied inversely with water hardness. Short term exposure to sub-lethal concentrations of mercury elicited significant hyperglycemia and glycogenolysis in liver and brain. Upon chronic exposure to 36.3 and 60.5 µg $HgCl_2$/l, there was hyperglycemia persisting up to four weeks. The glycogen levels in liver and brain remained low up to three weeks but an increase in myocardium glycogen was registered after 8 weeks exposure. Erythropenia and a fall in haemoglobin occurred following 1–3 week exposure but polycythemia was recorded after 6–8 weeks.

Deshmukh and Marathe (1979) reported that 96 h. LC_{50} values of copper and mercury in the fry and fingerlings of *L. rohita* and *Cyprinus carpio* showed that susceptibility varies with the size of fish, fry being more susceptible. Similarly, the 96 h LC_{50} values of 3 size groups of *Labeo raticulatus* were directly proportionate to the body weight with a significant correlation coefficient.

Diwan *et al.* (1979) studied the levels of blood glucose and tissue glycogen in two fish exposed to industrial effluents. The test fish were *H. fossilis* and *C. batrachus*. There was 92.2 per cent and 95.3 per cent increase in blood glucose level in *H. fossilis* and *C. batrachus* respectively after exposure to effluents. In *C. batrachus* a significant decrease in glycogen level was found in liver, muscles and gills, whereas in heart and kidney there was a slight increase. A significant decrease in glycogen level in *H. fossilis* was observed only in liver and gills.

The interaction between zinc and other metals has been investigated in fish by few workers (Brown and Dalton, 1970; Spehar, 1976; Broderius and Smith, 1979). Results from these experiments ranged from antagonistic to synergistic responses.

Singhal *et al.* (1973) reported that cadmium treatment decreased the level of hepatic glycogen. The concentration of food glucose and urea was significantly elevated by this heavy metal. Discontinuation of the heavy metal treatment for 28 days in rats previously infected with cadmium for 45 days failed to restore the observed biochemical alterations in hepatic and renal carbohydrate metabolism to control values.

Christensen (1972) determined that the plasma lactate dehydrogenase was unaffected by copper exposure in *Ictalurus nebulosus* while plasma glutamic oxaloacetate transminase (GOT) activity at low copper concentrations decreased, but high concentration of copper cause liver damage and leakage of GOT in the blood in white sucker (*Catastomus commerssonu*), caused increased activity of GOT.

Hewitt and Nicholas (1963) studied that the close relationship between enzymes and organ toxicity has made enzyme estimations important in the diagnosis of various diseases. Heavy metals are known to inhibit enzymes requiring sulfahydryl groups in the expression of activity.

By the survey of literature available in respect to the biochemical and histopathological changes due to heavy metals in fish, it is evident that still there is enough room for additional study in the mentioned fields as the information available is not comprehensive, particularly with reference to metal effects in combination in fresh water fish species.

References

Abhas, H.H., Zaghloul, K.H. and Mousa, M.A. 2003. Effect of same heavy metal pollutants on same biochemical and histopathological changes in blue tilapia, *Oreochromies aureuns. Egyptian J. Agril. Res.*, 80(3): 1395–1410.

Ahilan, B., Jeyaseelan, M.J. 2001. Effect of cobalt chloride and vitamin B_{12} on the growth and gonadal maturation of goldfish *Carassius auratus. Indian J. Fish.*, 48(4): 369–374.

Ahmad, S. and al-Ghais, S.M. 1996. Metal content in the tissue of *Lutjanus fulviflamma* (Smith, 1949) and *Epinepliehus tauvina* (Farskal, 1775) collected from Arabian Gulf. *Bull. Environ. Contam. Toxicol.*, 57: 957–963.

Beena, S. and Viswarajan, S. 1987. Effect of cadmium and mercury on hematological parameters of the fish, *Cyprinus carpio. Env. and Ecol.*, 5: 726–732.

Bhagwant, S. 2003. Bioaccumulation of copper and iron in edible fish species from Pourdre D'orbay, Mauritisu. *Poll. Res.*, 22(2): 293–297.

Bhatia, N.P. and Sandhu, G.S. 2004. Endosulfan induced morphological. changes in *Heteropneustes fossilis* (Bloch). *Proceedings of the National Workshop on Rational use of Water Resources for Aquaculture*, Hisar, March 18–19, p. 178–181.

Bhattacharya, T., Ray, A.K. and Bhattacharya, S. 1987. Blood glucose and hepatic glycogen interrelationship in *Channa punctatus:* A parameter of non-lethal toxicity bioassay with intestinal pollutants. *Indian J. Exp. Biol.*, 25: 539–541.

Bnenedetti, I., Albano, A.G. and Mola, L. 1989. Histomorphological changes in some organs of the brown bullhead, *Ictalurus nebulosus* Le Suettr, following short- and long-term exposure to copper. *J. Fish Biol.*, 34: 273–280.

Bonnel, J.A., Ross, J.H. and King, E. 1960. Renal lesion in experimental cadmium poisoning. *Dist. J. Ind. Med.*, 17: 69–80.

Borah, Sabita and Yadav, R.N.S. 1994. Endosulfan induced metabolic alterations in the muscle of a freshwater fish. *Heteropneustes fossilis* (Block). *Geobios.*, 21: 289–293.

Brafield, A.E. and Koodie, A.V. 1994. The effect of high dietary zinc on trypsin activity in carp, *Cyprinus carpio. J. Fish. Biol.*, 45(1): 169–172.

Broderius, S.J. and Smith, L.L. Jr. 1979. Lethal and sub-lethal effects of some mixtures of cyanide and hexavalent chromium, zinc or copper to the fathead minnow and rainbow trout (*Salmo gairdneri*). *J. Res. Board Can.*, 36: 164–172.

Brown, V.M. and Dalton, R.A. 1970. The acute lethal toxicity to rainbow trout mixtures of copper, phenol, zinc and nickel. *J. Fish. Biol.*, 2: 211–216.

Bull, K.R., Dearsley, A.F. and Jnskip, M.H. 1981. Studies of growth rates and mercury content of roach, perch and pike living in sewage effluent. *Env. Poll.*, A25: 229–240.

Carpene, E., Fedrizzi, G. and Giani, G. 1989. Concentrations of Cd, Zn and Cu in liver and bile of *Carassius auratus* (Italian). *Arch. Ital.*, 40(3): 197–202.

Chatterjee, S. and Bhattacharya, S. 1984. Response of climbing perch, *Anabas Testudineus* to industrial pollutants: Head kidney peroxidase, iodide peroxidase and blood thyroxine profiles. *Water, Air and Soil Poll.*, 25: 161–174.

Chaudhry, H.S. and Nath, Kedar. 1985. Nickel induced hyperglycemia in the fresh water fish, *Colisa fasciatus. Water, Air and Soil Poll.*, 24: 173–176.

Christensen, G.M., McKin, J.M., Brungs, W.A. and Hunt, E.P. 1972. Changes in the blood of the brown dullhead, *Ictalurus nebulosus* (Leswer) following short and long term exposure to copper (II). *Toxico. Appl. Pharm.*, 23: 417–427.

Das, S.K. and Kaviraj, A. 1990. Accumulation of cadmium in *Heteropneustes fossilis* and changes in haematological parameters. *J. Nat. Conserv.*, 2: 25–30.

Deb, S.C. and Santra, S.C. 1997. Bioaccumulation of metals in fishes: An *in vivo* experimental study of sewage fed ecosystem. *Journal Article ASEA*, Part 3.

Deshmukh, S.S. and Marathe, V.B. 1979. Size related toxicity of copper and mercury to *Lepisteus reticulates, Labeo rohita* and *Cyprinus carpio. Indian J. Exp. Biol.*, 18: 421–423.

Dimitrova, M. and Tishinova, V. 1989. Effect of sub-lethal concentration of cadmium on some biochemical processes in carp (*Cyprinus carpio*). 6th *International Trace Element Symposium*, 5: 1461–1464.

Dinodia, G.S., Gupta, R.K., Jain, K.L. and Yadava, N.K. 2003. Cadmium toxicity and its effect on the proteolytic enzyme activity in some fresh water carps. *Proceedings of 3rd Interaction Workshop*, December 17–18, pp. 195–199.

Diwan, A.D., Hingorani, H.G. and Naidu, N.C. 1979. Levels of blood glucose and tissue glycogen in two live fish exposed to industrial effluents. *Bull. Env. Contam. Toxicol.*, 21: 269–272.

Dua, A.K. 1995. Toxic effects of organochlorine pesticide, endosulfan on some tissues of *C. punctatus* (Bloch, 1793). *Ph.D. Thesis*, Punjab University, Chandigarh.

Dubois, M., Gilles, K.A., Hamilton, J.K., Robers, P.A. and Smith, F. 1956. Calorimetric method for determination of sugars and related compounds. *Anal. Chem.*, 28: 350–356.

Eisler, R. 1979. Behavioural responses of marine poikilotherms to pollutants. *Phil. Trans. R. Soc.*, 286(B): 507–521.

Espina, S., Salibian, A. and Diaz, F. 1999. Influence of Cadmium on the respiratory function of the grass carp *Ctenopharyngodon Idella. Water Air and Soil Poll.*, 119: 1–10.

Faraq, A.M., Suedkamp, M.J., Meyer, J.S., Banaws, R. and Woodward, D.F. 2002. Distribution of metals during digestion by cut-throat trout fed benthic invertebrates contaminated in the Clark Fork River, Montana and the Coeurd' Alene River, Idaho, U.S.A., and fed artificially contaminated Artemia. *J. Fish Biol.*, 56: 173–190.

Gadhia, Mohini. 1989. Histopathological lesions induced by rogor in the liver of a siluroid. *J. Inland Fish Soc.*, India, 21(1): 55–57.

Ghosh, B.B., Mukhopadhyay, M.K. and Bagchi, M.M. 1990. Status of organic pollution in the Mathabhanga Churni river system, West Bengal. *J. Inland Fish Soc.*, India. 22 (1&2): 16–21.

Ghosh, Santanu and Mukhopadhyay, M.K. 2000. Toxicity of five industrial metals on Gangetic catfish *Rita rita*. *Geobios.*, 27: 93–96.

Ghosh, Santanu, Das, A.K. and Vass, K.K. 2000. DDT, HCH and endosulfan residues in the lower stretch of river Ganga. *Geobios.*, 27: 161–164.

Gill, S., Tjendra, Pant, C. Jagdish and Tewari, Hema. 1988. Branchial and renal pathology in the fish exposed chronically to methoxy ethylmercuric chloride. *Bull. Environ. Contam. Toxicol.*, 41: 241–246.

Gill, T.S. and Pant, J.C. 1981. Effects of sub-lethal concentration of mercury in a teleost *Puntius conchonius:* biochemical and haematological responses. *Indian J. Exp. Biol.*, 19: 571–573.

Gill, T.S., Pant, S.C. and Pant, J. 1988. Gill, liver and kidney lesions associated with experimental exposures to carbaryland dimethoate in the fish *Puntius conchonius* (Ham). *Bull. Environ. Contam. Toxicol.*, 41: 71–78.

Giridhar, P. and Indira, P. 1997. Effect of an organophosphhorus Nuvan on total lipids and lipase activity of the fresh water fish *Labeo rohita*. *Indian J. Comparative Animal Physiol.*, 15: 37–40.

Golovanova, I.L., Kuzmina, V.V., Gobzhelian, T.E., Davlov, D.F. and Chuiko, G.M. 1999. *In vitro* effects of cadmium and DDVP dichlorovos on intestinal carbohydrase and protease activities in fresh water teleosts. *Camp. Bioch. Physiol.*, 122(1): 21–25.

Greig, R.A., Schurman, S., Preira, J. and Patricia, N. 1983. Metal concentration of silver, cadmium, copper and lead in livers of windopane flounder. *Bull. Environ. Contam. Toxicol.*, 31: 257–262.

Gupta, A.K. and Chakarbarti, P. 1993. Toxicity of zinc to fresh water teleosts, *Notopterus notopterus* and *Puntius javanicus. Freshwater Biol.*, 5(4): 359–363.

Gupta, A.K. and Soni, M.K. 1998. Dietary requirement of zinc for the fingerlings of *Cirrhinus mrigala. Geobios.*, 25: 261–265.

Gupta, A.K., Saini, C.B.B. and Rajbanshi, V,K. 1997. Influence of Mn on growth of *Cirrhinus mrigala* Fingerlings. *Geobios.*, 24: 219–222.

Hewitt, E.J. and Nicholas, D.J.D. 1963. Cations and anions inhibition and interactions in metabolism and enzyme activity. In: R.M., p. 311–319.

Hidalgo, M.C., Urea, E. and Sanz, S. 1999. Comparative study of digestive enzymes in fish with different nutritional habits, proteolytic and amylase activities. *Aquaculture*, 170(3–4): 267–283.

Hota, S. Arsenic toxicity to the brain, liver and intestine of a fresh water fish, *Channa punctatus. Geobios.*, 23: 154–156.

Jain, K.L. and Mittal, V. 2004. Heavy metal pollution in surface water bodies and its impact on fishes. In: *Proceedings of the National Workshop on Rational Use of Water Resources for Aquaculture*, Hisar, March, 18–19.

Jain, K.L., Sharma, M. 2003. Toxic effects of mercury and cobalt on biochemical composition of fresh water fish *C. mrigala*. In: *Proceedings of 3rd Interaction Workshop*, December, 17–18, pp. 203–207.

James, R., Sampath, K., Sivakumar, V. and Manthiramoorthy, S. 1991. Individual and combined effects of heavy metals on survival and biochemistry of *Oreochromis mossambicus. Indian J. Fish.*, 138(1): 49–54.

James, R., Sivakumar, V., Sampath, K. and Rajendran, P. 1991. Individual and combined effects of zinc, cadmium and copper on growth of *Oreochromis mossambicus. Indian J. Fish.*, 38(3): 198–200.

James, R., Sampath, K. and Ponmani, K. 1992. Effect of metal mixtures on activity of two respiratory enzymes and their recovery in *Oreochromis mossambicus* (Peters). *Indian J. Exp. Biol.*, 30: 496–499.

Jaunch, D. 1979. Gill lesions in cichlid fishes after intoxication with insecticide fenthion. *Experientia.*, 35: 371–372.

Jebakumar, S.R.D., Flora, S.D.J. and Ganesen, R.M. 1990. Effect of short-term sub-lethal exposure to cypermethlinon the organic constituent of freshwater fish. *J. Environ. Biol.*, 11: 203–209.

Jia, X.Y. 2002. Effect of four kinds of heavy metal on respiration intensity of juvenile *Misguxnus anguillieandatus. Journal of Zhejiang University Agriculture and Life Sci.*, 27 (5): 556–558.

Kanabur, V.V. and Sangli, A.B. 2000. Acute toxicity of resorchinol and nitrophenol on a freshwater fish, *Gambusia affinis* and their effects on oxygen uptake. *Geobios.*, 27: 188–190.

Kannupandi, T., Pasupathi, K. and Soundarapandian, P. 2001. Effect of endosulfan, HCH, copper and zinc on the phototactic and swimming rate of larvae of the mangrove crab *Macropthalmus erato. Indian J. Fish.*, 48(3): 313–322.

Kargin, R.1998. Metal concentration in tissues of the fresh water *Capoeta banoisi* from the seyham river (Turkey). *Bull. Environ. Contam. Toxicol.*, 60: 822–828.

Kaur, R. 1999. Fenvalerate induced changes in some tissues of air breathing cat fish *Hcteropneustus fossilis* (Bloch). *Ph.D. Thesis*, MDS University, Ajmer.

Kaviraj, A. 1989. Heavy metal concentrations of fish and prawn collected from *Hooghly estuary. Sci. Cult.*, 55: 257.

Kaviraj, A. and Ghosal, T.K. 1997. Effects of poultry litter on the chronic toxicity of cadmium to common carp, *Cyprinus carpio. Bio. Res. Tech.*, 60: 239–243.

Kaviraj, A. and Konar, S.K. 1983. Lethal effects of mixtures of mercury, chromium and cadmium on fish. *Environ. Ecol.*, 1: 153–158.

Khanee, K.M.A., Zazeemul, James, R., Sivakumar, K. and Muthuramalakshmi, M. 1991. Sub-lethal effects of mercury on growth and biochemical composition and their recovery in *Oreochromis mossambicus. Indian J. Fish.*, 38(2): 111–118.

Khanna, S.S. and Gill, T.S. 1975. Effect of cobalt salts on the glycemia and islet histology of *Channa punctatus* (Bloch). *Acta. Anat.*, 92: 194.

Kirunbagaran, R. and Joy, K.P. 1992. Toxic effects of mercury on testicular activity in the fresh water teleost, *Clarius batrachus* (L.). *J. Fish. Biol.*, 41: 305–315.

Kumar, S , Pant, S.C. 1981. Histopathological effects of acutely toxic levels of copper and zinc on gills, liver and kidney of *Puntius conchonius* (Ham). *Indian J. Exp. Biol.*, 19: 191–194.

Kureishy, T.W. and D'Silva, C. 1993. Uptake and loss of mercury, cadmium and lead in marine organism. *Indian J. Exp. Biol.*, 31: 373–379.

Lakota, S., Raszka, A., Kupczak, I., Hlond, S., Stefan, J. and Roskowski, J. 1978. The effect of methoxychlor and proposur on the health of carp fry (*Cyprinus carpio* L.). *Acta Hydrobiol.*, 20: 197–205.

Larsson, A., Bengtsson, B.E. and Svanberg, O. 1976. Effects of pollutants on aquatic organisms, A.P.M. Lockwood (Ed.) Soc. For Exptl. Biol. Seminar Series. Vol. 2, Cambridge Univ. Press, p. 35–43.

Lin, H.C., Hsu, S.C. and Hwang, P.P. 2000. Maternal transfer of cadmium tolerance in larval *Oreochromis mossambicus*. *J. Fish Biol.*, 57: 239–249.

Lowry, O.H., Rosebrough, N.J., Fan, A.L. and Randall, R.J. 1951. Protein measurement with folin-phenol reagent. *J. Biol. Chem.*, p. 193–265.

Metelev, V.V., Kanaev, A.I. and Dazasokhova, N.G. 1971. *Water Toxicology*, Amerind Publishing Co. Pvt. Ltd., New Delhi, pp. 93–97.

Midha, M. and Singhal, R.N. 1994. Effect of cobaltous chloride on the growth of *Labeo rohita*. *Geobios.*, 21: 297–302.

Mikryakov, V.R., Lapirova, T.B., Soltani, M., Marrin, A.S. and Vinogradov, G.A. 2002. Influence of sub-lethal concentration of some heavy metal salts (Hg, Cd and Cu) on the contents of lysozyme in tissues of sturgeon fingerlings, (*Acipenser baeri*). *Bull. European Assoc. Fish Pathologists*, 22: 15–20.

Mirenda, J.R. 1986. Acute Toxicity and Accumulation of Zinc in the Crayfish, *Orconectes virilis* (Hagen). *Bull. Environ. Contam. Toxicol.*, 37: 387–394.

Mohan, D. and Choudhary, A. 1991. Zinc accumulation in a few tissues of fish, *Puntius sophare* (Ham.) after sub-lethal exposure. *J. Natcon.*, 3: 205–209.

Motais, R., Isaia, J., Rankin, J.C. and Maetz, J. 1969. Adaptive changes of water permeability of the teleost gill epithelium in relation to external salinity. *J. Exp. Biol.*, 51: 529–546.

Mukhopadhyay, M.K. 1983a. Effects of copper on fish and aquatic ecosystem. *Environ. Ecol.*, 1: 53–59.

Mukhopadhyay, M.K. 1983b. Effects of zinc pollution on fish and aquatic ecosystem. *Environ. Ecol.*, 1: 89–96.

Mukhopadhyay, M.K. and Ghosh, B.B. and Bagchi, M.M. 1994. Toxicity of heavy metals to fish, prawn and fish food organisms of Hoogly estuarine system. *Geobios.*, 21: 13–17.

Mukhopadhyay, M.K. and Konar, S.K. 1984. Toxicity of cooper, zinc and iron to fish, plankton and warm. *Geobios.*, 11: 204–207.

Nath, Kedar and Kumar, Nishith. 1990. Gonadal histopathology following nickel intoxication in the giant gaurami *Colisa fasicatus* (Bloch and Schneider), a fresh water tropical Perch. *Bull. Environ. Contam. Toxicol.*, 45: 299–304.

Neil, J.H. 1957. Some effect of potassium cyanide on *Salvelinus fontinalis. Ontario Industrial Waste Conf.*, 4: 74–76.

Ney, J.J. and Martin, M.G. 1985. Influence of prefreezing on heavy metal concentrations in blue gill sunfish. 19(7): 905–907.

Obula, Reddy, K.P., Seshalatha, E., Muthuraj, D. and Neerja, P. 1999. Changes in certain metabolic components in fry in *Cyprinus carpio* on Ammonia stress. *J. Nat. Cenv.*, 11(2): 147–151.

Oladimeji, A.A., Qadri, S.U. and deFreitas, A.S.W. 1984. Long-term effects of Arsenic Accumulation in Rainbow Trout, *Salmo gairdneri. Bull. Environ. Contam. Toxicol.*, 32: 732–741.

Olasson, Per-erik, Larson, A. and House, C. 1988. Metallothionin and heavy metals in rainbow trout (*Salmo gairdneri*) during exposure to cadmium in water. *Marine Environ. Res.*, 24: 151–153.

Palaniappan, Pl. Pm., Karthikeyan, S. and Sabhanayakam, S. 2003. Studies on the effects of heavy metal nickel on gills of fingerlings of an edible fish *Cirrhinus mrigala. Poll. Res.*, 22(2): 247–250.

Pande, R. and Saxena, D.N. 1990. Effect of copper sulphate on histopathology of gill and liver in the fingerlings of an Indian major carp, *Labeo rohita*. *11th Ann. Sess. Acad. Environ. Biol.*, Aurangabad.

Pardo, R., Barrada Parozl, E. and Vega, M. 1990. Determination and speciation of heavy metals in sediments of the piswerga river. *Water Res.*, 24: 373–379.

Patil, G.P., Dhande, R.R. 2002. Effect of mercury and copper on ovarian recrudescence in the freshwater fish, *Channa punctatus*. *J. Ecotoxicology and Environmental Monitoring*, 11(2): 83–90.

Paulose, P.V. 1989. Histological changes in relation to accumulation and elimination of inorganic and methyl mercury in gills of *Labeo rohita* (Ham.). *Indian J. Exp. Biol.*, 27: 146–150.

Qayyum, M.A. and Shaffi, S.A. 1977. Changes in tissue glycogen of a fresh water cat fish *Heteropenustis fossilis* (Bloch) due to mercury intoxication. *Curr. Sci.*, 46: 652.

Radhakrishnaiah, K. 1988. Accumulation of copper in the organs of fresh water fish *Labeo rohita* (Ham.) on exposure to lethal and sub-lethal concentration of copper. *J. Environ. Biol.*, 9: 319–326.

Ram, Bhaskar, B. and Rao, K.S. 1987. Comparative haematology of ten species of marine fish from Vishakhapatnam coast. *J. Fish Biol.*, 30: 59–66.

Ram, R.N. and Sathyanesan, A.G. 1984. Mercuric chloride induced changes in the protein, lipid and cholesterol levels of the liver and ovary of the fish *Channa punctatus*. *Env. and Ecol.*, 4: 23–29.

Rangaswamy, C. and Padmanabha Naidu, B. 1999. Effect of endosulfan on oxygen equilibrium curves of blood of food fish *Oreochromis mossambicus* (Peters). *Indian J. Exp. Biol.*, 37: 166–168.

Rani, A. Usha. 1999. Pathological observations in gills of the fresh water teleost *Oreochromis mossambicus* (Tilapia) with reference to cadmium toxicity. *Indian J. Comp. Anim. Physiol.*, 17(1): 18–22.

Rao, A. Sreenivasa and Rao, P. Ramamohana. 2001 Study on the polycyclic aromatic hydrocarbons in tissues of fish from Kolleru lake. *Indian J. Fish.*, 48(3): 323–327.

Rema, L.P. and Philip, B. 1996. Metallothiolein on metallothionein like proteins and heavy metals toxicity in *Oreochromis mossambicus* (Peters). *Indian J. Exptl. Biol.*, 34: 527–530.

Renu, G., Singh, O.V., Tandon, S.N. and Mathur, R.P. 1991. A study of water quality and metal speciation of Yamuna river. *Asian Env.*, 13: 3–10.

Sahoo, P.K. and Mukherjee, S.C. 1999. Normal ranges for diagnostically important haematological parameters of laboratory-reared rohu (*Labeo rohita*) fingerlings. *Geobios.*, 26: 31–36.

Saikia, D.K., Mathur, R.P. and Srivastava, S.K. 1988. Heavy metals in water and sediments of upper Ganga. *Indian J. Environ. Hlth.*, 31: 11–17.

Sakthivel, M., Sampath, K. and Pandian, T.J. 1991. Sub-lethal effects of textile dye stuff effluent on selected oxidative enzymes and tissue respiration of *Cyprinus carpio*. *Indian J. Experimental Biol.*, 29: 979–981.

Sarkar, S.K. 1989. Evaluation of two heavy metals on the oxygen consumption of *Tilapia mossambica* (Peters. *Geobios.*, 16: 108–110.

Sastry, K.V. and Gupta, P.K. 1978. Effect of mercuric chloride on the digestive system of a teleost fish, *Channa punctatus*: A histopathological study. *Environ. Res.*, 16: 270–278.

Sastry, K.V. and Rao, D.R. 1981. Enzymological and biochemical changes produced by mercuric chloride in teleost fish, *Channa punctatus*. *Toxicol. Lett.*, 9: 321.

Sastry, K.V. and Rao, D.R. 1982. Chronic effects of mercuric chloride on the activities of some enzymes in certain tissues of the fresh water murrel, *Channa punctatus*. *Chemosphere.*, 11: 1203.

Sastry, K.V. and Shukla, V. 1990. Effects of cadmium on the intestinal absorption of some nutrients in the teleost fish *Channa punctatus*. *Pollut. Res.*, 9: 141–148.

Sastry, K.V. and Subhadra. 1982. Effect of cadmium on some aspects of carbohydrate metabolism in a freshwater cat fish. *Heteropneustes fossilis*. *Toxicol. Lett.*, 14: 45–55.

Sastry, K.V. and Subhadra. 1985. *In vivo* effects of cadmium on some enzyme activities in tissues of the fresh water catfish, *Heteropestes fossilis. Environ. Res.*, 36: 32–45.

Sastry, K.V. and Sunita. 1983. Alterations in the intestinal absorption of xylose induced by heavy metals in a fresh water teleost fish *Channa punctatus. Poll. Res.*, 2: 48.

Saxena, P.K., Singh, V.P., Kondal, J.K. and Soni, G.L. 1988. Effect of Malathion and Carbaryl on *in vitro* lipid and protein synthesis by liver of the fresh water teleost, *Channa punctatus* (Bl.). *Indian J. Experimental Biol.*, 26: 700–702.

Schultz, Randall and Hermanutz, Roger. 1990. Transfer of toxic concentrations of selenium from parent to progeny in fate head minnow *Pimephales promelas. Bull Environ. Contam. Toxicol.*, 45: 568–783.

Sehgal, R. and Saxena, A.B.N. 1986. Studies on the toxicities of cadmium and zinc to a freshwater teleost *Puntius ticto. J. Hydrobiol.*, 2: 33–35.

Sen, P.R. and Chatterjee, D.K. 1979. In: *Advances in Aquaculture*, (Eds.) T.V.R. Pillay and W.M.A. Dill. FAQ, England.

Seshasrinivas, V. and Rao, B.M. 1999. Chromium induced histological alterations in the gill of the freshwater teleost fish. *Labeo rohita* (Ham). *Indian J. Comp. Animal Physiol.*, 17(1): 31–33.

Sultana, R. and Rao, D.P. 1988. Bioaccumulation pattern of zinc, copper, lead and cadmium in grey mullet, *Mugil cephalus* (L), from harbour waters of Vishakhapatnam, India. *Bull. Environ. Contam. Toxicol.*, 60: 949–955.

Susan, T.A. and Tilak, K.S. 2003. Histopathological changes in the vital tissues of the fish *Cirrhinus mrigala* exposed to fenvalenate technical grade. *Poll. Res.*, 22(2): 179–184.

Thiruvalluvan, M., Manoharan, A.C. and Nagendran, N. 1997. Acute and genotoxic effect of cadmium and methyl parathion on *Cyprinus carpio* var. communis (Linn.). *Geobios.*, 24: 250–252.

Tiwari, A., Joshi, H.V., Raghunathan, C., Kumar, Sravan, V.G. and Kotiwas, Q.S. 2000. New record of *Mytilis virdis* Linn., its density, growth and accumulation of heavy metals on Saurashtra coast, Arabian sea. *Current Sci.*, 78(1): 97–101.

Tort, L. and Torres, P. 1988. The effect of sub-lethal concentrations of cadmium on haematological parameters in the dogfish, *Scyliorhinus canicula. J. Fish Biol.*, 32: 277–282.

Umminger, B.L. 1970. Physiological studies on supercooled fish *Fundulus heteroclitus*. III. Carbohydrate metabolism and survival at sub-zero temperature. *J. Expt. Zool.*, 17: 156–164.

Van Duijn, Jr. C. 1967. *Diseases of Fishes*. Iliffe Books Ltd., London.

Varghese, G., Naik, P.S. and Katdare, M. 1992. Respiratory responses and blood sugar level of the crab *Barytelphrsa cunicularis* (West wood), exposed to mercury, copper and zinc. *Indian J. Exp. Biol.*, 30: 308–312.

Vijayram, K., Geraldone, K., Varadarajan, T.S., George, J. and Longasnathan, P. 1989. Cadmium induced changes in the biochemistry of an air breathing fish *Anabas testudineus. J. Ecobiol.*, 1: 245–251.

Vincent, S., Ambrose, T. and Selvanyagam, M. 2002. Impact of cadmium on food utilization of the Indian major carp, *Catla catla. J. Env. Biol.*, 23(2): 209–212.

Visviki, I. and Rachlin, J.W. 1994. *Arch. Environ. Contam. Toxicol.*, 26(2): 154.

Wallace, Kendall, B. and Dargan, John, E. 1987. Intrinsic metabolic clearnace of parathion and paraoxon by livers from fish and rodents. *Toxicol. Appl. Pharmacol.*, 90: 235–242.

Whitton, B.A. and Say, J.J.. 1975. Heavy Metals. In: *River Ecology*, (Ed.) Whitton, B.A. Blackwell Scientific Publication, Oxford, p. 286–311.

Wise, M.L., Stiebel, C.L. and Grizzle, J.M. 1987. Acute toxicity of nitrofurazone to channel catfish *Ictalurus punctatus*, and Goldfish, *Carassius auratus. Bull. Environ. Contam. Toxicol.*, 38: 42–46.

Zhou, X.W., Zhu, G.N. and Sun, J.H. 2003. Effects of the interaction of heavy metals on the accumulation of copper in the tissues the fish (*Carassius auratus*). *J. Zhejiang University of Agriculture and Life Sci.*, 28 (4): 427–430.

Chapter 8

Impact of Seasonal Variation in Physico-chemical Features of Water on Inhabiting Fish Fauna

Md. Shamsul Haque, Kaulshalendra Kumar and Binay Kumar Singh

Department of Zoology, Government P.G. College, Shahdol – 484 001, M.P., India

ABSTRACT

The life of aquatic animals absolutely depends upon the physico-chemical conditions of the water. The suitable condition of water favours aquatic dwellers to perform their almost all vital phenomena in quite normal, coordinated and harmonious way. Any alterations in the hydro -biological parameters may affect the coordinated way of physiological actions. Observations reveal that present experimental pond is not quite suitable for fish culture as its physico-chemical features are unfavourable. Besides unfavourable physico-chemical features, the water also contains the mercury causing damage in physiology of fish.

Keywords: *Hydrogen ion concentration, Chemical oxygen demand, Biochemical oxygen demand, Dissolved oxygen, Transparency and Alkalinity.*

Introduction

Existence of life is indispensable to water. Water seems more significant for the water dwellers especially the fishes and others because it covers about 72 per cent of earth's surface and provides the most extensive medium for a majority of the animals. The physico-chemical conditions of water are responsible to provide the better survival to the inhabiting organisms to a greater extent. Extensive studies have been carried out by various workers like Arora (1966), Das and Shrivastava (1956), Das and Pandey (1978), Singh (1960), Mishra and Trivedi (1993), Tilak *et al.* (2001) and Zafar (1967 and 1968) on the hydro-biological features and physico-chemical nature of water in the different regions of the country. The contributions of Singh (1991), Khillare (1989) and Siddiqui (2001) are ever treated as milestone in the field of heavy metal salt (mercury) toxicology.

Materials and Methods

For the estimation of physico-chemical features of water, a pond was selected in the district Shahdol close to the downstream of effluent discharge center of Orient paper mills, Amlai. The water was collected in the middle of the month for a complete year. Water was analyzed by using standard methods as given by Chapman (1976), Adoni (1985) and NEERI. To avoid possible errors proper precautions were taken during the collection, transportation and analysis of water sample.

Results and Discussion

The data obtained through observations have been represented in Table 8.1. The temperature of water of ponds ranges from 18.8°C to 31.1°C. with minimum in December and maximum in May during the investigation period. It follows almost the trend of variation in atmospheric temperature.

The transparency of water was observed minimum (24.6 cm) in rainy season and maximum (45.5 cm) in the month of January. It reveals that water transparency varies with seasonal variations and it was moderate in winter but lower in monsoon.

The minimum value of dissolved oxygen (4.1 mg/l) was recorded in the month of May whereas the maximum value of dissolved oxygen (8.1 mg/l) was recorded in the month of December.

Table 8.1: Physico-chemical Features of Water Samples of a Pond of Shahdol District

Sl.No.	*Parameters*	*Average Value in Summer*	*Average Value in Winter*	*Average Value in Monsoon*
1.	Temperature	31.1°C	18.8°C	26.4°C
2.	Transparency	24.6 cm	45.5 cm	32.2 cm
3.	Dissolved oxygen	4.1 mg/l	8.1 mg/l	6.8 mg/l
4.	BOD	3.0 mg/l	1.8 mg/l	2.6 mg/l
5.	COD	80.5 mg/l	30.2 mg/l	26.2 mg/l
6.	Alkalinity	79.0 mg/l	150.0 mg/l	91.0 mg/l
7.	pH	7.35	7.57	7.42
8.	Dissolved solids	659.0 mg/l	442.0 mg/l	589.0 mg/l
9.	Calcium	39.0 mg/l	175.0 mg/l	89.0 mg/l
10.	Magnesium	116.0 mg/l	66.0 mg/l	17.0 mg/l
11.	Mercury	0.08 mg/l	0.03 mg/l	0.02 mg/l

Low dissolved oxygen content was probably due to the maximum water temperature and increased decomposition of dead aquatic vegetation in summer whereas the high dissolved oxygen content in winter was due to lower water temperature and increased phytoplankton density.

The minimum value of biochemical oxygen demand (BOD) was found to be 1.8 mg/l in winter season (December), and the maximum value of BOD was 3.0 mg/l in summer season (June).

The lower value of chemical oxygen demand (COD) was found to be 30.2 mg/l in the rainy season during July–August and the higher value was 80.5 mg/l in summer during April–May. Such variations in the value of COD showed a positive correlation with conductivity, free CO_2 and BOD and a negative correlation with pH and dissolved oxygen.

The minimum level of alkalinity of water was recorded to be 79.0 mg/l in the month of July while the maximum was 150 mg/l in the month of March. The alkalinity showed a positive correlation with conductivity, dissolved oxygen and chemical oxygen demand and negative correlation with pH, free CO_2 and BOD.

The minimum pH value was observed to be 7.35 in the month of June during monsoon while its maximum value was 7.57 in the month of September during onset of winter. The variable values of pH showed a negative correlation with conductivity.

The minimum value of dissolved solids was observed to be 442.0 mg/l during December–January while its maximum value was 659.0 mg/l during June–July.

An average concentration of mercury (0.08 mg/l) has also been observed in the water sample of the selected pond during the course of investigation.

The minimum value of calcium was observed to be 39.0 mg/l in the month of July while its maximum value was 175.0 mg/l in the month of December.

The minimum value of magnesium was observed to be 17.0 mg/l in the month of August while its maximum value was 116.0 mg/l in the month of May.

The minimum value of free carbon dioxide was found in the month of December while its maximum value was recorded in the month of June.

The impact of seasonal variation in physico-chemical factors of water on the inhabiting fish species in aquatic ecosystem have been established by Das *et al.* (1956) and Mishra *et al.* (1993). The findings of Khillari (1989), Singh (1991) and Siddiqui (2001) have also focused on the irregularities in the reproductive physiology of freshwater fish species due to heavy metals intoxication.

The change in physico-chemical factors of water usually drops fish species due to lack of ample availability of dissolved oxygen, biochemical oxygen and chemical oxygen in summer season. The demand of oxygen reached immensely at its peak when the ecological balances get disturbed due to intoxication by heavy metal salts. In this critical condition the fish species are bound to tolerate the toxic atmosphere till the death justifying that present experimental pond is not suitable for fish culture. Probably these alterations in physico-chemical nature of water are because of the effluents of Orient Paper Mills, Amlai, frequently reaching to the adjoining water bodies.

Acknowledgement

The senior Author is thankful to Dr. Bharat Sharan Singh for providing guidance and wise suggestions during this research work. Binay Kumar Singh is also grateful to UGC, CRO, Bhopal for providing the financial assistance vide sanction letter No. 4–001 (03) 2004 (MRP/CRO), Dated 27–03–2004.

References

Arora, H.C. 1966. Responses of rotifera to variations in some ecological factors, Proc. Ind. Acad.Sci., 58: 57–63.

Adoni, A.O. 1985. *Workbook on Limnology*. Indian Mob. Committee, Department of Environment, Govt. of India, pp. 216.

Chapman, S.B. 1976. *Method in Plant Ecology*. Blackwell-Scientific Publications, Oxford, London, Edinburgh, Melbourne, pp. 536.

Das, S.M. and V.K. Shrivastava. 1956. Some new observation on plankton from freshwater ponds and tanks of Lucknow. *Sci. Cult.*, 21(8): 466–467.

Das, S.M. and J. Pandey. 1978. Some physico-chemical and biological indicators of pollution in lake, Nainital, Kumaun, U.P. *Ind. J. Ecol.*, 5(1): 7–16.

Jhingaran, V.G. 1991. *Fish and Fisheries of India*, 3rd edn. Hindustan Pub. Corp., India, pp. 727.

Khillare, V.K. 1989. Effects of ammonium sulphate on oocytes development of the fish, *Puntius stigma* (Ham). In: *Nat. Symp. on Advances in Limnology and Conservation of Endangered Fish Species*, October 23–25, H.N.B. Univ. Srinagar.

Mishra, P.C. and R.K. Trivedi (Eds.) 1993. Ecology and Pollution of Indian Lakes and Reservoirs. Ashish, New Delhi, pp. 347.

Siddiqui, S. 2001. Toxicity of some heavy metal salts on hypothalamo-neurohypophyseal neurosecretory system in a freshwater teleost in relation to reproduction. *Ph.D. Thesis*, A.P.S. University, Rewa, M.P.

Singh, K.P. 1960. A new species of *oocystis* sp. Nor. From Nainital. *Curr. Sci.*, 29: 29–30.

Singh, B. 1991. Hypothalamo-neurohypophyseal system in a freshwater fish, *Danio equipinatus* (Ham) and neural control on reproduction. *Ph.D. Thesis*, A.P.S. University, Rewa, M.P.

Tilak, K.S., K. Veeraiah and S. Vijaylakshmi. 2001. Biochemical changes induced in freshwater fish *Labeo rohita* (Ham) exposed to pesticide mixture. *Asian J. of Microbial Biotech. and Env. Sci.*, 3(4): 315–319.

Zafar, A.R. 1968. Certain aspects of distribution pattern of phytoplankton in the lakes of Hyderabad. In: *Proc. Symp. Recent, Adv. in Tropical Ecology*, p. 368–375.

Chapter 9

Substrate Specificity in Relation to Spawning and Causes of Egg/Larval Mortality in Common Carp, *Cyprinus carpio*

Subash C. Gupta
Department of Environmental Sciences, University of Jammu, Jammu

ABSTRACT

The present communication is an attempt to identify substrates used for spawning in common Carp, *Cyprinus carpio* in order of preference and various causes of egg and larval mortality in the experimental pond of the Department of Botany, University of Jammu, Jammu. Different types of substrates provided in the pond included aquatic plants like *Eichhornia* sp., *Salvinia* sp., *Hydrilla* sp., *Potamogeton* sp., *Charra* sp. and *Aspidium* sp. (Fern). In addition to these, materials like polythene and jute were also tried for their possible role as substrates for spawning in *Cyprinus carpio*.

Studies conducted revealed that *Eichhornia* sp. acted as the most preferred and ideal substrate for egg deposition, followed by *Salvinia* sp. and *Hydrilla* sp., and it bas been extablished that *Cyprinus carpio* failed to breed in the absence of these spawning substrates.

Introduction

Suitable substrate is a pre-requisite for spawning in fishes and varies from fish to fish. Besides, fishes spawn at a particular site in a particular season which is ideally suited to them and this spawning represents optimum range along a gradient of different and interacting environmental factors, so that stresses on young ones are minimal and their optimum survival is ensured.

Different fishes prefer different substrates for spawning *i.e.*, they deposite/lay eggs in a specific habitat and prefer specific substrates and on this basis, they were classified by Nikolski (1963) into five categories *viz.*, Lithophils (spawning on stony grounds/beds), phytophils (spawning in plant covers), Psammophils (spawning in sand), Pelagophils (spawning in water column) and Ostracophils (spawning in the mantle cavity of molluscs). During the course of present study, spawning and various factors related to spawning of carp and their egg/larval mortality in the University pond have been studied in greater details, with the singular aim of identifying substrate, preferred most by carp for egg deposition. As carp eggs are adhesive and pelagic, the identification of ideal substrate for spawning assumes greater significance in carp culture so that the valuable eggs are not damaged. Since carp is a fast growing, hardy, delicious and an important food fish and thrives well both in lotic and lentic waters, these studies assumes greater importance in its culture.

Materials and Methods

As it is a known factor that carp fish needs some suitable substrate for egg deposition, various macrophytes alongwith other soft materials were provided to them in the pond few days before breeding starts. From the past experiences, it has been learnt that unless and until carp gets some soft material for egg deposition, they do not breed. Thus keeping this thing in mind, an experiment was carried out under yard conditions in a concrete pond (Figure 9.1) situated in the natural park of Department of Botany, University of Jammu, Jammu, where *Cyprinus carpio* has been introduced for experimental purposes. Various aquatic macrophytes (*viz., Eichhornia* sp., *Salvinia* sp., *Hydrilla* sp., *Potamogeton* sp., *Charra* sp., and *Aspidium* sp.) (Figure 9.1a), soft synthetic materials (*viz.*, polythene pieces in the form of clusters) and jute were provided to the fish in this pond, to test them for their role as substrate for spawning, and identify the most ideal/preferred substrate for egg deposition by Carp fish.

In order to identify the spawning grounds preferred by *Cyprinus carpio*, different breeding habitats were created in the pond (Mortishav, 1983) by providing different materials in the small breeding pockets/nurseries (Figure 9.2), constructed along the sides of the pond *viz.*,

1. Sand: Sandy bottom (Psammophilous) (Figure 9.2a)
2. Stones: Stony bottom (Lithophilous) (Figure 9.2b)
3. Plant covers: (Phytophilous) *viz.*, Algae, *Salvinia* (Figure 9.2c), *Eichhornia, Hydrilla* (Figure 9.2d), *Potamogeton, Charra, Aspidium* (Fern) etc.
4. Clay: Clayey bottom
5. Sand +stones
6. Clay + stones
7. Without any substrate (only water column): Pelagophilous

Figures 9.1–92c: Fish Pond with Breeding Pockets/Nurseries on the Sides and Different Substrates Provided in the Breeding Pockets for Carp Breeding

Figure 9.1: Fish Pond Showing Breeding Pockets/Nurseries on the Sides

Figure 9.1a: Various Macrophytes Provided in the Pond for Carp Breeding. Coloured plastic balls are tied to various macrophytes as floats and attracting/alluring objects for the brooders.

Figure 9.2: Breeding Pockets Covered with Wire Mesh to Protect Eggs and Larvae from Bird Predators and Jute Bags to Protect them from Direct Heat

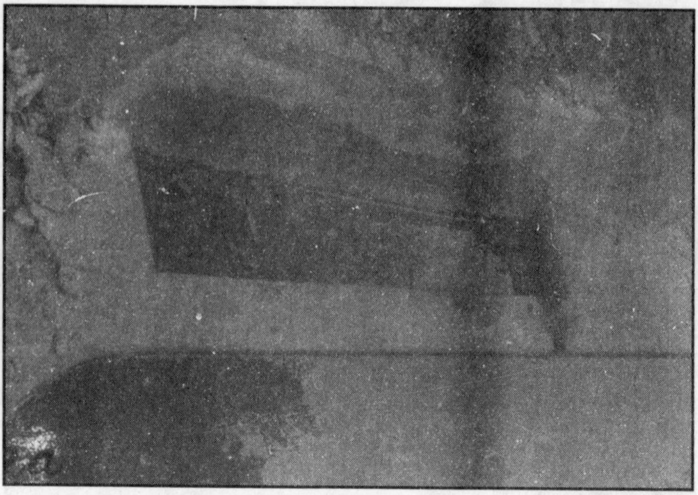

Figure 9.2a: Breeding Pockets with Psammophilous Habitat

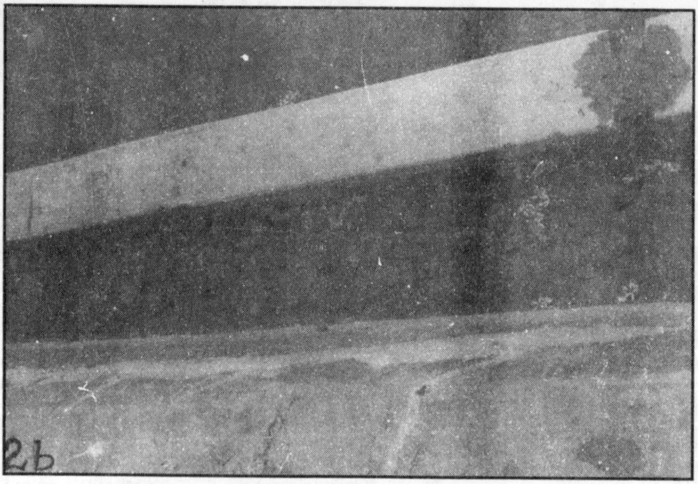

Figure 9.2b: Breeding Pockets with Lithophilous Habitat

Figure 9.2c: Breeding Pockets with Phytophilous (*Salvinia* sp.) Habitat

Observations and Discussion

Studies have been made under the following headings:

1. Substrate preference for spawning/egg deposition.
2. Egg/larval mortality.

Substrate Preference for Spawning

During the course of present investigations, *Eichhornia* roots (Figure 9.3) have proved to be the best substrate for spawning in *Cyprinus carpio*, followed by *Slavinia* (Figure 9.4) and *Hydrilla* (Figure 9.5) in order of their preference (Table 9.1). *Eichhornia* as a suitable substrate for egg laying in carp has also been reported by Jhingran (1975). In the absence of *Eichhornia*, carp exhibited preference for *Salvinia* and *Hydrilla* as substrate for spawning. *Aspidium* (Fern) and *Potamogeton* are the least preferred whereas *Charra* is completely rejected by fish for spawning.

The roots of macrophytes act as soft cushion enabling the fish to deposite eggs on them without any problem because while laying egg, the fish rubs its belly on the ground or any other substrate (which it likes for egg laying) and then deposites eggs on that particular substrate. Thus in order to project the belly from any injury and thus damage to the eggs while rubbing, carp requires and prefers soft cushion like substrate for spawning. Keeping in view the preference of carp for soft cushion like roots, other soft materials (like polythene and jute) were also provided to the fish to test, whether it likes soft materials other than roots of macrophytes. Among these, polythene paper (in clusters of small pieces) has been observed to be preferred by the fish over jute. However, this preference for polythene declined when *Eichhornia* was made available to carp in the pond. Thus from the studies it has been established that unless and until these aquatic plants are added to the pond, the carp fish didn't spawn either in the pond itself or in the nursery pockets made along the sides of the pond.

Table 9.1: Showing the Preference for Egg Deposition and their Mortality in *Cyprinus carpio*

Sl. No.	*Plant Species/ Other Soft Materials Provided as Substrate for Spawning*	*No. of Plant Species/Other Soft Materials Provided and Studied for Egg Deposition*	*No. of Eggs Attached to Plant Species/ Other Soft Materials Provided*		*No. of Eggs Died on Plant Species/Other Soft Materials Provided*	
			Total No.	*Average*	*Total No.*	*Average*
1.	*Salvinia* sp.	338	2236	6.61	298	0.88
2.	*Hydrilla* sp.	1380	5035	3.64	1124	0.81
3.	*Eichhornia* sp.	105	1865	17.76	261	2.48
4.	*Potamogeton* sp.	165	8	0.04	Nil	Nil
5.	*Charra* sp.	120	Nil	Nil	Nil	Nil
6.	Fern (*Aspidium* sp.)	113	10	0.08	Nil	Nil
7	Polythene	432	360	0.83	11	0.02
8.	Jute	48	53	1.10	2	0.04

Egg/Larval Mortality

On the basis of observations made during the course of present investigations on carp spawning, it has been found that on an average, mortality of eggs was high when deposited on roots of *Eichhornia* plant against low mortality on *Salvinia* and *Hydrilla* (Table 9.1). Eggs deposited on *Potamogeton* and Fern didn't show any mortality, which may be due to the fact that the number of eggs laid by carp on these plants is very less, as it is known factor that mortality of eggs is more where their deposition is more. Mortality was least in eggs deposited on pond and jute materials.

The differential mortality could be due to the difference in the number of eggs deposited on different substrate. The studies suggest that high mortality of eggs on *Eichhornia* roots is due to very high number of eggs occupying least area which disfavours their fertilization and hence survival. Carp larvae exhibited very high mortality in nursery pockets (made on the sides) of the pond following hatching and has been worked out to be mostly due to predation by:

1. Snakes
2. Frogs
3. Birds

Predation by Snake

During the course of present studies, the snake has frequently been found to visit the breeding/nursery pockets in which carp larvae were being reared. A number of times, the snake was found sitting either on the edge/sides of nursery pockets or in the pockets itself (Figure 9.6). Predation by snake on carp larvae was established from the absence of carp larvae in the nursery pockets, frequently visited by snake.

Further, it has been observed that the snake visits the breeding/ nursery pockets either in the early morning or in the late evening which actually are the feeding times for fish larvae–when the fish larvae are mostly seen moving on the surface for feeding purposes. Another possibility is that the snakes used to visit the pond or nursery pockets in the morning or in the evening to get rid of heat and to enjoy the cool breeze during hot summer months, but the absence of fish larvae from the pockets supports that these fish larvae might have been eaten by the snake.

Predation by Frog

Another important factor contributing towards larval mortality in nursery pockets has been established to be predation by frogs. Almost daily a number of occasions, frogs were observed sitting/ hiding under the plant covers/stones provided in these nursery pockets in which larvae were being reared, besides their presence in the main pond. The absence of fish larvae from the nursery pockets contributing towards predation by frogs has been confirmed, when frogs were caught from these nursery pockets and their stomachs were cut open/dissected and carp larvae found in their stomach (Figure 9.7). This confirms the mortality of carp larvae due to predation by frogs in the pond.

Predation by Birds

Mortality of carp larvae in nursery pockets has also been found to be the result of predation by birds. Almost daily, at times, the predators like Crane and Kingfisher were found feeding on eggs attached to the vegetation. The absence of carp larvae in these breeding pockets visited by the birds supports that these fish larvae must have been eaten by the birds. These predators were seen visiting the breeding pockets in the early morning or late evening (which are actually the feeding times for fish larvae) and feeding on the larvae.

Figures 9.2d–9.7: Various Breeding Substrates and Predation by Snakes and Frogs on Carp Larvae in the Breeding Pockits

Figure 9.2d: Breeding Pockets with Phytophilous Habitat (*Hydrilla* sp.)

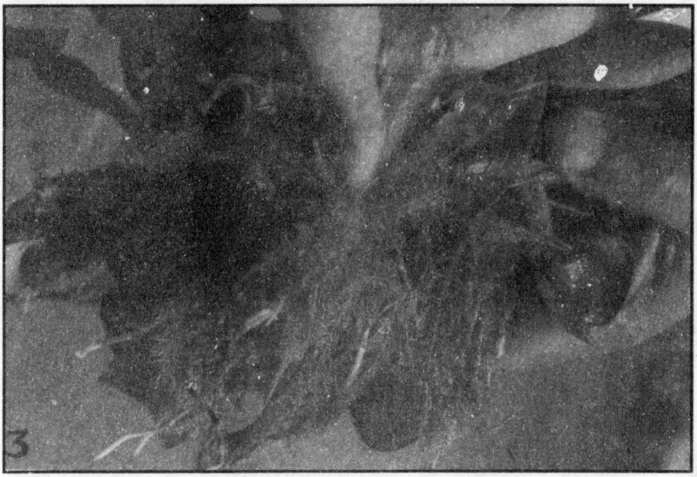

Figure 9.3: ***Eichhornia*** **Plant with Carp Eggs on its Roots**

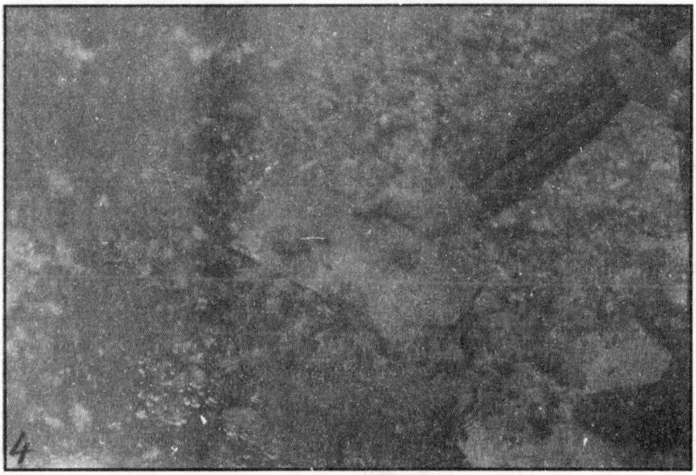

Figure 9.4: ***Salvinia*** **Plant with Carp Eggs**

Conclusion

During the course of investigation on carp spawning etc. in the experimental pond of the Department of Botany, University of Jammu, Jammu, an attempt was made to establish/identify type of substrate

Figure 9.5: *Hydrilla* Plant with Carp Eggs

Figure 9.6: A Snake Sitting Along the Breeding Pockets with its Head Inside the Pocket (Covered with Wire Mesh), in an Effort to Prey upon Carp Larvae

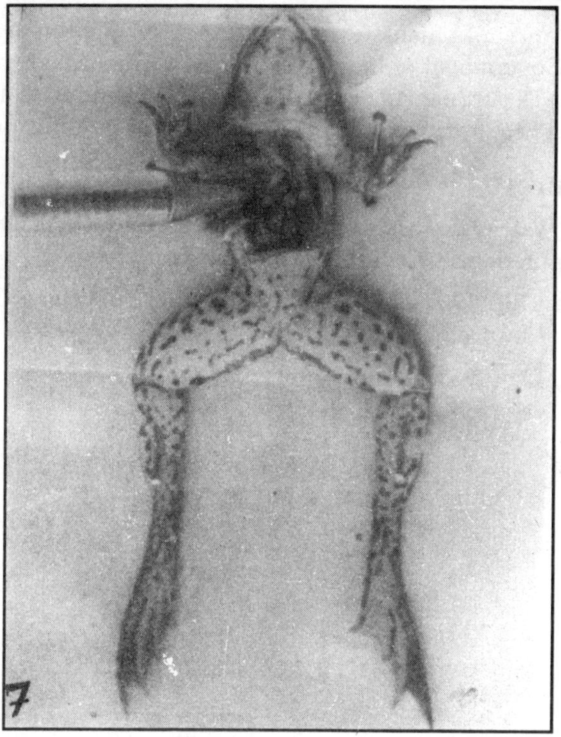

Figure 9.7: A Frog Captured from the Breeding Pocket, Dissected to Show Carp Larvae in its Stomach

preferred most by *Cyprinus carpio* for breeding. The various substrates provided to the fish for spawning purposes, comprised of *Salvinia* sp., *Hydrilla* sp., *Eichhornia* sp., *Potamogeton* sp., *Charra* sp., *Aspidium* (Fern) and other synthetic materials like polythene and jute etc. Of these, *Eichhornia* sp. has been found to be an ideal and preferred substrate for egg laying/deposition in carp, followed by *Salvinia, Hydrilla, Potamogeton, Aspidium,* jute and polythene in order of preference. *Charra* is totally rejected by the fish for egg deposition.

In addition, mortality of eggs and larvae have been attributed to predation by snakes, frogs and birds, causing decline in larval population in the nurseries.

Acknowledgements

The author is thankful to Department of Environment and Forests, Government of India, New Delhi for financial assistance and to the Department of Botany, University of Jammu, Jammu for providing necessary facilities to undertake these investigations.

References

Jhingran, V.G. 1975. *Fish and Fisheries of India*. Hindustan Publishing Corporation, India, 6-U.B. Jawahar Nagar, Delhi.

Mortishav, F.G. 1983. *Pond Fisheries*. Amerind Publishing Co. Pvt. Ltd., New Delhi.

Nikolsky, G.B. 1963. *The Ecology of Fishes*. Academic Press, London and New York, p. 153.

Chapter 10

Ecology and Natural History of Indian Crocodilians with Special Reference to Mugger Crocodile

V. Vijaya Kumar
Deputy Director, Gujarat Institute of Desert Ecology, Post Box No. 83, Bhuj-Kachchh, Gujarat, India

Introduction

Crocodilians are among the largest living reptiles, much larger than most lizards, snakes and turtles (Lang, 1989). The earliest known crocodilian is *Proterochampsa barrionuevoi*, which was discovered in Western Argentina in the middle Triassic period, while others were *Sphenosuchus* from South Africa and *Protosuchus* from Arizona, etc. Thus, about two hundred million years ago the most primitive crocodilian had emerged from some Thecodontia group including forms like phytosaurs (Singh, 1976), which was a secondary branch of dinosaurs during evolution. Therefore, the crocodilian forms existing now though differ considerably in their morphology, anatomy and physiology from their earliest representatives, they explicit living relatives of the extinct 'dinosaurs'. In view of this, crocodiles have high academic value in addition to being also an integral part of the natural ecosystem.

Crocodilian Species of the World

Crocodilians are found all over the world except in Europe (Singh, 1976), represented by 21 species, distributed in the tropical and sub-tropical waters of the world, all belonging to three families; Aligatoridae, Crocodilidae and Gavialidae. The Aligatoridae are distributed in South America, China and USA while Crocodilidae are in Asia, Africa, Australia, Pacific Islands and USA and Gavialidae in Asia, particularly in India and Nepal (Table 10.1).

Table 10.1: Distribution of Crocodilian Species in the World

Family	*Distribution*	*Genera*	*Species*
Aligatoridae	South America, China and USA	*Alligator*	*A. mississippiensis, A. sinensis*
		Caiman	*C. crocodilus, C. latirostris*
		Paleosuchus	*P. palpebrosus*
		Melanosuchus	*M. niger*
Crocodilidae	Asia, Africa, Australia and Pacific Islands, USA	*Crocodylus*	*C. palustris, C. porosus, C. acutus, C. mindorensis, C. intermedicus, C. moreletii, C. rhombifer, C. johnsoni, C. novaeguineae, C. siamensis, C. niloticus.*
		Osteolaemus	*O. tetraspis*
		Tomistoma	*T. schlegeli*
Gavialidae	India and Nepal	*Gavialis*	*G. gangeticus*

Crocodilians of Indian Sub-continent

Only three of the above said species of crocodilians occur in the Indian sub-continent. They are the fresh water crocodile, Mugger *Crocodylus palustris* and the Gharial *Gavialis gangeticus* and the Estuarine (Salt Water or Salty) Crocodile *Crocodylus porosus*.

Distribution of Indian Crocodilian in the World

Among the three Indian species, Mugger has the widest distribution range among the Asian fresh water crocodiles (Whitaker and Whitaker, 1989), which extended from the Sarbaz river in the South Eastern Iran to Assam in the east India, Nepal and also the tributaries of Ganges that cross the "terai" and the Himalayan foot hills in the north India and south up to Sri Lanka (Honegger, 1971). With only a single record of the Mugger at Thayetmyo, Burma, there are doubts on its distribution in Burma. The Estuarine Crocodile is distributed in India, South China, Indonesia, Philippines, Papua New Guinea and North Australia while the Gharial is distributed in India and Nepal (Singh, 1999). However, fossil evidences showed that the members of the family Gavialidae were also found in Tertiary deposits in Asia, North and South America and Europe (Smith, 1935). During the turn of the century, Gharials were also reported from the Indus river in Pakistan, the Ganges and some of its tributaries which flow through India, Bangladesh and Nepal, the Manas river (tributary of Brahmaputra river) in Bhutan and also in other Indian rivers like Mahanadi (Orissa), Jamuna (Uttar Pradesh) and Kosi (Bihar) (Shortt, 1921).

Distribution of Indian Crocodilians in India

In India, Mugger is distributed in the entire Indian sub-continent (Smith, 1927, 1935), the Estuarine crocodile is distributed in its best numbers in Bhitarkanika Sanctuary in Orissa, Sundarbans of West Bengal, and in Andaman and Nicobar Iislands, while Gharial is restricted to more inland northern Indian rivers like Chambal, Girwa, Mahanadi, Gangetic system and Brahmaputra.

The Gharial is distinct from the other two species by the presence of long slender snout with a pot like structure on the tip of the snout known as 'gada' there by the name Gharial while Mugger and Estuarine crocodile have short-broad snout. The Mugger differs from Estuarine species by the presence of post-occipital scutes, distinct nuchal scutes and body coloration. The Mugger grows up to a length of 4–5 m, Estuarine to 7 m while the Gharial reaches a length of 8 m (Bustard, 1974).

Threats to Crocodilians and Status of Indian Crocodilian Species

Until late 1950's, crocodiles were in large numbers in India and many parts of the world. The demand for its skin in the western market has resulted in an uncontrolled exploitation of crocodiles. The conservation of crocodiles therefore becomes necessary due to its vital ecological role and maintaining a healthy aquatic habitat aside from the obvious value of its skin, meat and other by-products. Apart from being the master predator in the aquatic habitats, it maintains genetic quality of fish species by feeding on weak and diseased fishes and controls predatory fish population by the habit of selective feeding. Thus the presence of crocodile actually helps to increase the yields of edible and economically important fish for man. In spite of all these, humans always view crocodile as an enemy to fishery, and tried to eliminate them from its habitats. In addition to these, the pressure for its skin and eggs for food, modern fishing techniques, destruction and shrinking of its habitats are the major reasons, that has caused local extinction of certain crocodilian species in some parts of the world.

In India, though the levels of exploitation have not been documented, there are reasons to believe that populations of all the three Indian crocodilian species were at a phase of intense exploitation for skins during the first two decades of the post-independence era, *i.e.*, 1947–1967. Similar was the case of these species distributed in the other parts of the world (Box 10.1).

BOX 10.1

Status of Indian Crocodilian Species in the Neighbouring Countries of India

Nepal: The Mugger is very rare in Nepal and the Gharial survives with only one known breeding population, in Royal Chitawan National Park.

Bhutan: There may have been few mugger in the cold Manas river, however, they may be extinct by now.

Pakistan: The status of Mugger and Gharial in Pakistan are rare and at the verge of extinction and little work has been done in the way of research, survey and rehabilitation. Habitat limitation is a key factor as more and more river courses and tank edges are settled and cultivated.

Bangladesh: The Mugger survives only at Bagerhat tank where there are perhaps two breeding pairs. The remnant Gharial population (about 20) was lost to India in the monsoon of 1991 when the Padma river shifted its course across the border, taking the Gharial with it. The Estuarine crocodile has extensive habitat of mangroves, Sunderbans, to multiply but unfortunately that population seems not to have recovered from overhunting. Law protects all the three species but without any specific rehabilitation there is little hope for significant recovery.

Burma: One single record of Mugger from Thayetmyo, Burma (Annandale, 1921) was reported however, the same was doubted by Smith (1921).

Sri Lanka: The Mugger is still reportedly common in protected areas such as Yella and Wilpattu National Parks. The Mugger present at Sri Lanka are kept under a separate sub-species known as *Crocodylus palustris kimpula* due its distinct behavior of man-eating, which is uncommon in Muggers (*Crocodylus palustris*) living in other parts of the world like India etc.

Increasing concern about the dwindling crocodilian population in India led the Government of India to enforce a ban on their killing and export, in 1958. However, the recovery of the wild populations from remnant surviving breeding adults has been made impossible by the direct and indirect effect of human activities, most notably those resulting in habitat loss.

The status of all three Indian crocodilian species was reported to be critical in its entire distribution range in the country (Biswas, 1970; Misra, 1969; Daniel, 1970; Daniel and Hussain, 1973; Whitaker, 1974; and FAO, 1974), which prompted the Government of India (GOI) to protect all three crocodilian species under the Schedule I of the Wildlife (Protection) Act, 1972. In addition to this, "species oriented conservation strategy" were also adopted to save the Indian crocodilian species since mid 1970s (Choudhury, 1999) with the technical assistance from Food and Agriculture Organization (FAO) and United Nations Development Programme (UNDP).

Mugger Crocodile *Crocodylus palustris*

Mugger or marsh crocodile or swamp crocodile is locally known as *'makara'* (derived from Sanskrit) which belongs to the order Crocodilidae of the group Craniata (Box 10.2). This species is highly adaptable and occupies a variety of habitats including hill streams, large man-made reservoirs, annual tanks, large rivers, small pools in the wilderness and also irrigation channels.

BOX 10.2
Nomenclature

Phylum	Chordata
Group	Craniata
Sub-phylum	Gnathostomata
Class	Reptilia
Sub-class	Diapsida
Super Order	Archosauria
Order	Crocodilia
Genus	*Crocodylus*
Species	*Palustris*

Total length of the Mugger hatchlings are 25–30.5 cm while the average adult male measures 3–3.5 m and female is 2–2.5 m. The age at sexual maturity is 6 years 8 months for female and 8–10 years for male (Whitaker and Whitaker, 1989). Potential life span in the wild is not known, except for some documentation done based on captive animals. However, such records are far from doubts because, it is often impossible to verify the age of animal captured and transferred to captive centers. It is also possible that a crocodile lives longer in the wild than in the restricted and artificial conditions. However, Mugger lives about 70 years, while Gharial and Estuarine live about 100 years or over.

Mugger is regarded as endangered and has been exterminated in most of its range. It is rare in Iran, and near extinction in Pakistan (Webb, 1978). It appears in the IUCN Red Data Book and Appendix I of the CITES. This chapter provides a detailed on the ecological and biological aspects of the Indian Mugger crocodiles based on the field study results conducted between 1986 to till date.

Legal Protection to Mugger in India and Neighbouring Countries

Legal protection for crocodiles is a recent phenomenon. In the past, the crocodiles were protected under religious sentiments. Mugger figures in Hindu mythology and local myths of many groups and certain tribal people in India. Mugger was regarded as the vehicle for goddess of Kodiyar (Goddess of Royal People of part of Gujarat), rain God (Varuna), river Goddess (Gangadevi) and also totem of the Mogri people in part of Gujarat State. Still recent decades crocodiles used to inhabit temple tanks in Kerala, Gujarat etc. However, these types of protections are at very local level but in reality crocodiles in many places were either simply ignored by national law or listed as pest or vermin species whose destruction was encouraged at all times and by any means. In spite of all these, Mugger were protected by the national laws of the respective countries in which they occur (Table 10.2).

Crocodile Conservation Project in India

The greatly depleted status of Indian crocodilians were highlighted in the international level for the first time in 1971 at the General Assembly Meeting of the International Union for Conservation of Nature and Natural Resources (IUCN) and the

IUCN/SSC (Species Survival Commission)/CSG (Crocodile Specialist Group) meeting. As a follow up, the Government of India approached UNDP for the development of crocodile farming in India. Consequent to this, FAO/UNDP team visited India to assess the countrywide crocodile distribution and population. They estimated crocodile population to be approximately 60 Gharials, 200 Estuarine crocodiles and about 1000 isolated Mugger (FAO, 1974). Based on the FAO/UNDP recommendation, the Crocodile Conservation Programme (CCP) was launched in India during the year 1975 (Bustard, 1975) in the State of Orissa, due to its special significance, *i.e.*, the only State in the country having all the three Indian crocodilian species.

Table 10.2: Legal Status of Mugger in Different Countries (Klemm and Navid, 1989)

Sl.No.	*Name of the Country/Region*	*Law for Protection*
1.	Bangladesh	Wildlife Preservation Act, 1973
2.	Baluchistan (Pakistan)	Wildlife Protection Act, 1974
3.	India	Wildlife Protection Act, 1972
4.	Iran	Game Laws and Regulations–Undated
5.	Islamabad (Pakistan)	Wildlife Protection, Preservation, Conservation and Management Ordinance, 1979
6.	Nepal	National Parks and Wildlife Protection Rules, 1974
7.	North West Frontier Province (Pakistan)	Wildlife (Protection, Preservation and Management) Act, 1975
8.	Pakistan	State Act (Hunting Under the Control of State Jurisdiction)
9.	Jammu and Kashmir (India)	Jammu and Kashmir Protection Act, 1975
10.	Sind (Pakistan)	Sind Wildlife Protection Ordinance, 1972
11.	Sri Lanka	Fauna and Flora Protection Ordinance, 1938

The main objective of the CCP was to build up the depleted crocodile populations in suitable habitats by restocking of captive crocodiles (FAO, 1974; Bustard, 1980) into specially created or already existing protected areas and create data base for all three species. The project also had a view to utilize the resource in a sustainable manner after the re-establishment of populations in its former wild habitats. In 1975, captive breeding and rearing stations were established in Orissa (Bustard, 1975; Singh, 1978). The number of rearing centers increased to twenty seven in twelve different States/ Union territories (Table 10.3) of the country (Singh, 1999).

For captive breeding, wild-laid crocodile eggs were collected and incubated in hatchery in the rearing centers. The young ones are released into the wild after attaining a size of about 1.2 m, as this size has very less natural enemies and thus enhance the survival in the wild (Choudhury and Bustard, 1982). Between 1977 and 1992, in total 3559 Gharial, 1384 Estuarine crocodile and 1193 Mugger have been released in the suitable habitats in many parts of the country (IUCN/SSC/CSG, 1993). However, the monitoring of the released stocks was reported only in few places like Chambal (Madhya Pradesh) for Gharial, Bhitarkanika (Orissa) for Estuarine crocodile, and at Ethipothalla, Sivaram, Manjira (Andhra Pradesh), Gir (Gujarat) and Similipal (Orissa) for Mugger crocodiles (IUCN/ SSC/CSG, 1993). For individual identification, which helps in the monitoring of the released crocodiles, a special marking system known as 'scute clipping' (clipping or cutting of selected tail scutes) was done for all the released stocks, which helped in assessing the growth rate (after the release), nesting details of the released stocks.

Since the inception of the Crocodile Conservation Project in 1975, the survival of the endangered Indian crocodiles has been assured in India. By reintroduction, Gharial has re-established itself in the Mahanadi, Chambal and Ganges river systems. The Estuarine crocodile is safe in its mangrove habitats of Sunderban in West Bengal, Bhitarkanika in Orissa and the Andaman Islands. The Mugger crocodile is now no longer in the endangered list of IUCN Red Data Book (1982) (Choudhury and Choudhury, 1986). All the three species are enjoying protection at 34 protected areas covering an area of 20,399 km^2.

Life History of Mugger

The author carried out a study in Andhra Pradesh (between 1986 and 1990) which was funded by the Wildlife Institute of India, Dehra Dun and in Gujarat (1995–1997), funded by the Fauna and Flora Preservation Society, UK.

Table 10.3: State-wise List of Crocodile Rearing Centers and Release Locations in India

State	*Some Rearing Centers*	*Few Important Release Sites*
Andaman Island	Haddo Mini Zoo	Lohabarack Crocodile Sanctuary
Andhra Pradesh	Vishakhapatnam, Warangal, Hyderabad, Nagarjunasagar-Sri Sailam	Siwaram WLS, Manjira WLS, Pakhal WLS, Kinnersani WLS, Nagarjunasagar-Srisailam WLS
Bihar	Hazaribagh	
Gujarat	Gandhinagar, Sasan Gir	Ranjit Sagar, Sasan Gir, Ajawa Sarovar, Narmada Dam, Karjan Dam
Kerala	Neyyar Dam, Perruvanmoozhy	Neyyar WLS, Parambikulam WLS
Madhya Pradesh	Deori	National Chambal WLS, Shivpuri NP, Son WLS Ken WLS, Vanvihar NP
Maharashtra	Sanjay Gandhi, Tadoba	Gugamal WLS, Melghat WLS
Orissa	Satkoshia, Nandan Kanan, Dangmal, Ramatirtha	Satkosia WLS, Basipalli WLS, Hadgarh WLS, Bhitarkanika WLS, Similipal TR. Chandaka WLS
Rajasthan	Kota	National Chambal WLS, Vanvihar WLS, Ranthambhore TR, Jawahar Sagar WLS
Tamil Nadu	Hoganekkal, Amaravati, Sathanur, Mudumalai, Chennai	Krishnagiri Reservoir, Hoganekkal Falls, Sathanur Reservoir, Amaravati Reservoir, Mudumalai WLS, Kalakad-Mundanthurai WLS, Annaimalai WLS
Uttar Pradesh	Katamiyaghat, Lucknow	River Ganga, Sarda, Rapti Gharga Champal etc., Corbett NP, Katemiyaghat WLS, Pinahat Ghat, Sundarban TR. Sajnakhali WLS, Gorumara NP
West Bengal	Bhagabatpur	Sunderban TR, Sainakhali WLS, Gorumara NP

Population Status

As per estimation (IUCN/SSC/CSG, 1993) the Mugger population in the country was roughly 3000–5000 in the wild and 12000 in the captivity. Surprisingly, these estimations were not on the basis of authenticated survey data and such data on the population was not available for any of the States of the country (except for Tamil Nadu and Orissa). Keeping this in view, the author surveyed the Mugger population in the entire Gujarat State based on which the population has been described.

All the eighteen districts of the Gujarat State were surveyed (Gandhinagar is considered as part of Ahmedabad district), of which, the reports on the existence of Mugger were obtained from 14 districts (Table 10.4). However, the direct sightings of Mugger were recorded only from 6 districts, Amreli, Junagadh (Saurashtra region), Kachchh (Kachchh region), Kheda, Panchmahal and Vadodara (South Gujarat region), indirect evidences confirmed the presence of very few Mugger in Bharuch and Surendranagar districts. In the whole State, a total of 149 water bodies, which include 14 major dams, 69 medium dams and 66 other water bodies that includes minor dams, lakes, village tanks and 42 locations (each location is about 5–10 km stretch) along 21 rivers, were surveyed.

The existence of numerous pressures to the Mugger habitats (water bodies) was also noted. Out of 149 water bodies surveyed, 68 per cent were used for irrigation (high level of water fluctuation during summer), 5 per cent for generating hydropower, 33 per cent for fishing and 21 per cent for domestic/drinking water purposes. Further, 33 per cent were reported to be dry during the drought years, which is a common phenomenon in Kachchh and Saurashtra regions of the State.

The total number of Mugger recorded in Gujarat was 429, which included 368 adults (above 6 feet), 44 sub-adults (4 to 6 feet), 6 juveniles (1.5 to 4 feet) and 11 hatchlings, recorded from 25 water bodies and 6 rivers of the State. However, as per the release records, 1,013 Mugger were released and according to local reports, at least 1,653 Muggers would possibly be present in the State (Vijaya Kumar, 1997). The low sightings of Mugger (429) question the survival of most of the released stocks. Out of 973 Mugger released in Gir Forest from the Sasan rearing centre, 66 per cent (645) were released during winter (November to February) months, as it being the best season

Table 10.4: Status of Mugger in Different Districts of the Gujarat State

Regions of the Gujarat State	Districts	Number of Mugger		
		Released[1]	Reported[2]	Recorded/ Sighted
Kachchh	Kachchh		186	94
Saurashtra	Amreli		7	5
	Bhavnagar		2	0
	Jamnagar	13**	18	0
	Junagadh	973	1272	306
	Rajkot		1	0
	Surendranagar		8	0
South Gujarat	Bharuch	19*	51	0
	Kheda		11	8
	Panchmahal		50	2
	Vadodara	8	37	14
	Surat		?	0
North Gujarat	Banaskantha		7	0
	Mahesana		3	0
Total Number of Mugger		**1013**	**1653**	**429**

GSFD: Gujarat Sate Forest Department; *: Wild Caught Released; **: 12 from Sasan Rearing Center and 1 is Wild Caught

[1]: By GSFD and Others Agencies; [2]: No of Mugger Reported by the Local and GSFD etc, which include Released Stock.

for the release of captive reared Mugger into wild habitats due to two major reasons. Firstly the food requirement of Mugger during winter is very low and secondly the animal will get enough time to settle in the new environment before the onset of monsoon. However, the above view can prove dangerous if the release is done in the existing adult's territories (Vijaya Kumar, 1993) which, leads to conflicts, due to territorial behaviour, courtship and mating, nest and hatchling protection of the adult Mugger's existing in the release site. These conflicts lead to either cannibalism (Vijaya kumar and Choudhury, 1994) or emigration of the released stocks. While emigrating to far off areas, the small mugger might fall prey to other animals like Panther (Pitman, 1942; Thomas and Scott, 1949; Attwell,

1957), Lion (Vijaya Kumar and Choudhury, 1994; Vijaya Kumar, 1997) etc. or ran over by vehicles when crossing roads. Such incidents are also recorded in Kachchh district of Gujarat (Vijaya Kumar, 1997). This could be the reason for the low sightings of the released Mugger at various water bodies of the Gir.

Further, 16 per cent of Mugger which were released in the Gir during monsoon (July to October) may have been either taken away far off areas by the heavy monsoon currents because of their less adaptability to overcome the monsoon currents. The similar incidents of Mugger dispersal by monsoon currents were reported by Ahmed, 1945; Vijaya Kumar, 1993; Singh, 1982, 1983 and of other crocodilian species (Bara, 1975; Joanen and McNease, 1972). The above clearly suggests that most of the released stocks in the Gir forest either died or migrated to longer distances. However, the sightings of 16 Mugger in the down stream of Singoda river, three in Rawal river, six in Hiran river, one in Mobnesh dam and one in Barmund tank, located far from the release site, but in and around the Gir forest indicated the survival of the some of the released Mugger and also long distance dispersal of the stocks from the release site.

The population structure of Mugger mentioned above is based, on size rather than age structure. Due to occurrence of high mortality among hatchlings their per cent may vary greatly within a given population in a short period of time. Hence, the data on hatchlings were eliminated from the population analysis. The percentage of adults to the non-hatchlings in Gujarat State accounted for 88 per cent during the surveys in Gujarat. An interesting observation was, presence of more number of adults in the natural environment than the other size classes like sub-adults, juveniles, yearlings etc. The per cent sightings of juvenile and sub-adults were high (22.9 per cent) in the Kachchh region whereas it decreased to 9.3 and 8.3 per cent in the Saurashtra and South Gujarat regions respectively (Table 10.5). The presence of 12 per cent (N = 50) of sub-adults and juveniles in the State indicates a low natural recruitment and this could be due to high loss in the egg and hatchling stages. Predation, habitat alterations and environmental factors are the major causes for the loss of eggs during the post hatching periods (Thorbjarnarsan, 1989) and hatchlings, which tend to reduce the number of hatchling reaching the adult stage in the wild. This may be true in Gujarat State, where high fluctuation of water was noted in the irrigation dams like Hiran, Singoda etc. (Gir), where during summer water was released for irrigation. As a result, the nest made during February/March would become gradually far from the water edge (over 2 km was recorded), which made it difficult for the nest guarding females to protect its nest from the predators. As a result, out of 52 nests recorded in the Gujarat State, 62 per cent were lost due to animal predation (31 nest–60 per cent) and human destruction (1 nest–2 per cent). The animal predators destroying eggs include Jackal, Fox, Common Mangoose, Wild Boar (Vijaya Kumar, 1993) and *Varanus* (Choudhury and Bustard, 1979) while the predator on hatchlings include Turtle (Wait Ku Wait, 1982), Great Horned Owl, Grey heron, Crested Serpent Eagle, Short Toed Eagle, Brahminy Kite (Vijaya Kumar, 1993) and Night Heron (Alvarez Del Toro, 1974). Other enemies of small crocodiles include sub-adult crocodiles (Messel *et al.*, 1984 and Reddy, 1978) while *Panthera* spp. (Lion etc.) are enemies for both sub-adult and adult crocodiles (Alvarez Del Toro, 1974 and Vijaya Kumar and Choudhury, 1994). With the increasing body size, the predators of Mugger diminish rapidly. Other than man, large crocodile have very few potential predators (Panther-Pitman, 1913; Lion-Vijaya Kumar and Choudhury, 1994; Vijaya Kumar, 1997). The above clearly highlights the large number of predators dependent on small-sized crocodiles, which could be the reason for the presence of more number of adults in the natural environment.

Table 10.5: Percent Sightings of Small Size Class Mugger

Region	*Total Number of Non-Hatchlings (Adult, Sub-adult, Juvenile)*	*Total Number of Sub-adults and Juveniles*	*Percentage of Sub-adults and Juveniles*
Kachchh	83	19	22.9
Saurashtra	311	29	9.3
South Gujarat	24	2	8.3
North Gujarat	0	0	0
Gujarat State	418	50	11.9

Further, high crocodile density at places like Hiran, Prakshar etc. might have triggered density dependent population control, which may have led to complete exclusion of smaller individuals from the main population or mortality via mechanisms such as cannibalism. Schmidt (1924) reported similar observation at Hondrus, where he recorded a high crocodile density with a total absence of small juveniles. The juvenile mortality becomes a bottleneck to population growth (Hutton, 1984). Added, the spatial segregation of crocodiles by size class may represent a difference in the physical habitat requirements by different sized crocodile or it may reflect social factors and agnostic encounters between size classes (Messel *et al.*, 1984). These could be some of the possible reasons for the low sighting of small size class Mugger in Gujarat State.

Captive Mugger in Gujarat

As part of CCP, crocodile rearing centres were established in the Gujarat State with a purpose to increase the hatching success and survival rate. Over 80 per cent hatching success and 90 per cent survival was achieved in the captive centres (Choudhury and Choudhury, 1986) against 2 per cent survival in the wild (Singh, 1985). Along with the two rearing centres (Sasan and Gandhinagar) of the Gujarat State, eight more zoos were also involved in the captive breeding of Mugger. Total number of Mugger recorded in the rearing centres and zoos during 1995-96 was 289 (all size classes). Due to lack of identification of appropriate reintroduction sites the stock rates increased in the captive centres which led to excess of expenditure for maintenance. As a result, three zoos had stopped the captive breeding while others were also gradually towards the same path.

Habitat Preference

Site for basking, courtship and mating, nesting and rearing the young ones, are the four major requirements essential for the survival of Mugger in the natural environment, which leads to select a suitable area (microhabitat) for their life activities. Certain microhabitat features needed to determine also strongly influence the distribution of crocodiles in an area. The shoreline vegetation, which includes *Typha* reed beds etc. forms an important habitat for hatchlings, yearlings and juveniles (Figure 10.1) which offers cover from predator and also protect them from being washed away by water currents and winds etc. In addition, the prey species of the young Mugger (fish fingerlings, frogs, insects etc.) are mostly restricted to the shoreline habitats. Since the habitat provides food and protection

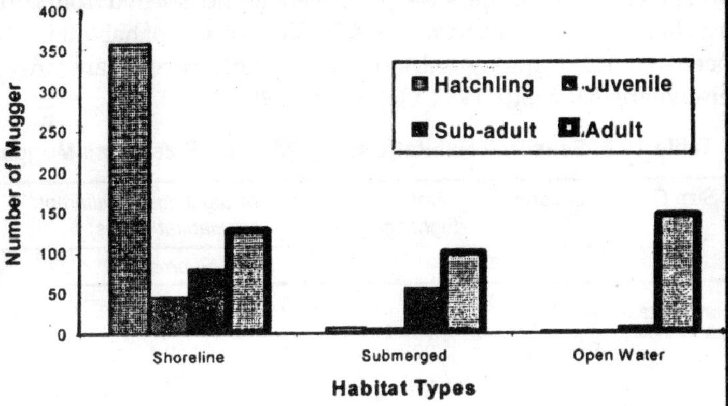

Figure 10.1: Habitat Preference by Different Size Class Mugger

for the young ones, it is highly preferred. Therefore, hatchlings, yearlings and juveniles are restricted more to the shoreline while, sub-adults frequents at shoreline and submerged vegetation area which is adjacent to the shoreline habitat. The adults being master predator have very few enemies and being opportunistic feeders use open water, submerged vegetation and shoreline habitats of the water body. Shoreline including the banks of the water body forms an important habitat for basking, part of normal and daily activity in all the size class crocodiles (Station and Dixon, 1975; Maness, 1976 and Whitaker and Whitaker, 1984).

The hatchlings were mainly restricted to shoreline habitat in all three seasons (91.7–100 per cent) followed by juveniles (90 per cent). Most of the sub-adults were recorded along the shoreline during winter (72 per cent) and monsoon (76.2 per cent) while in summer (43.6 per cent) they used the submerged vegetation habitat and were rarely (6.9 per cent) recorded in open water areas. Sub-adults respond more to monsoon currents which is similar to the sub-adult Alligators (Chabreck, 1965). The adults were recorded in all three habitats in all the seasons (Vijay Kumar, 1993). The adults used open water area was less (30.4 per cent) during winter mainly for basking and maximum usage was during summer (42.2 per cent) mainly in search of food. Further, as adults always chase sub-adults and juveniles, therefore, sub-adults were found isolated from adults and hatchlings (Vijaya Kumar, 1993). Similar type of habitat use has been recorded among sub-adults of other crocodilians species elsewhere (Mazzotti, 1983; Gapy *et al.*, 1981, 1985).

Table 10.6: Seasonal Habitat Use by Different Size Class Mugger

Size Class	Seasons	No. of Sightings	% of Sightings in Different Habitat Types		
			Shoreline	Submerged	Open Water
Hatchling	Winter	12	91.7.	0.0	8.3
	Summer	293	97.9	2.1	0.0
	Monsoon	61	100.0	0.0	0.0
Juvenile	Winter	20	90.0	10.0	0.0
	Summer	20	90.0	5.0	5.0
	Monsoon	10	90.0	10.0	0.0
Sub-adult	Winter	18	72.2	27.8	0.0
	Summer	101	49.5	43.6	6.9
	Monsoon	21	76.2	23.8	0.0
Adult	Winter	112	43.8	25.9	30.4
	Summer	206	30.6	27.2	42.2
	Monsoon	53	28.3	26.4	45.3

Breeding

There are few *in situ* studies (Schultze, 1914; Dharmakumarsinghji, 1947; Whitaker and Whitaker, 1977; Choudhury, 1981; Singh, 1979, 1982) on the nesting of Mugger but the nesting success and hatchling survival in the wild was studied only after 1986 (Vijaya Kumar, 1993, 1997). Box 10.3 shows the nesting details of other Indian crocodilians.

Nest site selection is governed by three factors; sufficient depth of soil to provide a pit where the eggs can be deposited, nearby shade in which the guarding mother can retire during the heat of the day and access to permanent water body. The adult showed territorial behavior during October and mating was observed in November/December (Whitaker and Whitaker, 1984, 1989 and Vijaya Kumar, 1993) in shallow water close to the bank.

BOX 10.3
Nesting Details of Estuarine and Gharial Species

The nesting season differs from species to species; it is between March and April for Gharial and May to June for Estuarine crocodile. The Gharial is a hole nester and lay eggs in a pit digged along the bank of the water body. The Estuarine crocodile is a mound nester, which construct nest using leaves, twigs and soil above the ground, generally in open areas located amid of the mangroves. The Gharial select a high steep sandy bank for nesting while Mugger lay eggs either in sandy or in muddy banks.

The number of eggs per clutch size also varies: 10–97 for Gharial, 10–75 for Estuarine crocodile. The incubation period extends from 75–80 days for both Gharial and Estuarine crocodile.

The nesting activity began in the month of March in Andhra Pradesh whereas it is in April in Gujarat. Before laying the eggs, the female dig a number of trail nests, which appears to be for selecting suitable nesting site and substrate. Miller (1985) reported that the nest location is very important for hatching success. The hatching success was high in the nests located in the sunny area (89.2–100 per cent), which decreased to 61.5 per cent in partly shady and 42.5 per cent in shady area.

Similarly the nesting substrate is also equally important. Maximum hatching success (91.5 per cent) was recorded in sandy soils, which decreased to 48.4 per cent in clayee soil. This is because the sand has larger particle, lower mean water content and high oxygen diffusion capacity (Lutz and Dunbar-Cooper, 1984). The gaseous environment can affect embryonic development (Packard and Packard, 1984). Miller (1985) reported that temperature, hydric environment and gaseous exchange synergistically get to influence embryonic metabolism. Therefore the trail nests were made to find suitable nesting environment as well as to divert the nesting predators like fox, jackal, mangoose etc.

The nests were recorded from a minimum of 0.7 m to a maximum of 10 m from water edge (n = 23 nests) in Andhra Pradesh whereas it ranged between 0.6 and 146 m for 52 nests in Gujarat. This gives protection to the nest from unusual flood, and maintains the nest temperature, and also easy access to the nest guarding female. If suitable nesting sites were not available, the nests were placed even away from the water edge as seen in Gujarat (146 m). In drought prone areas like Saurashtra and Kachchh region of Gujarat State, Mugger are forced to nest at the mouth of the burrows (tunnel) where it spends the summer periods. This enables the female to protect the nest easily, though in some cases it may be far from water (importance of tunnel are given separately). In Gujarat, 22 nests (42 per cent) out of 52 were recorded in the burrows while it was two (9 per cent) out of 23 nests in Andhra Pradesh.

The clutch size ranged between 14 to 46 in Gujarat and 17 to 25 (n = 23 nests) in Andhra Pradesh. The incubation period varies with incubation temperature which is influenced by the atmospheric temperature. A minimum of 50 days and a maximum of 83 days of incubation period were recorded in various Mugger habitats of Andhra Pradesh (Vijaya Kumar, 1993). Whitaker and Lang (1985) reported that at nest temperature of 28°C the incubation period prolonged up to 90–101 days and at 30°C it came down to 82 days and at 33°C it was about 63 days. The incubation temperature also plays an important role in sex determination of Muggers (Lang and Andrews, 1989).

During the entire incubation period, the nesting female remains in attendance either lying on the nest or nearby cover to protect the nest from the predators. At the completion of the incubation period, the young make a "croak" (an unusual call of the young when it is inside the egg), and this is the signal for the mother to uncover the nest to liberate her offspring. Therefore, towards the end of incubation period the mother remains at the nest for longer time (Lang, 1986; Whitaker and Whitaker, 1989; Vijay Kumar, 1993). Hatchlings emerged out from late May to Mid June in Andhra Pradesh while it was from June to July in Gujarat. The yearly nest data between 1986 and 1990 showed that the number of nest in an area varied in certain years. This is because, some females do not nest every year (Messel *et al.*, 1978–86) and some wild females may nest twice in anyone season (Webb *et al.*, 1983; Whitaker and Whitaker, 1984, 1989). It is presumed that either food supply or the condition of the female along with the environmental situation that determine the female to nest once or twice or not at all. This may be an adaptation to prevent loss of eggs during drought periods in arid and semi-arid areas like Kachchh and Saurashtra.

Out of 23 nests in Andhra Pradesh, two were damaged by flood while other nests hatched successfully, with a hatching success of

60.1 to 100 per cent. Interestingly, in Andhra Pradesh, the rivers in which Mugger were present are perennial and with minimum fluctuation of water level during the breeding season. In Gujarat there are no perennial rivers in Kachchh and Saurashtra regions and the reservoirs/tanks of these regions are facing heavy pressure due to irrigation demands. Mugger living in such reservoirs/water bodies and rivers faces heavy fluctuation of water level, which is one of the major causes for egg predations. The female as per the existing situation choose the nesting site (distance) at the time of egg laying in February and March, however, irrigation demands from March lower the water table and thereby increase the nest distance to over 2 km from the water edge. In such cases, the nest falls under predation due to lack of protection and in some cases the nesting female falls under prey to larger carnivores like lion etc. while undertaking such land movements to guard their nests. This is reported in India and elsewhere (Gorzula, 1978; Whitaker and Whitaker, 1977; Choudhury and Bustard, 1979; Madem, 1981, 1983; Hutton and Child, 1989 and Thorbjarnarsan, 1989).

The embryos within a single clutch have been reported to be at different stages of development (Pooley, 1969 and Hutton, 1984). In Andhra Pradesh, a single egg from a nest hatched 40 hours later than other eggs. After hatching, the hatchlings stay together forming discrete pods (Gorsula, 1978; Medem, 1981; Whitaker and Whitaker, 1989), which remain together for years if the area provides suitable protection from enemies (Rodda, 1984; Thorpjarnarsan, 1984, 1988) and other requirements. Two peaks were observed in the loss or mortality of the hatchings in Andhra Pradesh, one immediately after hatching and the other during flood periods. When the hatchlings stay together in areas having poor cover they easily fall under predation. Further, hatchlings can not withstand the monsoon currents and may be killed by heavy currents or washed off to far areas. The estimated survival of wild hatchlings in different Mugger habitats at the end of one year in Andhra Pradesh was 0–2.9 (Vijaya Kumar, 1993) with the dispersal extending from 30 to 1100 m within the same period.

Conflicts with Human and Livestock

Crocodile's grotesque appearance, crawling habits and the capacity to remain submerged in water have made it a much-feared animal, though they are much-maligned animals. Even though reports on man eating by Mugger exist since the beginning of the century (Pitman, 1913; Short, 1921; Battye, 1944), generally it has been stated that, attacks on human are probably a mistaken identity (Whitaker and Whitaker, 1983). The earlier reports mentioned that the Mugger eats the remains of the dead bodies of human which were thrown in the rivers like Ganga. However, during the recent decades, the conflicts between human and Mugger have increased considerably due to limited resource availability and excess pressure on the resources. In Andhra Pradesh, one incident of human attack was recorded at Ethipothalla area while occasional livestock attacks were recorded from other areas.

In Gujarat, totally 10 incidents of human attack by Mugger was reported (3 in Kachchh, 3 in Saurashtra and 4 in South Gujarat) during the period between 1991 and 1996, of which nine occurred between March and July when the water level was very low. Further it was also understood that adult males were involved in these attacks. Except for two, in all the other incidents the person survived. Attacks on livestock were reported from 11 locations, in Kachchh, Vadodara and Junagadh districts of the Gujarat State. However, in Kheda district of Gujarat where Mugger and livestock share the common water body no such attacks were reported. Communal use of village pond by Mugger has been reported (Singh, 1983). The recent increase of conflicts in Kachchh and Saurashtra region of Gujarat is mainly due the reduced resource availability and high pressure on the available resources.

Factors Affecting the Sighting of Mugger

There are several factors such as phase of the moon, cloud cover, water depth, air and water temperature which presumably affect the sighting of the crocodile during night spot light surveys. The biotic disturbances and air and water temperatures are the factors that affect the sightings of the crocodiles during the day-time. The Figure 10.2 shows that 26–28°C is the ideal temperature for Muggers to float on the surface of water during night periods. Further, the sub-adults were noted only above 24°C indicating that they preferred warmer temperatures. Therefore, while estimating the population of Mugger crocodile, the survey should be planned in the above water temperature regimes so that maximum number of crocodile can be recorded.

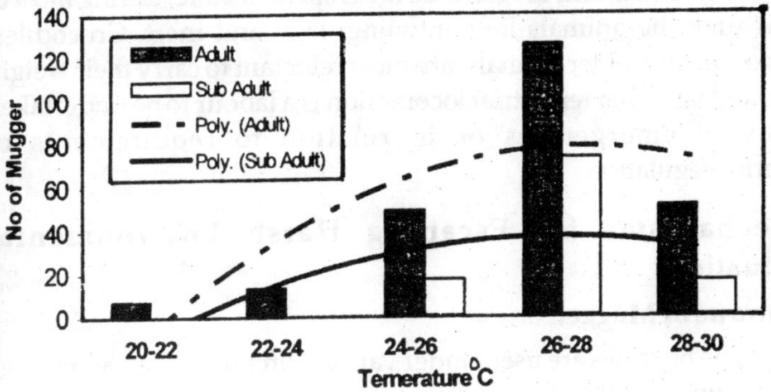

Figure 10.2: Water Temperature During Nights and the Number of Mugger Recorded

Basking behaviour in relation to air temperature was studied in Andhra Pradesh. Basking commenced at 18–20°C and was prevalent between 22–24°C air temperature, thereafter decrease gradually as the air temperature increase to 28°C during which no Mugger was seen basking in the wild. During winter they are found maximum on land during day while in summers they are in the water for maximum period. Out of 82.5 hrs observation during the months of January and February, the adult spend 52.5 per cent of the time for basking on land while in summer it spends 17.4 per cent (out of 134.5 hrs observation between March and June). They have two basking peaks during day; 0900–1100 and 1500–1800 hrs, which is similar to that reported for other crocodilians elsewhere (Maness, 1976; Station and Dixon, 1975 and Ayarzaguena, 1983).

Being an ectothermic animal, the ambient temperature is very important for regulating body temperature and thereby maintain the physiological activities like digestion, metabolism etc. Diegenbach (1975) reported that the rate of digestion is three times faster at 30°C than at 15°C in Caimen crocodiles. Coulson and Hernandeaz (1979) accounted that at 28°C the metabolic rate of Alligators increases three times following feeding. Early morning basking evidently plays an important role in the restoration of heat lost during the night. When they are lying ashore or partly submerged by day, crocodiles are commonly seen with the jaws held widely agape. The exposure of the moist mucosa does provide, through evaporation, an important accessory cooling mechanism.

Although about half of crocodile life is spend ashore, during most of this time the animals lie sprawling, lazy and inert. Crocodiles, especially the older animals, are most reluctant to carry their weight on land, and that terrestrial locomotion is a labour to be undertaken only in emergencies or in relation to requirements of thermoregulation.

Mechanisms for Escaping Harsh Environmental Situations

Burrows of Mugger

The burrows are used under various circumstances; as a refuge for resting and thermoregulation, an aestivation site for prolonged droughts or protection from natural predators or man or used for nesting purposes. Burrowing seems to be a survival tactic in Mugger to withstand the drought conditions, which are a standard feature of the dry season in many parts of its drier range. It seems likely that Mugger group together in a single tunnel as observed in Nile Crocodiles (Guggisber, 1972). Further, the lowering of water table in high crocodile density areas leads to competition for food, space etc. In such areas burrowing helps them to escape from such unfavourable situations.

In total 114 Mugger burrows were recorded in Gujarat State. Of which Muggers were present in 5 burrows and 22 were used for nesting. Many of the burrows were constantly used by Mugger, which was evident from the presence of spoor marks. During summer at Gir forest, the burrows were observed at three different heights; 5–10 m up the bank, at the water level and at the middle of the bank (between top of the bank and water table). It was presumed that the burrow on top may have been used during monsoon when the water body was full, the middle one have been used during winter when the water level retreats, whereas the bottom ones may have been used in summer when the water body have minimum level of water (Vijaya Kumar, 1997). It was interesting to note that the temperature at 3 m depth of a burrow at Hiran dam of the Gir forest fluctuated between 19.2°C to 19.8°C in a 24 hour cycle while the temperature outside the burrow ranged between 12°C to 43°C. The low and constant temperature inside the burrow helps the mugger in reducing their metabolic rate via body temperature and thereby reduces the energy expenditure on maintenance. Being an ectothermic animal, crocodile's metabolic rate increases at higher body temperature

(Lewis and Gatten, 1985). Hence inside the burrow a Mugger can live for longer periods without food.

The monsoon water spread of Hiran dam (Gir) is about 3.35 km^2 and the density of Mugger (201) in this reservoir is 57/km^2, whereas in summer the 201 Muggers recorded are forced to adjust within 0.83 km^2 of the water spread due to release of water for irrigation and evaporation etc. Interestingly, the recorded home range of adult Mugger ranged between 0.78 to 3.54 km^2 (Vijaya Kumar and Choudhury, 1992; Vijaya Kumar, 1993). With this home range size, the Mugger experiences difficulties even with the water level of the reservoir being full during the monsoon. These difficulties related with the habitat availability and shortage/competition for food, forces the Mugger to develop certain escape mechanism like burrowing which help them in escaping from the above difficulties. This could probably be the reason for the maximum record of 58 burrows at Hiran dam in the Gir forest (Vijaya Kumar, 1997).

Overland Movement

Overland travel is a likely mode of colonization particularly by sub adult and juvenile Mugger. This is well documented in Mugger, which normally occurs when water dries during summer (Ahmed, 1945). When moving on land the crocodile has three distinct gaits; high walk, belly run and gallop. Of which, high walk is used generally during unharried overland travel, immediate reaction to disturbance is to move towards the water with high speed by belly run while gallop is very rarely used.

When the usual habitat is dry these movements sometimes take the animal far from the water. Under prolonged droughts and at times can even survive through prolonged aestivation. In their tropical state, crocodiles can be clubbed, one at a time, and put up no fight.

Food and Feeding Habits

Dietary habit is one of the key aspects in assessing the role of crocodilians in the ecology of inland waters. Such studies not only provide a general picture of predator prey relations, but also offered comparison between food habits of crocodilians of different size classes. There are few published reports on the food habits of Mugger (Pitman, 1913; D'Abreu, 1915; McCann, 1935; Abdulali, 1938; Lewis, 1940), mostly on the basis of stomach content analysis of dead animals. Food habits can also be studied by removing the stomach contents from crocodiles without sacrificing the animal (Tayler *et al.*, 1978). The killing or capturing of Mugger for stomach content analysis in India is difficult, since they are protected under Schedule I of the Wildlife (Protection) Act, 1972. Therefore food habits of the Mugger was studied through scat analysis, a standard method suggested to derive the crocodilian food habit (Whitaker and Whitaker, 1989; Vijaya Kumar, 1993; Vijaya Kumar *et al.*, 1995).

Total 289 scat samples were collected from the wild, of which 216 were of adults and 73 belonged to sub-adults. Table 10.8 shows the proportional occurrence of different categories of food items eaten by sub-adult and adult Muggers. It also reveals an ontogenetic change in feeding pattern, with the larger individuals exhibiting a preference for large vertebrate prey. The sub-adults restrict themselves to smaller prey like small fish, arthropods, and amphibians. Adult Mugger though primarily piscivorous, feed on a variety of diets. The scat samples of adults also had remnants of mammals, reptiles and birds. The frequency occurrence of mammals in the diet increased with the size of the Mugger, which was evident in increase from 6.85 per cent in sub-adults to 44.4 per cent in adults. Evidence of predation on domestic livestock was found in 16 samples (Vijaya Kumar *et al.*, 1995). The bird prey of adults included Cormorant, Coot and Cattle egret while sub-adults fed on Coot and Purple Moorhen.

Table 10.8: Frequency Occurrence of Food Items from Scat Samples of Sub-adult and Adult Mugger

Food Items	*Sub-adults (n = 73)*		*Adults (n = 216)*	
	Frequency	*% Frequency*	*Frequency*	*% Frequency*
Mammals	5	6.85	96	44.44
Birds	1	1.37	9	4.16
Reptiles	2	2.74	11	5.09
Amphibians	13	24.65	26	12.03
Fish	49	67.12	148	68.52
Arthropods	43	58.90	23	10.64
Molluscs	4	5.47	4	1.85
Plants	0	0.00	6	2.77
Non-food Material	0	0.00	3	1.38
Unidentifiable Debris	6	8.22	11	5.09

There is seasonal variation in diet of Mugger. Fish formed the main part of diet in all the three seasons (Table 10.9). Mammals were taken more frequently during winter which gradually decreased in summer and monsoon. Basking an important activity during winter (Vijaya Kumar, 1993), leads to prolonged time spent on land during the day which resulted in increased feeding on land animals, such as mammals. Further, Mc Nease and Joanen (1981) reported that although excellent growth of crocodiles was achieved on a diet of fish, the diet did not result in high reproductive success. It appears from the above that mammalian diet was required for enhancing reproductive performance of the crocodilians (Vijaya Kumar *et al.*, 1995). Bird formed part of diet during winter and summer months. This may be due to the high availability of wintering birds in the area (Vijaya Kumar and Choudhury, 1995a, 1995b). Amphibians were eaten during the summer and winter, whereas frequency of arthropod was found to be relatively high during the winter and low during the summer. Fish formed the major constituent of the diet of sub-adults only during the summer. Singh (1983) reported that intake of fish in sub-adults was reduced to less than 1 per cent of body weight while it's increase during summer, was suspected to be due to lowered water level, which made the capture easy. Therefore, on the ecological side, crocodiles are not detrimental to fishery interests; on the other hand, their presence is directly or indirectly beneficial to the industry. Large proportions of fish prey are either of secondary value or of predatory species. The small crocodiles feed on enemies of fish, fry and eggs while the adults take the toll of Otter, Cormorants, Darters, Storks etc. Crocodiles are also important scavengers as they feed on sick and dead fishes and keep the aquatic environment healthy and clean.

Crocodile acquire stones before reaching maturity, which are deliberately swallowed by the animal to serve physical and physiological function such as digestion, by the triturating of food, serve to relieve the pangs of hunger during periods of enforced starvation, hydrostatic function etc. A 3.42 m Mugger contained about 1 kg of stones (Simcox, 1905; D' Abreu, 1915). The same was reported on other species of crocodiles by McCann (1940). Since there was no postmortem report of Mugger available during the study period this was not recorded during the study.

Table 10.9: Seasonal Changes in the Dietary Constituents of Sub-adult and Adult Mugger

Food Items	*Sub-adults (n = 73)*			*Adults (n = 216)*		
	Winter (n = 23)	*Summer (n = 43)*	*Monsoon (n = 7)*	*Winter (n = 75)*	*Summer (n = 120)*	*Monsoon (n = 21)*
	% Frequency	*% Frequency*	*% Frequency*	*% Frequency*	*% Frequency*	*% Frequency*
Fish	69.56	72.09	28.57	61.33	70.00	85.71
Mammals	21.74	0.00	0.00	61.33	40.83	4.76
Birds	4.35	0.00	0.00	4.00	5.00	0.00
Arthropods	95.65	44.10	28.57	21.33	4.17	9.52
Amphibians	10.67	15.00	0.00	10.67	15.00	0.00

Current Disputes in Mugger Conservation and Management

Except for few States, there is paucity of systematic survey results on the population of Mugger in the entire country. As a result, no information on the rate of increase or decline in Mugger population was available from most of the release locations in the country, which otherwise is, very important for the managers and planners for proper planning of the future conservation and management of the species and their habitats.

Due to efficient captive breeding programs, stocks of Mugger have piled up in all the rearing /captive breeding centres and presently facing a crisis of overcrowding and inadequate care. Due to this captive breeding of Mugger has been currently stopped in many zoos, rearing centers etc.

Further, Indian Government has also adopted a policy prohibiting any possibilities of commercial use of captive-bred crocodiles. The non-identification of new release locations other than Gir Protected Area in Gujarat has led to repeated stocking of Mugger in the same location, resulted in severe problem for the survival of released stocks.

The water bodies where Mugger have been released are subjected to severe water fluctuation between monsoon (wet) and summer (dry) seasons due to use of water for irrigation purposes, which has led to unwanted effects such as cannibalism, emigration and predation during dry periods (Vijaya Kumar and Choudhury, 1994).

Thus, the main objective of the conservation project to build up the depleted population, suffered due to non-identification of new suitable Mugger habitats.

Past debates on crocodile conservation and management has focused largely on the population within the Protected Areas, where favourable habitats with protection from interference, very low human and livestock population density exists. These features add up to conservation of Mugger population in Protected Area, which is very different from the population that exists outside the protected areas. A good number of Mugger have been recorded outside the Protected Area in Kachchh, Saurashtra and South Gujarat region of the Gujarat and Manjira and Godawari rivers of Andhra Pradesh

States. Further, crocodile conservation is not a simple case of preservation. Most of the species, which remain endangered, can never regain their historical numbers because their habitats have disappeared. Thus, there have also been major changes over the years in the types of threats to Muggers in the State. The early actions were almost entirely concerned with limiting the killing of Mugger, which either had been, or were at the verge of becoming seriously depleted by human exploitation, when Mugger habitats were still relatively undisturbed. During the recent past the most invidious threat to Mugger is not excessive killing, but disruption of their habitat. Further regional variation in environment and climate with in a State also needs to be considered, because Kachchh region entirely falls under arid tract where droughts are a recurring phenomenon whereas South Gujarat region is a heavy rainfall tract in Gujarat. Under these circumstances it is not an easy task to simply transfer a set of management plan right across the State. Instead, it is essential to develop a management strategy specific to each region (*e.g.* Kachchh, Saurashtra etc.), which is sensitive to regional variation in habitat quality and conservation potential, in addition to the fact that the information base is lacking. The management plan should also emphasize the most pressing problems and concentrate on resources in areas of maximum benefit to state-wise crocodile conservation. More particularly the management program recognized that sharing water ways by people and crocodiles brings conflicts which have to be resolved, because, such conflicts create a significant impact on the survival of Mugger in the unprotected areas. Hence, managing the people/crocodile interface is pivotal to any conservation program that can bring a social dimension to the problem which some of the earlier debate have ignored or trivialized. Hence, concrete measures must be taken to avoid disastrous consequences of human deaths resulting from crocodile attack where sharing of common water bodies by Mugger and human exists. Further, the success of these programs is mainly based on the support and active participation of the local people. To achieve this, an education program designed to eliminate apprehensive beliefs about dangerous interactions between people and crocodiles will be necessary throughout, particularly in areas where the restocking of crocodiles are contemplated. However, high priority should be given to the interest and need of the local people and measures should be taken to arrange alternate resources where possible.

As part of the public safety campaign within the State, the 'problem' crocodiles from unprotected areas need to be identified and removed from the wild and assigned to captive centres which would enhance the genetic viability of the captive stock. Such animals are broadly defined as those attacking human and livestock or residing in human habitations, where their presence constitutes a risk to people and livestock.

Added, a complete protection to crocodiles throughout Gujarat State is presently an impractical consideration, because, most of them are found in remote areas where little or no law-enforcement capabilities exist. Hence, under this condition, the recovery of Mugger in many parts of its historical range will be very difficult, because of heavy human usage, the antipathy of man towards crocodiles, as well as human and livestock population expansion and other associated factors that cause habitat destruction are still not controlled.

Agriculture puts an enormous pressure through direct competition with crocodiles in Gujarat State. Use of reservoir water for irrigation purposes during the period between February and June leads to a decrease in water level, which is the period for egg laying and hatching and it causes tremendous nest losses. There is no denying of the paramount importance of the agricultural development and food production in the State for the ever-increasing human population. However, it should be also noted that surface water irrigation is the oldest form of irrigation and it requires large quantities of water, as the losses in the form of infiltration, evaporation etc. are enormous. Hence, the use of large quantities of water from the Mugger habitat is to be regulated during the breeding season by the application of modern techniques like sprinkler, drip irrigation etc. water use for agriculture can be moderated and thereby, save the Mugger and its environment.

It is very important to avoid restocking in high crocodile density areas, because, high crocodile densities and limited extent of available habitat results in a greater degree of overlap in habitat use by different size class, thereby creating conflicts, which may lead to cannibalism and out migration of smaller size classes. Based on the hydrological characteristics of the water body (dry and wet periods), the number of stocks for restocking has to be regularised. Therefore, it is suggested to stop further restocking within the Gir Sanctuary.

Monitoring the major Mugger population area is very essential to understand the effects of the implementation of the project. The formulation of state level Crocodile Monitoring Committee (CMC) comprising subject experts, Forest Department officials and NGO's would be of great help for this purpose. This would enhance the information base, help in prioritising site specific management issues and evolve appropriate strategies.

References

Abdulali, H. 1938. The food of mugger (*Crocodylus palustris*). *J. Bombay Nat. Hist. Soc.*, 40: 336.

Acharya, H. 1949. List of reptiles from Gujarat state. *Prakaruti* (A Gujarati Journal), 8: 158–178.

Ahmed, Q. 1945. Syncope in a crocodile. *J. Bombay Nat. Hist. Soc.*, 45: 429.

Alvarez Del Toro, M. 1974. *Los crocodilia de Mexico.* D.F. Institute Mexicano de Recursos Nalorales Renovalbles, A.C.

Attwell, R.I.G. 1957. Unpubl report. Game and Tsetse Control Dept. Northern Rhodesia. In Transactions of the Zoological Society of London by Cott, H.B. 29 (4), Publ.

Ayarzaguena, J. 1983. Ecologia del Caimen anteojus obaba (*Caimen crocodilus*) en losiianos de Apure (Venezuela). *Donana Acta Vertebrata,* 10: 1–136.

Battye, R.K.M. 1944. Crocodile bellowing. *J. Bombay Nat. Hist. Soc.*, 45(1): 93–94.

Bara, M.O. 1975. *American Alligator Investigations, Final Study Report 1970–1975.* Wildlife and Marine Resources Department, South Carolina.

Biswas, S. 1970. Proposal for the protection of marsh crocodile. *Indian Forester,* 96(9): 704.

Bustard, H.R. 1974. *India: A Preliminary Survey of the Prospects of Crocodile Farming.* FO/FO: IND/71/033, Rome, p. 1–66.

Bustard, H.R. 1975. Crocodile conservation in India. *Tiger Paper,* 2(4): 17.

Bustard, H.R. 1980. The Government of India Crocodile project. *Cheetal,* 22(1/2): 11–16.

Chavan, S.A. 1979. Distribution and conservation of crocodiles in and around Gir Wildlife Sanctuary. *Indian J. Forestry*, 2(3) 266–268.

Choudhury, B.C. 1981. Mugger (*Crocodylus palustris*) released in Andhra Pradesh and Tamil Nadu. *Hamadryad*, 6(1): 10–11.

Choudhury, B.C and H.R. Bustard. 1979. The status of mugger in Tamil Nadu. In: *Proc. First Indian Crocodilian Researcher's Symposium*, Katerniyaghat, U.P. India, 21–25 January.

Choudhury, B.C and H.R. Bustard. 1982. Restocking mugger crocodile *Crocodylus palustris* (Lesson) in Andhra Pradesh: Evaluation of a pilot release. *J. Bombay Nat. Hist. Soc.*, 79(2): 275–289.

Choudhury, B.C. and S. Choudhury. 1986. Lesson from crocodile reintroduction project in India. *Indian Forester*, 12(10): 881–890.

Chabreck, R.H. 1965. Methods of capturing, marking and sexing alligators. In: *Proc. 17th Ann. Conf. SE Assoc. Game Fish Comm.*, 17: 47–50.

Coulson, R.A. and T. Hernandeaz. 1979. Increase in the metabolic rate of alligator fed proteins or amino acids. *J. Nutr.* 109: 538–550.

D'Abreu. 1915. Notes on the mugger (*Crocodylus palustris*) contents of their stomach, folklure etc. *J. Bombay Nat. Hist. Soc.*, 23(4): 780.

Daniel, J.C. 1970. A review of the present status and position of endangered species of Indian reptiles. In: *Proc. 11th Tech. Meeting*, IUNC Publ., NS, 18: 75–76.

Daniel, J.C. and S.A. Hussain. 1973. The record (?) salt water crocodile (*Crocodylus porosus*). *J. Bombay Nat. Hist. Soc.*, 71(2): 309–312.

Daniel, J.C. and E.M. Shull. 1963. List of the reptiles and amphibians of Surat, Dangs, South Gujarat. *J. Bombay Nat. Hist. Soc.*, 60: 737–743.

Dharmakumarsinghji, K.S. 1947. Mating and parental instinct of a marsh crocodile (*Crocodylus palustris*). *J. Bombay Nat. Hist. Soc.*, 47(1): 175.

Diegenbach, C.O. 1975. Gastric function in *Caimen Crocodylus* (Crocodilia–Reptilia) II. Effects of temperature on pH and proteolysis. *Comp. Biochem. Physiol.*, 51A: 267–274.

Gaby, R., M.P. Mc Mahon, B.A. Bonsack, W.N. Gillies and J. Gleman. 1981. The population of American crocodile *Crocodylus acutus* (Reptilia–Crocodilidae) at the turkey point power plant site. *Coral Gables*. Metcaff and Eddy. Inc.

Gaby, R., F. Mozzotti, W.N. Gillies and J.R. Wilcox. 1985. Ecology of a population of *Crocodylus acutus* at a power point site in Florida. *J. Herpetol.*, 19: 189–195.

Gorzula, S.J. 1978. An ecological study of *Caimen crocodilus* inhabiting savanna lagoons in the Venezuelan Guayana. *Oecologia* (Berl.), 35: 21–34.

Guggisber, C.A.W. 1972. *Crocodiles: Their Natural History, Folklore and Conservation*. C.A.W. Guggisberg, Newton Abbot, David and Charles, pp. 195.

FAO. 1974. *India: A Preliminary Survey of the Prospects for Crocodile Farming*. UNDP/FAO, FO: IND/71/033, pp. 66.

Honegger, R.E. 1971. *The Status of Four Threatened Crocodilian Species of Asia*. IUCN Pub., 32: 42–50.

Hutton, J.M and G.T.F. Child. 1989. Crocodile management in Zimbabwe. In: *Crocodiles*. IUCN Publ. (NS), Gland, p. 62–79.

Hutton, J. 1984. Home range, territoriality and dispersal in the Nile crocodile at Nyezl, Zimbabwe. Field Do.

IUCN/SSC/Crocodile Specialist Group. 1993. Crocodile conservation and management in India. In: *Report of CSG Workshop held at Madras Crocodile Bank*, Madras, India, 1st–3rd March. Pub. by Africa Resources Trust, Zimbabwae, pp. 27.

Joseph, A., E. Chauhan, K. Khanna, and R.V.J.M. Taker. 1975. Marsh Crocodile (*Crocodylus palustris*) in the Gir. *J. Bombay Nat. Hist. Soc.*, 72(3): 862–863.

Joanen, T. and L. Mc Nease. 1972. A telemetry study of adult male alligators of Rockfeller refuge. In: *Louisiana Proc. of 26th Ann. Conf. S.E. Assoc. Game and Fish Comm.*, 32: 182–186.

Klemm, C. and D. Navid. 1989. Crocodilians and law. In: *Crocodiles: Their Ecology, Management and Conservation*. IUCN Publ., NS, Gland, Switzerland.

Lang, J.W. 1986. Male parental care in mugger crocodiles. *National Geographic Res.*, 2(4): 519–525.

Lang, J.W. and H. Andrews. 1989. Sex determination and sex ratios in *Crocodylus palustris*. *American Zoologist*, 29(3): 935–952.

Lewis, E.S. 1940. Crocodile and wild boar. *J. Bombay Nat. Hist. Soc.*, 41: 903.

Lewis, I.Y. and R.E. Gatten. 1985. Aerobic metabolism of American Alligators *Alligator mississippiensis* under standard conditions and during voluntary activity. *Comp. Biochem. Physiol.*, 80A: 441–447.

Lutz, P and A. Dunbar-Cooper. 1984. The nest environment of American crocodile (*Crocodylus acutes*). *Copeia*, p. 153–161.

Madem, F. 1981. Los crocodylia de sur America. Vol. I: Los crocodilia de colombia colciensians, Bogota, Colombia.

Madem, F. 1983. Los crocodylia de sur America. Vol. II: Colciensians, Bogota, Colombia.

Maness, S.J. 1976. Contribution and conocimiento del ciclo de vida del *Caimen crocodilus* Parte I Epoca de sequia. II Seminario de chiguires Y Babas. Programe Y Rasumenson Moracay, Venezuela, Conicit.

Mc Cann, G. 1935. The mugger *Crocodylus palustris* (Lesson) feeding on large water beetles. *J. Bombay Nat. Hist. Soc.*, 38(2): 409.

Mc Cann, G. 1940. A reptile and amphibian miscellany. *J. Bombay Nat. Hist. Soc.*, 41(4): 742–747.

Me Cann, C. 1938. The reptiles and amphibia of Kutch state. *J. Bombay Nat. Hist. Soc.*, 40: 425–427.

Mc Nease, L. and T. Joanen, 1981. Nutrition of Alligators. In: *Alligator Production Conference*, February 12th. Gainesville, Florida, p. 3–4.

Messel, H., Wells, A.G. and W.J. Green. 1978–86. Surveys of tidal river systems in Northern Australia and their crocodile populations. *Monographs*, p. 1–19.

Messel H., G.C. Vorlicek, W.J. Green and I.C. Onlay. 1984. The continuing and mysterious disappearance of a major fraction of sub-adult *Crocodylus porosus* from tidal water ways in Northern Australia. In: *Crocodile Proc. of the 6th IUCN/SSC/CSG Meeting*.

Miller, J.D. 1985. Criteria for staging reptilian embryos. In: *Biology of Australian Frogs and Reptiles*, (Eds.) G. Grigg, R. Shine and H. Ehmann. Survey Beatty and Son, Sydney, p. 305–310.

Misra, R.N. 1969. The endangered crocodile in India. *Cbeetal*, 12(1): 143–148.

Mazzotti, F.J. 1983. The ecology of *Crocodylus acutus* in Florida. *Ph.D. Thesis*. Pennsylvania State University.

Oza, G.M. 1975. Conservation of the crocodile in the Sayaji Sarovar lake, Baroda, India. *Biol. Conservation*, 7: 235–236.

Packard, G.C. and M.J. Packard. 1984. Coupling of physiology of embryonic turtles to the hydric environment. In: *Respiration and Metabolism of Embryonic Vertebrates*, (Eds.) R.S. Seymour and W. Junk. Dordrecht, Netherlands, p. 99–119.

Pitman, C.R. 1913. The contents of marsh crocodile's stomach. *J. Bombay Nat. Hist. Soc.*, 22(3): 641.

Pooley, A.C. 1969. Preliminary studies on the breeding of the Nile crocodile *Crocodylus niloticus* in Zululand. *Lammergeyer*, 10: 22–24.

Rashid, M.A. 1978. A note on the status and management of crocodile sin Gujarat state. Unpubl. Report to IUCN/SSC/CSG.

Reddy, P.S. 1978. Crocodile breeding in Indira Gandhi Zoological park. *Wildlife Club Newsletter*, Dehra Dun, 6(4): 68–69.

Rodda, G.H. 1984. Movement of American crocodile in Gaton Lake, Panama. *Herpetologica*, 40: 441–451.

Schmidt, K.P. 1924. Notes on Central American Crocodiles. Field Mus. nat. 1–list. *Zool. Ser.*, 12: 77– 96.

Schultze, W. 1914. Notes on the nesting place of *Crocodylus palustris* Lesson. *Philippines J. of Sci.*, 9: 313–315.

Short, W.H.O. 1921. A few hints on crocodile shooting. *J. Bombay Nat. Hist. Soc.*, 28: 76–84.

Smith, M.A. 1935. *Fauna of British India*, Vol. I. Taylor and Francis Ltd. London.

Smith, M.A. 1927. The distribution of the mugger. *J. Bombay Nat. Hist. Soc.*, 81(2): 22–25.

Simcox, A.H.A. 1905. The crocodile, its food and muscular vitality. *J. Bombay Nat. Hist. Soc.*, 16: 375–376.

Singh, L.A.K. 1976. Back to the wild. *Cheetal*, 18(1–2): 22–23.

Singh, L.A.K. 1978. Crocodile projects in Orissa. *Indian Forester*, 104(12): 823–829.

Singh, L.A.K. 1979. Sexual attraction of wild mugger (*Crocodylus palustris* Lesson) towards captive mugger. *J. Bombay Nat. Hist. Soc.*, 76(1): 167–172.

Singh, L.A.K. 1982. Indian mugger, *Crocodylus palustris* Lesson (Reptilia, Crocodilia): Observation on the behaviour of a female from nature. *J. Bombay Nat. Hist. Soc.*, 80 (3): 499–506.

Singh, L.A.K. 1983. Communal use of village pond by the mugger crocodile (*Crocodylus palustris*), cattle and man. *J. Bombay Nat. Hist. Soc.*, 80(3): 642–645.

Singh, L.A.K. 1985. Population trend of Gharial in National Chambal Sanctuary with a notes on radio tracking. Study report. Crocodile Research Center of Wildlife Institute of India, Hyderabad.

Singh, L.A.K. 1999. Significance and achievements of the Indian crocodile conservation project. *Envis*, Wildlife Institute of India, Dehra Dun, 2(1): 10–16.

Station, M.A. and J.R. Dixon. 1975. Studies on the dry season biology of the Spectacled Caimen *Caimen crocodylus crocodylus* from the Venezuelan Llanos. *Mem. Soc. Cienc. Nat. La Salle*, 35: 237–265.

Taylcr, J.A., G.J.W. Webb and W.E. Magnusson. 1978. Methods of obtaining stomach contents from the live crocodilians (Reptilia–Crocodilidae). *J. Herpetol.*, 12(3): 415–417.

Thorpjanarsan, J.B. 1989. Ecology of American crocodile *Crocodylus acutus*. In: *Crocodiles*. IUCN Publ., NS, Gland, p. 228 –258

Thorpjarnarsan, J.P. 1984. Status and ecology of American crocodile *Crocodylus acutus* in Haiti. *Bull. Florida State Museum*, 33(1): 1–86.

Thomas, H.B. and R. Scott. 1949. Uganda Oxford press. In: *Transactions of the Zoological Society of London*, (Ed.) Cott, H.B., 29(4).

Vijaya Kumar, V. and B.C. Choudhury. 1995a. A report on waterfowl study in Manjira Wildlife Sanctuary. *Pavo*, 32(1): 47–57.

Vijaya Kumar, V. and B.C. Choudhury. 1995b. Studies on the waterfowl of Manjira Wildlife Sanctuary Medak District, Andhra Pradesh. *Mayura*, Vol. 12, p. 20–33.

Vijaya Kumar, V. and B.C. Choudhary. 1994. Mugger Crocodile Conservation in Gujarat. IUCN/SSC/Crocodile Specialist Group, Florida, U.S.A. *Newslatter*, 13(1): 6–7.

Vijaya Kumar, V. 1993. Studies on the Indian mugger Crocodile *Crocodylus palustris* (Lesson) in Andhra Pradesh, India. *Ph.D. Thesis*, Saurashtra University, Rajkot, pp. 162.

Vijaya Kumar, V. 1994. Studies on the Indian Mugger. *IUCN/SSC/ Crocodile Specialist Group Newsletter*, 13(3): 6–7.

Vijaya Kumar, V. 1995. Studies of the Indian mugger crocodile *Crocodylus palustris* (Lesson) in Andhra Pradesh, India. *Zoo's Print*, 10(7): 53–54.

Vijaya Kumar, V. 1996. Diet of Muggers. *IUCN/SSC/Crocodile Specialist Group, Florida. Newsletter*, 14(4): 6–7.

Vijaya Kumar, V. and B.C. Choudhury. 1992. Home range of introduced muggers. *IUCN/SSC/Crocodile Specialist Group, Florida, USA, Newsletter*, 11(2): 9.

Vijaya Kumar, V. 1997. Evaluation of Restocked Mugger Crocodiles and its Implication in Long-term Conservation and Management of the Species in Gujarat, India. Final Study Report, Fauna and Flora Preservation Society, London, U.K., pp. 82.

Vijaya Kumar, V., B.C. Choudhury and V.C. Soni. 1995. Dietary habits of the Indian mugger crocodile (*Crocodylus palustris*) in Andhra Pradesh, South India. *Hamadryad*, 20: 8–12.

Vijaya Kumar, V., R. Vyas and B.C. Choudhury. 1999. Status of Mugger and its conservation problems in Gujarat. In: Wildlife and Protected Areas. *Envis.*, Wildlife Institute of India, Dehra Dun, 2(1): 69–76.

Vyas, R. 1994. Mugger crocodiles in the vicinity of human settlements in South Gujarat. *Hamadryad*, 19: 94–96.

Wait Ku Wait, W.E. 1982. Investigations in the breeding biology of Selendar Snouted Crocodile *Crocodylus cataphractus*. Paper contributed to the symposium on crocodile conservation and utilization, at 6th working meeting of the CSG Victoria Falls, Zimbabwe.

Webb, G.J. and H. Missel. 1978. Movement and dispersal pattern of *Crocodylusporosus* in some rivers of Arnchem Land, Northern Australia. *Aust. Wild. Res.*, 5: 263–283.

Webb G.J. *et al.* 1983. *Crocodylus johnstoni* in the McKinlay river area. N.T. IV. A demonstration of homing. *Aust. Wildl. Res.*, 403–406.

Whitaker, R. and Z. Whitaker. 1976. Collection and hatching of marsh crocodile (*Crocodylus palustris*). *J. Bombay Nat. Hist. Soc.*, 73(2): 403–407.

Whitaker, R. and Z. Whitaker, 1977. Notes on the natural history of *Crocodylus palustris*. *J. Bombay Nat. Hist. Soc.*, 74(2). 358–360.

Whitaker, R. and Z. Whitaker, 1984. Reproductive biology of the mugger (*Crocodylus palustris*). *J. Bombay Nat. Hist. Soc.*, 81(2): 297–317.

Whitaker, R. 1978. Notes on the status of Gir crocodiles. *J. Bombay Nat. Hist. Soc.*, 75(1): 224–227.

Whitaker, R. 1974. Notes on behaviour, ecology and present status of marsh crocodile (*Crocodylus palustris*) in south India. Publ. Madras Snake Park Trust, pp. 22.

Whitaker, R. and Z. Whitaker 1989. Ecology of the Mugger crocodile. In: *Crocodiles*, IUCN Publ., NS, pp. 276 –296.

Chapter 11

Habitatic Features, Relative Abundance, Sex-ratio and Population Decline of Peafowl *Pavo cristatus* in Channagiri Area of Karnataka, India

S. Basavarajappa

Department of Zoology, University of Mysore, Manasagangotri, Mysore – 570 007, Karnataka, India

ABSTRACT

Indian peafowl (*P. cristatus*) density, relative abundance of the sexes and trend in population decline was studied in three different habitats for three consecutive years (1998–2000) in the Channagiri area, Karnataka. The data revealed that the distribution of peafowls depends on the availability of the preferred vegetation in the respective habitats. Further, the density of peafowl varied significantly ($F = 78.5$, $P > 0.05$) in these habitats. The peafowl population consists of adult (Cock : peahen = 1 : 4.9), sub-adults (Juveniles), and their relative abundance varied significantly ($P > 0.05$) between the habitat III compared to habitats II and I. The steady decline of peafowl during 1998–2000 in habitats I and II was due to increased habitat destruction, encroachment and other associated disturbances followed by hunting. Considering the high degree of human interference, it is recommended that habitat destruction and human associated disturbances be minimized to conserve the existing population of peafowl in their preferred habitats.

***Keywords**: Peafowl density, Sex-ratio, Population decline, Habitats, P. cristatus.*

Introduction

Indian crested peafowl *Pavo cristatus* is a gallinaceous bird (Order: Galliformes; Family: Phasianidae) widely distributed in the diversified agro-climatic zones of the Indian subcontinent. In brilliancy of plumage the peafowls stand unrivalled among the feathered race. Since time immemorial, it has been known that India was the cradle of the peafowls (Cuvier, 1993). Being the national bird of India (Saharia, 1996), it is considered as an important species because of its ecological, economical, ethical, medicinal and scientific values (Ali and Ripley, 1983; Ensminger, 1992).

The *P. cristatus* in Peninsular India is confronted with hunting pressure and increased human interference (Thirumurthi *et al.*, 1993). Its status is alarming in different agro-ecosystems (Basavarajappa, 2001). In many parts of India a considerable number of peafowls are dying due to various reasons (Table 11.1). Several workers (Schaller, 1964; Sharma, 1972 and 1979; Trivedi and Johnsing, 1996; Yasmin,

Table 11.1: Some Reports on Peafowls Deaths Due to Various Reasons in Some Parts of India

Sl.No.	*Date*	*No. of Peafowls Died*	*Reason*	*Place*	*Reference*
1.	31–3–1986	17	Poaching	Kanarhiya Village, Indore, M.P.	Bird Watcher
2.	Jan. to Dec. 2000	21	High Temp.	Rajasthan	Prajavani
3.	16–2–2001	6	Food poisoning	Phokanpur, Delhi	Janavahini
4.	Jan. to April, 2001	129	High Temp. and poaching	Rajasthan	Prajavani
5.	30–4–2001	11	Food poisoning	Jaipur out skrit, Rajasthan	Prajavani
6.	3–05–2001	07	High Temp.	Jaipur dist. Rajasthan	–do–
7.	4–5–2001	19	High Temp.	Rajasthan	–do–
8.	May, 2001	5	Hunting	Belalagere, Karnataka	Unpublished data
9.	13–5–2001	11	Not known	Luhaari, Ujjaini Dist. M.P.	Prajavani
10.	June, 2001	4	Hunting	Near by Shanthi-sagar, Karnataka	Unpublished data

1997; Yasmin and Yahya, 1996; Parasharya and Mukherjee, 1999) have reported on various aspects of peafowl exclusively from central and northern part of India. There are no reports on the status of *P. cristatus* in its wild state in the maidan areas of central part of South India. *P. cristatus* although not threatened (Saharia, 1996) in its wild state, is fast declining in its natural-cum-agro-ecosystems. Further, their previous abundance as well as the preferred habitat at the agro-ecosystems has shrunk greatly (Basavarajappa, 2001 a and b).

Despite the strict protective enactment on killing of peafowls under the Indian Wildlife (Protection) Act (1972), the population in agro-ecosystems are disappearing due to habitat destruction (Saharia, 1996; Basavarajappa, 2001a). The information on the population dynamics of Indian crested peafowl and its decline in maidan areas of central part of South India is relatively meager. Therefore, the present study was conducted to estimate the variation in population size and sex-age composition of peafowl.

Study Area

Physiographically, the study area (Channagiri–a part of Davanagere district) occupies a central position in the state of Karnataka (Figure 11.1), and constitutes the southern part of the Deccan Peninsula. It lies in between 13°27′–14° 39′ N latitudes and 74°38′–76°40′ E longitudes with an altitude of 671 meters above msl (Anonymous 1998). The area receives 871.8 mm average annual rainfall. The climate is dry with 266.6 days sunshine. The mean temperature is 32.5°C. and the relative humidity is 70 per cent. The landscape consists of vast stretches of plains with scattered vegetation of small hillocks. The natural vegetation is confined to South West part of Channagiri taluka and it is diversified with dry deciduous tree species. The remaining area of the land surface is with scattered horticultural gardens, bushy vegetation particularly at the valley of small and uneven hillocks and food crops at the plain (Maidan) land. A brief description of habitat features of the study area is detailed in Table 11.2.

Methodology

The study was carried out from 1998 to 2000 in different habitat types of Channagiri taluka of Davanagere district. There were three different habitats: (I) Plain area of 966.81 hectares (at Belalagere), (II) Uneven land and small hills with sparse vegetation of 300 hectares

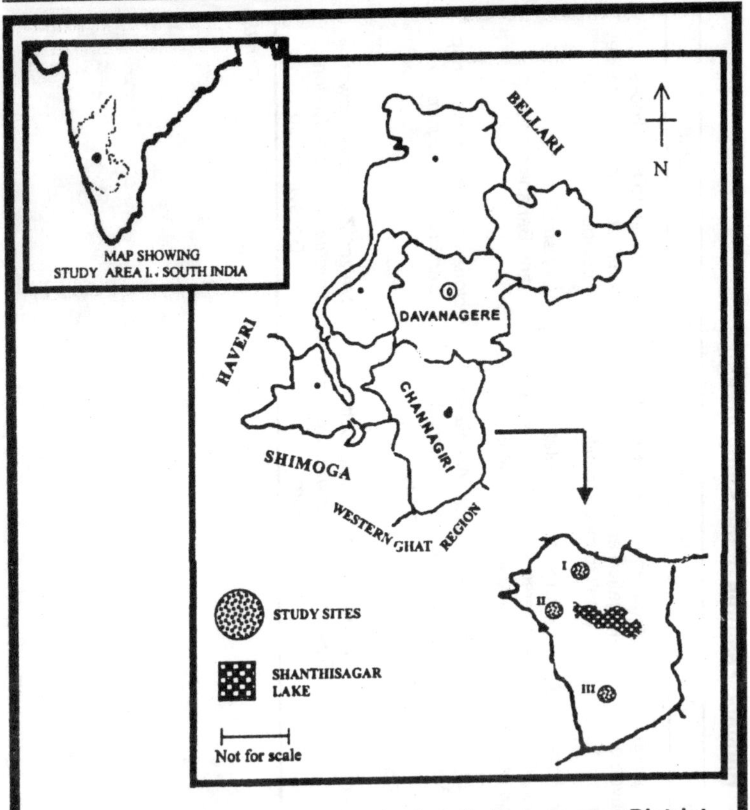

Figure 11.1: Map Showing Study Area in Davanagere District

(nearby Shanthisagar-second biggest lake in south India) and (III) Hillocks with dry deciduous vegetation of 600 hectares (at Joldhal) identified at random within a radius of 50 km to study the peafowl density, sex ratio and status of population or variation in size of population at these habitats (Figure 11.1). In each habitat, three permanent observation sites (2 km^2 each) were selected and which are 18 to 25 km apart form one another. In each site regular observations were made on group size of peafowl and sex-age composition of peafowl from early morning before sunrise (5.30 AM) to 10.00 AM and evening (5.00 PM) to late evening (7.00 PM) after sunset at an interval of fifteen days. Constant vigil was made by using 7 × 50 binoculars to observe from a distance to see their natural activities without disturbance. The peafowl was sampled through

Table 11.2: Habitatic Features of the Study Area

Sl.No.	*Characters*	*Habitat I*	*Habitat II*	*Habitat III*
1.	Topography	Plain land with irrigation, intensively cultivable low land	Uneven land surface with irrigation and rain fed land, small to medium hillocks	Medium to big hillocks, uncultivable land with dense deciduous vegetation
2.	Dominant cultivable vegetation	Scattered horticultural gardens *viz.*, Coconut, Banana, Areca and Mango	With sparse vegetation, scattered Areca, Banana and Mango plantation	Very small patch of Areca and Banana plantation
3.	Major tree species	*Cocus nucifera, Mangifera indica, Syzygium cumini, Tamarindus indica, Acacia* sp., *Pangamia pinnata, Ficus* sp.	*Mangifera indica, Acacia* sp., *Eucalyptus* sp., *Tamarindus indica*	*Tectona grandis, Terminalia* sp., *Bamboo* sp., *Sygium* sp.
4.	Major crops	Paddy-*Oryza sativa*	Paddy: *Oryza sativa,* Cotton: *Gossypium* sp., Jowar: *Sorghum vulgare*	Maize: *Zea mays,* Jowar: *Sorghum vulgare,* Cotton: *Gossypium* sp.

Contd...

Table 11.2–Contd...

Sl.No.	*Characters*	*Habitat I*	*Habitat II*	*Habitat III*
5.	Water source	All the year round, dominated by irrigated fields, canals, brooks and streams (with favourable wet land patches) near by roost sites	Sunflower: *Helianthus* sp., Maize; *Zea mays* all the year round, partially dominated with irrigated fields, canals and 2nd biggest lake Santhisagar is present nearby, very few patches of wet lands (¼ to ½ km away from roost sites)	Seasonal Brooks located 1 to 3 km away from roost sites
6.	Source of food for peafowls	Pest and non-pest insects, food grains	Arthropods, food grains	Arthropods, other invertebrates
7.	Predation by wild/ domesticated animals	Fox and stray dogs	Fox; wolf and panther	Panther, wolf and wild dogs
8.	Human interference (a) Human settlements (b) Cultivation (C) Cattles foraging (d) Encroachment of wet/waste land	 Nearby (100 meters to 3 km) intensive Moderate Maximum	 Far away (5 to 8 km) Moderate Normal Maximum	 Very far away (10 to 12 km) Seasonal Normal

transect surveys (Jaswinder *et al.*, 1993), using point census to determine the number of individuals which exist and, line transect was chosen in each habitat to census peafowl (Rathinasabapathy and Asokan, 1993). The probability of the observed sex-ratio was analysed as per Graeme (1978). The peafowls including adults (Cock and Peahen) and sub-adults (Juveniles) was made as per Sharma (1979) and Cuvier (1993). Recorded data was analyzed statistically by following ANOVA (Saha, 1992). The relative abundance of male/female, and sex-ratio was calculated (Wesley, 1993). Friedman's two way analysis of variance (X^2-test) was used to test the statistically significant differences in peafowl population between different habitats. The peafowl population was calculated as per Thirumurthi *et al.* (1993).

Results and Discussion

Habitatic Features

The details about the different habitatic conditions of the study area is predicted in Table 11.2.

Population Density

The population of *P. cristatus* under three different habitats of Channagiri is shown in Table 11.3. It is evident from the data that, the habitat III (near Joldhal) had highest (190 to 195) number of peafowls, followed by the habitat I (located at the vicinity of Belalagere) which recorded lowest number (65 to 96) of peafowls (Table 11.3). Thus, the population density is high (192.7/km^2) in habitat III compared to habitats II and I. The density of peafowl was highest in habitat III because it had dense vegetation cover and more number of tall tree species (Table 11.2) for resting and roosting of peafowls (Basavarajappa, 2001b).

In habitat II about 111.3 peafowls were recorded per km^2 area. The area is favoured with Santhisagar lake for continuous water source and located nearby rainfed croplands. Moreover, small to medium hillocks with scattered huge rocks here and there, and the few tall trees like Mango, *Acacia*, *Eucalyptus* and *Tamarindus* sp. (Table 11.2), might have provided resting and roosting sites to these birds to stay. However, during the crop season, there is a regular stream of workers into irrigated and rainfed crop fields. The increased human activity due to intensive cultivation could have disturbed the peafowl (Thirumurthi *et al.*, 1993). Thus, the density is less in habitat II compared to habitat III. However, the density of peafowls is further reduced to 77.3/km² in habitat I. Though there is good supply of food and water in habitat I (Table 11.2) the population density is less probably because of the intensive cultivation, cattle grazing, increased encroachment and high degree of human interference prevailing here throughout the year. Hence, the distribution of these birds depended on the occurrence of specific habitat with respect to the availability of the preferred vegetation in different habitats (Table 11.2).

Table 11.3: Population Density (No. per km²) of Peafowls in Channagiri, (N = 72)

Year	*Population of Peafowls in Different Habitats*			*Total*	*Mean*
	I	*II*	*III*		
1988	96	125	195	416	138.7
1999	71	112	193	376	125.3
2000	65	103	190	358	199.2
Mean±SD	77.3±16.44	111.3±11.04	192.7±2.54	383.3±29.69	127.8±9.93

Note: Each value is a mean of 24 observations.

Sex-Age Composition

Pavo cristatus is a polygamous bird, the cock live in the company of 4 or 5 peahens (Cuvier, 1993). Occurrence of male (Cock) and female (Peahen) fowl population varied much in different habitats (Table 11.4). The total cock density was in the range of 8 to 12 (habitat I), 12 to 16 (habitat II) and 19 to 20 (habitat III) during the years 1998 to 2000 respectively. However, the peahen density was 38 to 45; 59 to 68 and 88 to 89 in habitats I, II and III during the study period in respective habitats. Thus, the male ($F = 22.25$, $P > 0.05$) and female ($F = 137.69$, $P > 0.05$) population significantly differed between the habitats during the years 1998 to 2000. Further, the cock-peahen density was high (16–67/km²) in 1998 followed by 13–64/km² in the year 1999 and 2000. It could depict that the population has high reproductive potential in this area. However, it needs further studies to resolve probability of breeding potential. Such type of work is under progress and results of those observations will be published in near future.

Table 11.4: Occurrence of Male/Female Peafowls Density in Channagiri

Year	*Sex*	*Habitats*			*Total*	*Mean*
		I	*II*	*III*		
1998	Male	12	16	20	48	16.0
	Female	45	68	88	201	67.0
1999	Male	08	12	20	40	13.0
	Female	38	65	89	192	64.0
2000	Male	08	12	19	39	13.0
	Female	38	59	88	185	64.0

Note: Each value is a mean of 24 observations.

Sex-ratio

The relative abundance of the sexes in different habitats during 1998 to 2000 is shown in Figure 11.2. During majority of our observations single male was sighted with 4 females. Thus the sex ratio was in the range of 1 : 4.32 to 1 : 4.92 (Table 11.5).

Table 11.5: Sex-ratio (Male : Female) of Peafowls in Channagiri (Based on Table 11.3)

Year	*Habitat I*	*Habitat II*	*Habitat III*
1998	1 : 3.75	1 : 4.25	1 : 4.40
1999	1 : 4.75	1 : 5.41	1 : 4.45
2000	1 : 4.70	1 : 4.91	1 : 4.63
Mean	1 : 4.32	1 : 4.91	1 : 4.90

Juveniles Population

During the observations several number of immature peachicks of age less than 2 years were seen (Table 11.6). Until second year, the male and female have the same plumage (Cuvier, 1993) and, it is difficult to distinguish the immature peachicks as male or female (Sharma, 1979). Thus, all the immature peachicks were together called as juveniles (Thirumurthi *et al.*, 1993) during the present investigation. The juveniles population includes the sub-adult males, immature males, sub-adult females (Sharma, 1979) and, the total

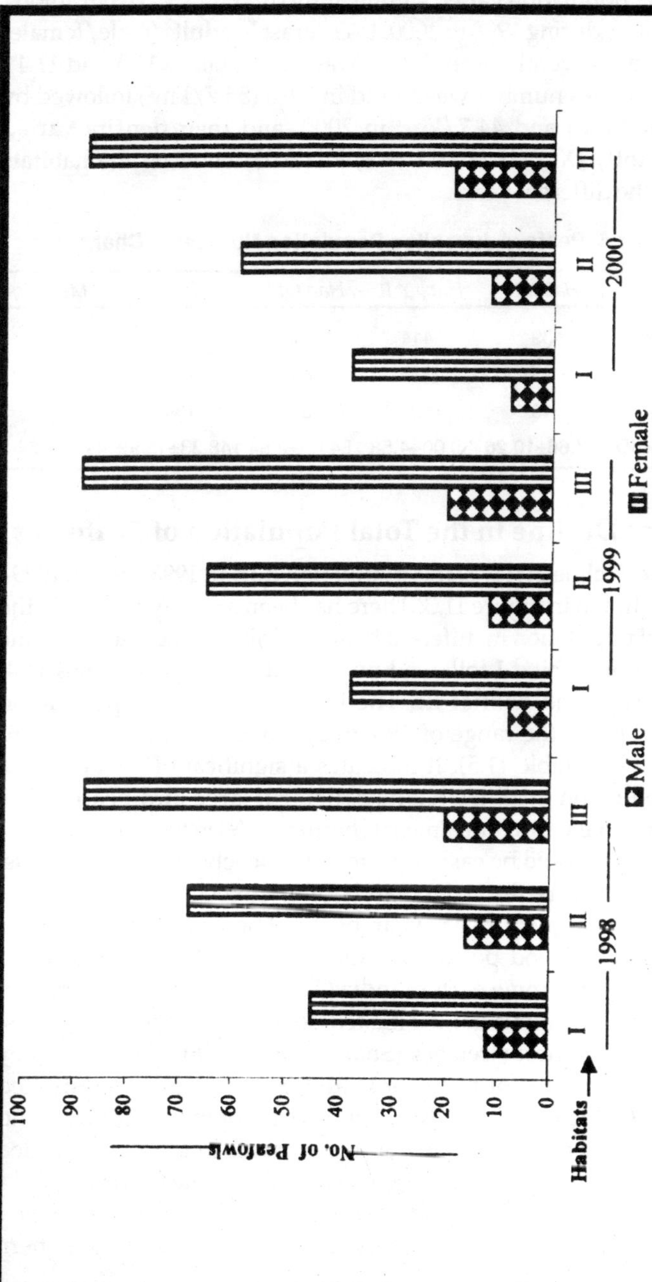

Figure 11.2: Relative Abundance of the Sexes in the Peafowl *Pavo cristatus*

juveniles population varied significantly (F = 65.27, P > 0.05) among the habitats during 1998 to 2000. In contrast to adults (male/female) the average juveniles population was 148.3 (Tables 11.3 and 11.4). High juveniles number was found in 1998 ($55.7/km^2$) followed by 1999 ($48/km^2$) and $44.7/km^2$ in 2000, and their density varied significantly (X^2 = 18.00, d.f. = 8, P > 0.05) between the habitats during the different years.

Table 11.6: Peafowl Juveniles Population Number in Channagiri

Year	*Habitat I*	*Habitat II*	*Habitat III*	*Total*	*Mean*
1998	39	41	87	167	55.7
1999	25	35	84	144	48.0
2000	19	32	83	134	44.7
Mean±SD	27.66±10.26	36.00±4.58	84.66±2.94	148.33±16.92	49.5±5.64

Percent Decline in the Total Population of Peafowl's

The decline in *P. cristatus* population during 1998–99 and 1999–2000 is shown in Figure 11.3. There has been a steady decline in the peafowl population in different habitats. Initially the rate of decline was high in habitat I followed by habitat II (Tables 11.3 and 11.4; Figure 11.3) and habitat III. The juveniles and cock population decline was in the range of 19.4 to 21.1 and 17.9 to 25.9 per cent respectively (Table 11.5). It indicates a significant (F = 14.04, P > 0.05) reduction in all the habitats. Being a large bird peacock has very little chance of concealment (Sharma, 1979) in the open cropping fields. These could be easily spotted and poached in their habitats. Perhaps, it may be the major reason for more cock decline in various parts of the country. Every year, peafowl become victim to man's activities (Ex. Food poisoning) and natural calamities (Ex. High temperature) in some parts of India (Table 11.1). Peachicks are very active by nature but get exhausted soon. They take enough rest frequently in short stretches (Sharma, 1972). During resting they have more chance of being predated by wild and domesticated animals (Table 11.2). Moreover, intensive cropping in the plains and increased encroachment of wet/waste lands might have hurdled the breeding activity of peafowl's. Thus, the prime factor that put a dent in the adults and sub-adults population is human interference in habitat I and II. Further, the extent of wet/waste land use pattern for agriculture has increased and worsened the natural habitat conditions of these birds. Thus, alteration and destruction of habitats by human could have drastic effects on these species. Due to all these reasons, it is suspected that in the near future, the crested peafowl *Pavo cristatus* may be pushed into the category of endangered birds at the maidan area of central part of southern India. Therefore, a deliberate attempt should be made to resolve the depletion of this species from poaching (Sharma, 1979), habitat destruction (Saharia, 1996) and human interference's (Basavarajappa, 2001). Such protective measures could help in conserving the viable population of peafowls in their natural abodes.

Acknowledgements

I thank Dr. B.B. Hosetti, the Chairman, Department of Applied Zoology, Kuvempu University, B.R. Project for his critical corrections, advise and necessary facilities extended during this work. I am thankful to Prof. Asad Rahmani, Director, BNHS, Mumbai. I am grateful to Prof. G. Sashidharaprasad, Vice-Chancellor, University of Mysore and N.D. Tiwari, Registrar, University of Mysore, Mysore

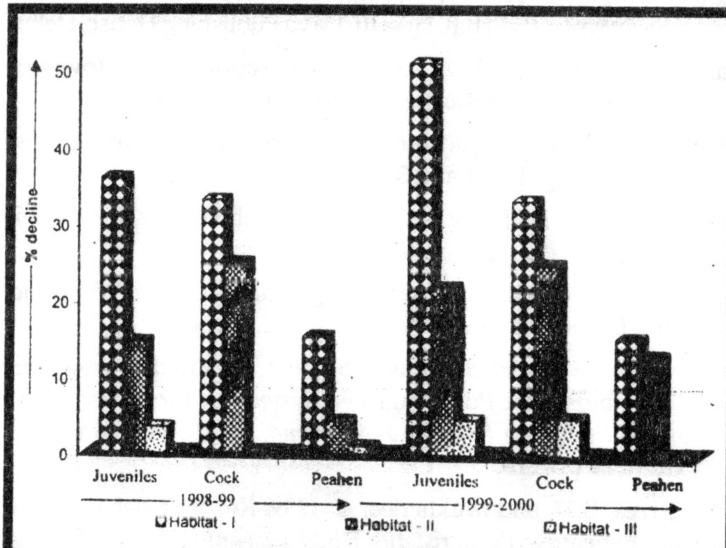

Figure 11.3: Status of Peafowl Population in Different Habitats During 1998–2000

for their encouragement. I am thankful to the Chairman, DOS in Zoology, University of Mysore, Manasagangotri, Mysore for the constant support and help extended during the preparation of this manuscript. I am also thankful to Prof. V. Vasudev, Department of Applied Zoology, Kuvempu University, B.R. Project for encouragement and Mangesh Ghodke, Assistant Librarian, for the timely supply of reference material during the preparation of this manuscript. Thanks are also due to late. Hanumanthappa and Sri Doddappa, Belalagere for their help during this investigation.

References

Anonymous. 1998. *Davanagere District Gazetteer*. Government Press, Bangalore, p. 1–20.

Ali, S. and Ripley, S.D. 1983. *Handbook of the Birds of India and Pakistan*. Compact Edn. Oxford University Press, New Delhi, p. 1–70.

Basavarajappa, S. 2001a. Studies on habitatic features of Indian peafowl *Pavo cristatus* under diversified conditions of Karbataka, India. In: *Proc. 88th ISC.*, New Delhi, pp. 29.

Basavarajappa, S. 2001b. Habitat degradation of peafowls. In: *Glimpse of Biodiversity*, (Ed.) B.B. Hosetti. Daya Publishing House, Delhi.

Basavarajappa, S. 2001. A preliminary Report on Peafowls in Channagiri area of Karnataka, India. *Birdwatchers*.

Cuvier, B. 1993. *Encyclopaedia of Life Sciences*. Anmol Publications, New Delhi, p. 12–14 and 35–152.

Ensminger. 1992. *Poultry Science*. Interstate Publications, Inc., Delhi, p. 381–382.

Graeme, C. 1978. *Analysis of Vertebrate Populations*. John Wiley and Sons, Chinchestor, New York, Brisbane, Toronto.

Jasaswinder, S.S. and Santhus, P.S. 1993. Population trends in Columbids in the Punjab agro-ecosystem. In: *Bird Conservation Strategies for the Nineties and Beyond*, (Ed.) Abraham *et al.*, Mumbai, pp. 216.

Parasharya, B.M. and Mukherjee, A. 1999. Roosting behaviour of Indian peafowl *Pavo cristatus*. *J. Bombay Natural History Society*, Mumbai, 96(3). 471–472.

Rathinasabapathy, B. and Asokan, S. 1993. Avifaunal diversity in different vegetation types of Eastern Ghat of Andhra Pradesh.

In: *Bird Conservation Strategies for the Nineties and Beyond*, (Ed.) Abraham *et al.*, Mumbai, pp. 65.

Saharia, V.B. 1996. *Wildlife in India*. Nataraj Publishers, Dehra Dun, p. 11–19.

Saha, T.K. 1992. *Biostatastics in Theory and Practice*. Emkay Publications, Delhi, p. 55–67.

Schaller, G.B. 1964. Notes on the behavior of the peafowl. *Birdwatchers*, 4(9): 1–2.

Sharma, I.K. 1972. Ecological study of rest and roost behaviour of the peafowl *Pavo cristatus*. *Birdwatchers*, 12(6): 5–8.

Sharma, I.K. 1979. Ecological aspects of population trends of the peafowl *Pavo cristatus* at Jodhpur. *Pavo*, 17(1 and 2): 50–53.

Thirumurthi, S., Annamalali, R. and Gunasekharan, V. 1993. Effects of Industrialisation on the Populations of Peafowl, *Pavo cristatus* Linn. In: *Bird Conservation Strategies for the Nineties and Beyond*, (Ed.) Abraham *et al.*, Mumbai, pp. 77.

Trivedi, P. and Johsingh, A.J.T. 1996. Roost selection by Indian Peafowl (*Pavo cristatus*) in Gir forest, India. *J. Bombay Natural History Society*, Mumbai, 93: 25–29.

Yasmin, S. 1977. Group size and composition of Indian Peafowl (*Pavo cristatus*) in an agro-ecosystem at Aligarh, Uttar Pradesh. *J. Bombay Natural History Society*, Mumbai, 94: 478–482.

Yasmin, S. and Yahya, H.S.A. 1996. Correlates of mating success in Indian Peafowl. *The Auk*, 1132: 490–492.

Wesley, D.H. 1993. Breeding behaviour sequential Polyandry and population decline in *Rostratula benghalensis*. In: *Bird Conservation Strategies for the Nineties and Beyond*, (Ed.) Abraham *et al.* Mumbai, pp. 166.

Chapter 12

Diversity, Status and Feeding Ecology of Avifauna of Ramnagar Wildlife Sanctuary, Jammu

Deep N. Sahi and Bharti Sharma

Department of Zoology, University of Jammu, Jammu – 180 006

ABSTRACT

An avifaunal survey was carried out in Ramnagar Wildlife Sanctuary, Jammu from January 2002 to May 2003. A total of 69 species belonging to 13 orders and 29 families were recorded. The present communication deals with their status vis-à-vis resident/migratory, feeding guilds, habitat and abundance. The study area was divided into three habitats *i.e.*, aquatic habitat (AqH), scrubby habitat (SH) and mixed semi-evergreen habitat (MSeF). Results indicate that the avifauna at MSeF is more diverse than in the other two habitats. Five major food categories were considered, namely insectivorous, granivorous, carnivorous, frugivorous and omnivorous. The overall highest proportion observed is of insectivorous birds followed by carnivorous. Both of these food categories were further subdivided into feeding guilds for more specific approach towards the feeding behaviour of species.

Keywords: *Diversity, Status, Feeding ecology, Avifauna.*

Introduction

India is one of the twelve megadiversity countries in the world. This classification is based on terrestrial species especially mammals, birds and butterflies. Thus, avian diversity is one of the important components of faunal diversity. Barring a few, most of the birds are useful for mankind. Birds play a useful role in the control of insect pests of agricultural crops; as predators of rodents; as scavengers; as seed dispensers and as flower pollinating agents, thus forming an important component in natural ecosystem.

Ali (1941) estimated 8600 species, the world over. Woodcock (1980) has estimated nearly 1250 species of birds from the Indian subcontinent. Grimmet *et al.* (1998) documented 1300 species of birds in Indian subcontinent.

While many common species are spread over large areas of the Indian subcontinent, others may be limited to a region or habitat. Since the subcontinent has a dense human population, the birds which adapt themselves to human habitat flourish well.

Ornithological studies in India have been documented from Jerdon's (1862–64) pioneering investigations to Ali and Ripley (1968–1974) authoritative handbook. However, it was not possible for them to record precise distribution pattern of the birds of a particular locality and the local adaptations of the birds as a result of interaction with the prevailing environment of a place.

It appears from previous records that our knowledge about birds is still patchy with regard to their ecological behaviour and natural history. In Jammu particularly, no such studies have been carried out barring a few stray records here and there in the works of Whistler (1928); Ali and Ripley (1968–1974; 1983); Grimmet *et al.* (1998) and Alfred *et al.* (2001).

Though the Indian subcontinent is gifted with a rich diversity of bird life and an equally diverse set of habitat types, our knowledge on the subject is still elementary (Ali and Ripley, 1983). Moreover, species richness and community structure of birds vary from region to region (Recher, 1969; Pearson, 1975 and Karr, 1976) as well as within a region, as abiotic and biotic factors vary from habitat to habitat. Several studies have identified the factors responsible for variations in avifauna from habitat to habitat outside India (Anderson, 1970 and Beedy, 1981) and within India (Johnsing *et al.*, 1987; Katti, 1989 and Rai, 1991). These studies also emphasize the value of avifaunal studies in quantifying and monitoring forest degradation.

Needless to mention that for several reasons detailed documentation of biodiversity of this vulnerable and important state of J&K is lacking as compared to other Indian states. Therefore, there is an imperative need for uncovering the avian diversity. To make a beginning in this direction, Ramnagar Wildlife Sanctuary in district Jammu has been chosen to carry out status survey.

Location of Study Area

The study site, Ramnagar Wildlife Sanctuary is located in district Jammu, J&K state, between 32°44′.37″ to 32½47′.32″ north latitude and 74°50′.45″ to 74°53′.37″ east longitude at an elevation of 340 mts. The climatic conditions in the study area are dry sub-humid having four distinct seasons *i.e.*, winter (mid-November to mid-March), summer (mid-March to end of June), monsoon (July to September), autumn (October to mid-November). The average rainfall is 115 cm of which 80 per cent occurs during monsoon.

Flora

Predominant plant species in the study area are *Acacia arabica, A. catechu, Zizyphus marlitiama, Morus nigra, Dalbergia sisso, Cassia fistula, Ficus bengalensis, Ageratum conizoides, Adhathoda vesica, Calotrophis procera, Lantana camera, Cuscuta reflexa, Abrus precatorius, Dudonia viscosa* etc.

The different types of habitat surveyed are:

1. Aquatic Habitat (AqH) in the form of natural ponds, artificial waterhole and run-off water.
2. Scrubby Habitat (SH).
3. Mixed Semi-evergreen Habitat (MSeF).

Methodology

The study area was surveyed for recording avifaunal diversity by applying line-transect method (Sale and Berkmuller, 1988). Census was carried out once in a week starting from January 2002 to May 2003. The following assumptions were made while doing the census. Observer always covered the distance of 4 km within a fixed duration of 120 minutes, thus covering 2 km/hr and this speed was maintained throughout the census. In order to maintain uniformity all surveys were conducted from 6.00 a.m. to 11.00 a.m. in the morning and 4.30 p.m. to 6.30 p.m. in the evening during summers and 7.30 a.m. to 11.30 a.m. and 3.30 p.m. to 5.30 p.m. during winters. All the birds are detected and correctly identified.

Binoculars (Bushnell 7 × 50 U.S.A. made) were used to record the observations from the distance to avoid any disturbance to the birds. Care was taken that the sun was always behind the observer so that the plumage patterns of the birds could be distinguished.

The identification of birds in the field is based on Ali (1941) and Ali and Ripley (1983). Confirmation of the species was done with the help of Grimmet *et al.* (1998).

Observations and Discussion

A systematic list of 69 bird species belonging to 13 orders and 29 families along with their status vis-à-vis resident/migratory, feeding guild, habitat and abundance has been presented (Table 12.1). The list shows that out of the total bird species recorded, 53 species are resident to the study area, 15 migratory and status of one species is unknown. Migratory birds were further categorized as passage migrants, winter migrants and summer migrants. Of the 15 species of migrants, 2 species are passage migrants, 9 winter migrants and 4 summer migrants.

The study area, despite its small size, appears to support an extremely rich and diverse bird community. Of the total birds (*i.e.*, 1300 spp.) recorded from Indian subcontinent by Grimmet *et al.* (1998), avifauna of Ramnagar Wildlife Sanctuary, Jammu represents 5.3 per cent. The observed bird diversity in the relatively small study area (31.50 km^2) underlies the importance of this area for biodiversity conservation.

In the present communication, the birds recorded from Ramnagar Wildlife Sanctuary are classified in five major food categories *viz.*, insectivorous (22 species), granivorous (7 species), carnivorous (10 species), frugivorous (8 species) and omnivorous (3 species) (Figure 12.1).

Besides this, remaining 19 species are recorded to utilise more than one feeding categories/guilds (Figure 12.2).

A view of above data depicts that the overall highest proportion is of insectivorous birds followed by carnivorous. Both of these food

Table 12.1: Avifaunal Inventory of Ramnagar Wildlife Sanctuary with Habitat and Feeding Ecology

Sl.No.	*Common Name*	*Scientific Name*	*Status*	*Feeding Guild*	*Habitat*	*Abundance*
	Class: Aves **Order**: Podicipediformes **Family**: Podicipitidae					
1.	Little Grebe	*Podiceps* (=*Tachybaptus*) *ruficollis capensis* Salvadori	Rst	DC	AqH	C
	Order: Ciconiiformes **Family**: Ardeidae					
2.	Indian Pond Heron	*Ardeola grayii grayii* Skyes	Rst	WC	AqH	C
3.	Cattle Egret	*Bubulcus ibis coromandus* Boddaert	Rst	WC/UI	AqH/MSeF	C
4.	Little Egret	*Egretta garzetta garzetta* Linnaeus	Rst	WC	AqH	F
	Order: Falconiformes **Family**: Accipitridae					
5.	Pariah Kite	*Milvus migrans govinda* Skyes	Rst	ATC	MSeF	C
6.	Indian Shikra	*Accipiter badius dussumieri* Temminck	Rst	ATC	MSeF	F
	Order: Galliformes **Family**: Phasianidae					
7.	Indian Red Jungle Fowl	*Gallus gallus murghi* Robinson and Kloss	Rst	GI	SH	C

Contd...

Table 12.1–Contd...

Sl.No.	Common Name	Scientific Name	Status	Feeding Guild	Habitat	Abundance
	Order: Gruiformes **Family**: Rallidae					
8.	Indian White Breasted Waterhen	*Amaurornis phoenicurus phoenicurus* (Pennant)	Rst	WC/SIP	AqH/SH/MSeF	C
9.	Indian Moorhen	*Gallinula chloropus indica* Blyth	Rst	WC/SIP	AqH/SH/MSeF	F
	Order: Charadriiformes					
	Family: Charadriidae					
10.	Red Wattled Lapwing	*Vanellus indicus indicus* (Boddaert)	Rst	TI/SIP	AqH/MSeF	C
11.	Indian Little Ringed Plover	*Charadrius dubius jerdoni* (Legge)	Rst	SIP	AqH	O
	Sub-family: Scolopacinae					
12.	Common Sandpiper	*Tringa* (=*Actitis*) *hypoleucos* Linnaeus	WM	SIP	AqH	F
13.	Green Shank	*T. nebularia* (Gunnerus)	WM	SIP	AqH	O
	Order: Columbiformes **Family**: Columbidae					
14.	Indian Blue Rock Pigeon	*Columba livia intermedia* Strickland	Rst	G	MSeF	C
15.	Indian Ring Dove	*Streptopelia decaocto decaocto* (Frivaldszky)	Rst	G	SH/MSeF	C
16.	Indian Spotted Dove	*S. chinensis suratensis* Gmelin	Rst	G	SH/MSeF	C
17.	Little Brown Dove	*S. senegalensis cambayensis* Gmelin	Rst	G	SH/MSeF	C
18.	Indian Red Turtle Dove	*S. tranquebarica tranquebarica* (Hermann)	Rst	G	SH/MSeF	R

Contd...

Table 12.1–Contd...

Sl.No.	*Common Name*	*Scientific Name*	*Status*	*Feeding Guild*	*Habitat*	*Abundance*
	Order: Psittaciformes **Family**: Psittacidae					
19.	Large Indian Parakeet	*Psittacula eupatria nipalensis* (*Hodgson*)	Rst	F	MSeF	F
20.	Rose Ringed Parakeet	*P. krameri manillensis* Bechstein	Rst	F	MSeF	C
21.	Blossom Headed Parakeet	*P. cyanocephala* (Forester)	Rst	F	MSeF	O
	Order: Cuculiformes **Family**: Cuculidae					
22.	Indian Koel	*Eudynamys scolopacea scolopacea* (Linnaeus)	Rst	F	MSeF	C
	Order: Strigiformes **Family**: Strigidae **Sub-family**: Striginae					
23.	Northern Spotted Owlet	*Athene brama indica* Franklin	Rst	ATC	MSeF	O
24.	Punjab Collared Scops Owl	*Otus bakkamoena plumipes* (Hume)	Rst	ATC	MSeF	R
	Order: Coraciiformes **Family**: Alcedinidae					
25.	Indian Pied Kingfisher	*Ceryle rudius leucomelanura* Reichenbach	Rst	AAqC	AqH	F
26.	White Breasted Kingfisher	*Halcyon smyrnensis smyrnensis* (Linnaeus)	Rst	ATC/AAqC	AqH/MSeF	C

Contd...

Table 12.1–Contd...

Sl.No.	Common Name	Scientific Name	Status	Feeding Guild	Habitat	Abundance
	Family: Meropidae					
27.	Indian Small Green Bee-eater	*Merops orientalis orientalis* Latham	Rst	AI	MSeF	C
28.	Blue Cheeked Bee-eater	*M. superciliosus persicus* Pallas	PM	AI	MSeF	R
29.	Kashmir Roller	*Coracias garrulus semenowi* (Loudon & Ischudi)	SM	ATC	MSeF	R
	Family: Upupidae					
30.	European Hoopoe	*Upupa epops epops* Linnaeus	Rst	G/UI	MSeF	C
	Family: Bucerotidae					
31.	Common Grey Hornbill	*Tockus* (=*Ocyceros*) *birostris* (Scopoli)	Rst	F	MSeF	F
	Order: Piciformes **Family**: Capitonidae					
32.	Large Green Barbet	*Megalaima zeylanica zeylanica* (Gmelin)	Rst	F	MSeF	C
33.	Copper Smith	*M. haemacephala indica* (Latham)	Rst	F	MSeF	C
	Family: Picidae **Sub-family**: Picinae					
34.	Lesser Golden-backed Woodpecker	*Dinopium benghalense benghalense* (Linnaeus)	Rst	T/BF	MSeF	O
35.	Maharatta Woodpecker	*Picoides maharattensis maharattensis* (Latham)	SM	T/BF	MSeF	R

Contd...

Table 12.1–Contd...

Sl.No.	*Common Name*	*Scientific Name*	*Status*	*Feeding Guild*	*Habitat*	*Abundance*
	Order: Passeriformes					
	Family: Hirundinidae					
36.	Kashmir House Martin	*Delichon urbica cashmeriensis* (Gould)	WM	AI	AqH	F
	Family: Lanidae					
37.	Rufous-backed Shrike	*Lanius scahach erythronotus* (Vigors)	Rst	ATC	MSeF	F
	Family: Oriolidae					
38.	Indian Golden Oriole	*Oriolus oriolus kundoo* (Skyes)	SM	F	MSeF	O
	Family: Dicruridae					
39.	Black Drongo	*Dicrurus adsimilis* (=*macrocercus*) *albirictus* (Hodgson)	Rst	AI	MSeF	C
	Family: Sturnidae					
40.	Indian Myna	*Acridotheres tristis tristis* (Linnaeus)	Rst	G,F,I	SH/MSeF	C
41.	Bank Myna	*A. ginginianus* (Latham)	Rst	G,F,I	SH/MSeF	C
42.	Northern Jungle Myna	*A. fuscus fuscus* (Wagler)	Rst	G,F,I	SH/MSeF	R
43.	Brahminy Myna	*Sturnus pagodarum* (Gmelin)	Rst	G,F,I	SH/MSeF	C
44.	Indian Pied Myna	*Sturnus contra contra* Linnaeus	Rst	G,F,I	SH/MSeF	F
45.	Starling	*S. vulgaris indicus* Blyth	WM	G,F,I	SH/MSeF	O

Contd...

Table 12.1–Contd...

Sl.No.	*Common Name*	*Scientific Name*	*Status*	*Feeding Guild*	*Habitat*	*Abundance*
	Family: Corvidae					
46.	House Crow	*Corvus splendens splendens* Vieillot	Rst	O	MSeF	C
47.	Jungle Crow	*C. macrorhynchos culminatus* Skyes	WM	O	MSeF	R
48.	Northerneastern Tree Pie	*Dendrocitta vagabunda* (Blyth)	Rst	O	MSeF	O
	Family: Campehagidae					
49.	Small Minivet	*Pericrocotus cinnamomeus* Baker	WM	CI	MSeF	F
	Family: Pycnonotidae					
50.	Red Vented Bulbul	*Pycnonotus cafer cafer* (Linnaeus)	Rst	F,I	SH/MSeF	C
51.	White-cheeked Bulbul	*P. leucogenys leucogenys* (Gray)	Rst	F,I	SH/MSeF	C
	Family: Muscicapidae **Sub-family**: Timaliinae					
52.	Jungle Babbler	*Turdoides striatus somervillei* (Skyes)	Rst	UI	SH/MSeF	C
53.	Western-spotted Babbler	*Pellorneum ruficeps punctatum* (Gould)	?	UI	SH	R
	Sub-family: Monarchinae					
54.	Paradise Flycatcher	*Terpsiphone paradisi paradisi* (Linnaeus)	SM	AI	MSeF	O
55.	Verditer Flycatcher	*Muscicapa (=Eumyias) thalassina thalassina* Swainson	PM	AI	MSeF	R
56.	Indian Tailor Bird	*Orthotomus sutorius guzuratus* (Latham)	Rst	CI/TI	SH/MSeF	F

Contd...

Table 12.1–Contd...

Sl.No.	Common Name	Scientific Name	Status	Feeding Guild	Habitat	Abundance
	Sub-family: Turdinae					
57.	Indian Magpie Robin	*Copsychus saularis saularis* (Linnaeus)	Rst	UI	SH	C
58.	Kashmir Black Redstart	*Phoenicurus ochruros pheonicuroides* (Moore)	WM	TI	MSeF	O
59.	Brown Rock Chat	*Cercomela fusca* (Blyth)	Rst	TI	MSeF	F
60.	Pied Bush Chat	*Saxicola caprata bicolor* (Skyes)	Rst	TI	SH	O
61.	Indian Robin	*Saxicoloides fulicata cambaiensis* (Latham)	Rst	TI	SH	F
	Family: Paridae **Sub-family**: Parinae					
62.	Indian White Wagtail	*Motacilla alba dukhunensis* Skyes	WM	SIP	AqH	F
63.	Large Pied Wagtail	*M. maderaspatensis* Gmelin	Rst	SIP	AqH	F
	Family: Nectariniidae					
64.	Purple Sunbird	*Nectarinia asiatica asiatica* (Latham)	Rst	N,I	MSeF	C
65.	Yellow-backed Sunbird	*Aethopyga siparaja seheriae* (Tickell)	WM	N,I	MSeF	C
	Family: Zosteropidae					
66.	Indian White Eye	*Zosterops palpebrosa palpebrosa* (Temminck)	Rst	CI	MSeF	F

Contd...

Table 12.1–Contd...

Sl.No.	Common Name	Scientific Name	Status	Feeding Guild	Habitat	Abundance
	Family: Ploceidae **Sub-family**: Passerinae					
67.	Indian House Sparrow	*Passer domesticus indicus* (Jardine and Selby)	Rst	G,I	SH/MSeF	C
68.	Spotted Munia	*Lonchura punctulata punctulata* (Linnaeus)	Rst	G	SH	R
69.	Baya Weaver Bird	*Ploceus philippinus philippinus* (Linnaeus)	Rst	G	SH	O

Feeding Category/Guilds

C: Carnivore; AAqC: Arboreal aquatic carnivore; ATC: Arboreal terrestrial carnivore; DC: Diving carnivore; WC: Wading carnivore; F: Frugivore; FI: Frugivore insectivore; G: Granivore; GFI: Granivore frugivore insectivore; GI: Granivore insectivore; I: Insectivore; AI: Aerial insectivore; CI: Canopy insectivore; SIP: Shore insect prober; T/BF: Trunk or bark feeder; TI: Terrestrial insectivore; UI: Understorey insectivore. NI: Nectarivore insectivore. O: Omnivore.

Habitat

AqH: Aquatic habitat; MSeF: Mixed semi-evergreen forest; SH: Scrubby habitat.

For recording the abundance of avifauna during the survey of study area based on sight records, the terminology of Khan (2002) was followed:

C–Common: means it can invariably be seen in that habitat where it occurs with the proviso of course that the season is also appropriate.

F–Frequent: means that even visiting appropriate habitat, it will not be seen or heard invariably, perhaps only in one visit out of three.

O–Occasional: means seen or heard only in one visit out of six.

R–Rare: means even less likelihood of occurrence.

Besides this, for assigning status, depending upon whether the species of bird is sighted in all the months/seasons of year or only in particular season/some months of the year and absent in others from the study area, it was referred as Resident (Rst) or Migrant (M), respectively. Migrant category is further differentiated into:

SM–Summer Migrants: Those which visit the study area in summers.

WM–Winter Migrants: Those which visit the study area in winters.

PM–Passage Migrants: Those which visit the study area just for few hours or few days.

? –Status undefined.

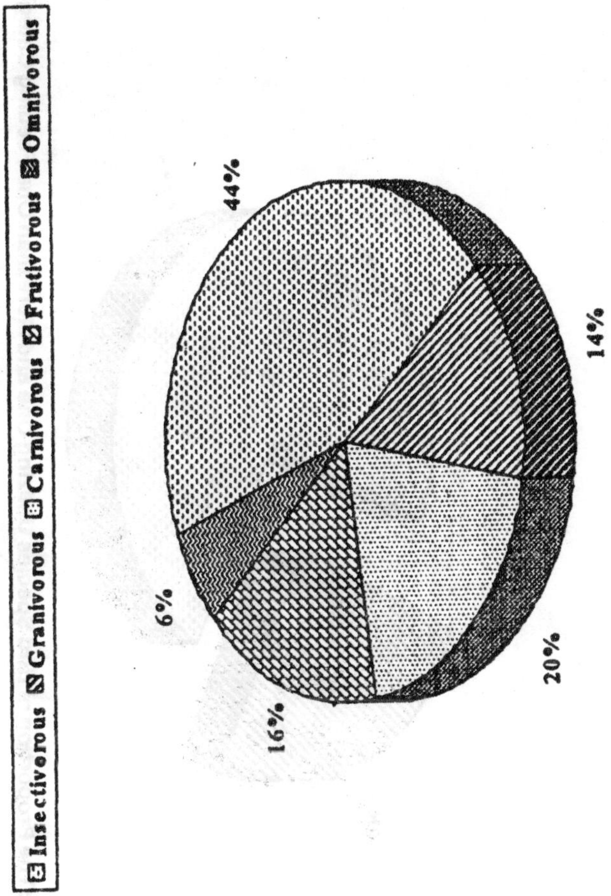

Figure 12.1: The Proportion of Bird Species in Various Feeding Categories

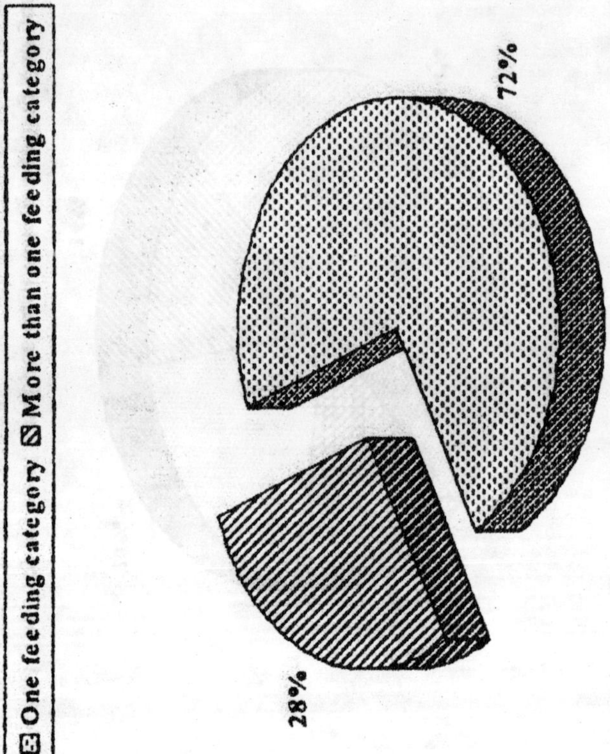

Figure 12.2: The Proportion of Bird Species Utilizing One and More than One Feeding Category

categories was further subdivided into feeding guilds. Among insectivores, 6 species are of aerial insectivore, 2 species of canopy insectivore, 5 of shore insect prober, 2 of trunk/bark feeder, 4 of terrestrial insectivore and 3 of understorey insectivore (Table 12.2 and Figure 12.3).

Table 12.2: The Proportion of Bird Species in Various Insectivorous Guilds

Guilds	*Number of Birds*	*Habitat Utilized*			
		AqH	*SH*	*MseF*	*SH/MseF*
Aerial insectivore	6	1	–	5	–
Canopy insectivore	2	–	–	2	–
Shore insect prober	5	5	–	–	–
Trunk/bark feeder	2	–	–	2	–
Terrestrial insectivore	4	–	2	2	–
Understorey insectivore	3	–	2	–	1
Total	22				

Carnivore category comprises 1 species of arboreal aquatic carnivore, 6 species of arboreal terrestrial carnivore, 1 species of diving carnivore and 2 species of wading carnivore (Table 12.3 and Figure 12.4).

Table 12.3: The Proportion of Bird Species in Various Carnivorous Guilds

Guilds	*Number of Birds*	*Habitat Utilized*		
		AqH	*SH*	*MSeF*
Arboreal aquatic carnivore	1	1	–	–
Arboreal terrestrial carnivore	6	–	–	6
Diving carnivore	1	1	–	–
Wading carnivore	2	2	–	–
Total	**10**	**4**		**6**

Food defined guild structure of the birds observed (Table 12.1), highlights the efficient distribution of food resources among different species in the present area. The feeding guilds of bird categories are directly related to the structural adaptations of each species (*i.e.,*

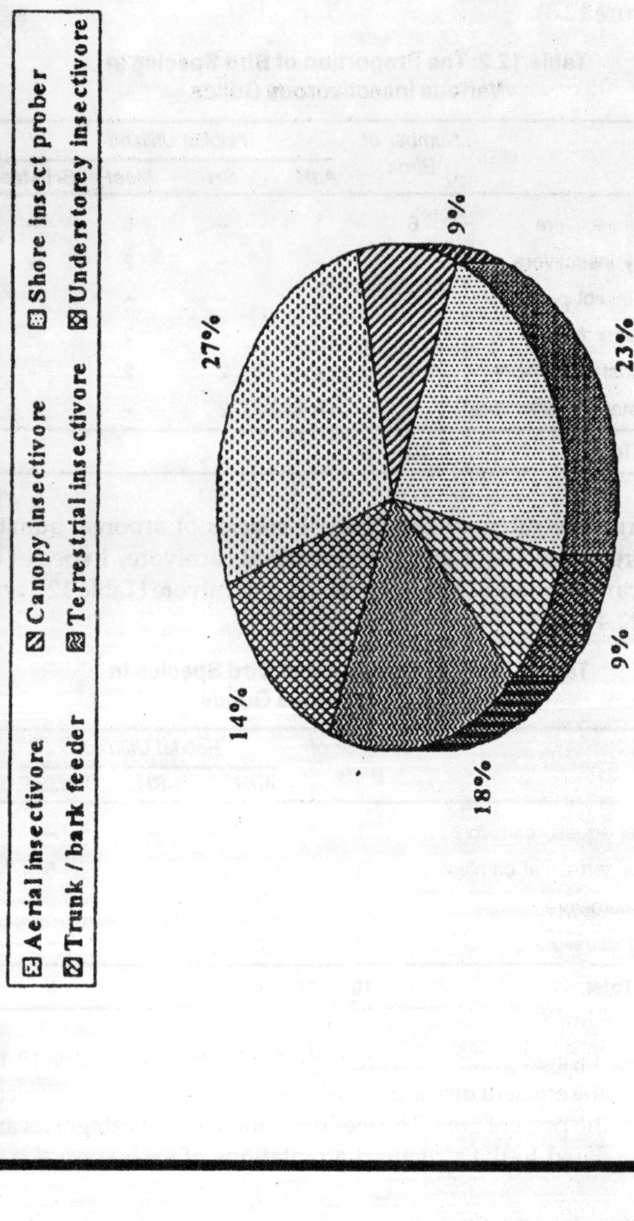

Figure 12.3: The Proportion of Bird Species in Various Insectivorous Guilds

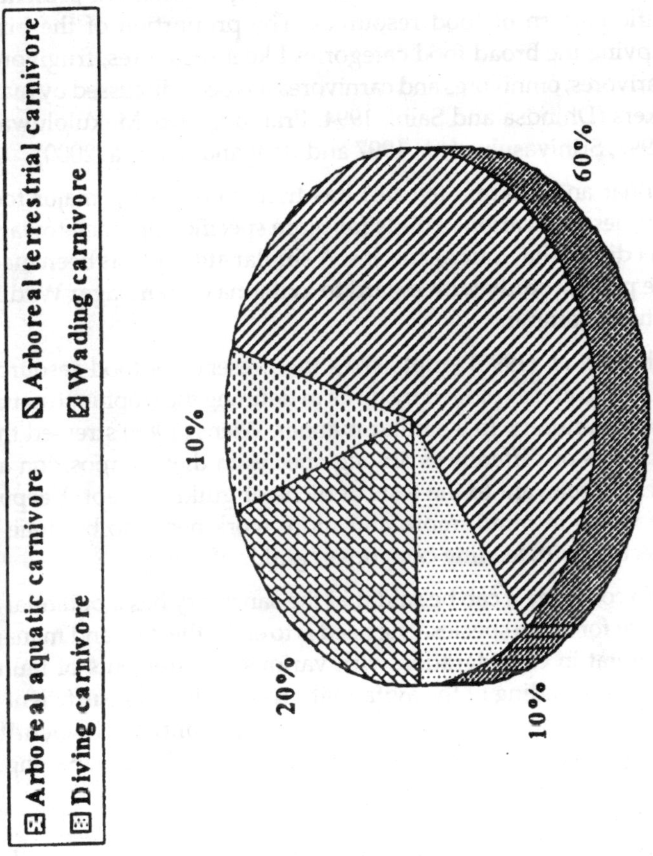

Figure 12.4: The Proportion of Bird Species in Various Carnivorous Guilds

structure of bill, legs and feet etc.). The types of bird present is a reflection of the food sources available.

The number of birds occupying the various habitats is highest at MSeF indicating that the mixed vegetation is good and suitable habitat for birds feeding as well as nesting.

The bird community in the light of trophic relations presents a specific pattern of food resources. The proportion of the birds occupying the broad food categories like insectivores, frugivores, nectarivores, omnivores and carnivores has been discussed by many workers (Dhindsa and Saini, 1994; Pramod, 1995; Makuloluwa *et al.*, 1997; Srinivasulu *et al.*, 1997 and Bhat and Sharma, 2000).

Bhat and Sharma (*op cit.*) has tried to break-up major food categories into various guilds for more specific approach towards the feeding behaviour of species. A similar attempt has been made in the present study with regard to avifauna of Ramnagar Wildlife Sanctuary, Jammu.

Karr *et al.* (1990) emphasizes the presence of food resources available to and exploited by birds in defining the trophic structure of the community. In a similar manner, Wiens (1989) stressed that similarities or differences among species in diet composition are especially relevant to the tests of niche or guild concept. Keeping these views in mind, more and more work needs to be done in connection with feeding categories/guilds of aves.

To conclude it can be said that the sanctuary has a potential as a habitat for avian species. The need is to enlist the data and manage the habitat in consideration with various requirements of fauna. Our understanding of the avifaunal diversity is still insufficient to guarantee proper conservation and only continued scientific research can throw light on the improved methods for managing it.

References

Alfred, J.R.B., Kumar, A., Tak, P. C. and Sati, J.P. 2001. *Waterbirds of Northern India*. Zoological Survey of India.

Ali, S. 1941. *Book of Indian Birds*. Bombay Natural History Society.

Ali, S. and Ripley, S.D. 1968–74. *The Handbook of the Birds of India and Pakistan*. Ten Volumes. Oxford University Press, Bombay.

Ali, S. and Ripley, S.D. 1983. *Compact Handbook of Birds of India and Pakistan*. Oxford University Press, Bombay.

Anderson, S.H. 1970. The avifaunal composition of Oregon White Oak Stands. *The Condor*, 72: 417–423.

Beedy, E.C. 1981. Bird communities and forest structure in the Sierra Nevada of California. *The Condor*, 83: 97–105.

Bhatt, D. and Sharma, R. 2000. Diversity, status and feeding ecology of avifauna in Motichur area of Rajaji National Park, India. *Ann. For.*, 8(2): 171–191.

Dhindsa, M.S. and Saini, H.K. 1994. Agriculture ornithology: An Indian perspective. *J. Biosci.*, 19(4): 391–402.

Grimmet, R., Inskipp, C. and Inskipp, T. 1998. *Birds of Indian Subcontinent*. Oxford University Press, Delhi.

Jerdon, T. 1862–64. *The Birds of India*, 2 Vols in 3 Parts, published by the author, Kolkata.

Johnsing, A.J.T., Martin, N.H., Balasingh, J. and Chelladurai, V. 1987. Vegetation and avifauna in a thorn scrub habitat in south India. *Tropical Ecol.*, 28: 22–34.

Karr, J.R. 1976. Within and between habitat avian diversity in African and neotropical lowland habitats. *Ecological Monographs*, 46: 457–481.

Karr, J.R., Robinson, S.K., Blake, J.G. and Berregard, R.O. 1990. The bird communities of four neotopical forest. In: *Four Neotropical Forest*, (Ed.) A. Gentry. Yale University Press, New Haven, CT, p. 237–269.

Katti, M.V. 1989. Bird communities of Lower Dachigam Valley, Kashmir. *M.Sc.* (Wildlife Science) Dissertation. Saurashtra University, Rajkot.

Majumdar, N. and Brahmachari, G.K. 1987. Prime avian predators controlling insect and rodent pests of paddy in India; Management of their econiches: Its feasibility and some suggestions. *Tiger Paper*, 14(2): 16–22.

Makuloluwa, S., Alagoda, T. and Santiapillai, C. 1997. Avifaunal diversity in the highlands of Sri Lanka. *Tiger Paper*, 24(3): 17–21.

Mason, C.M. and Maxwell-Lefroy, H. 1912. The food of birds in India. *Mem. Agr. Dept. India, Entomological Series*, 3: 1–371.

Parmod, P. 1995. Ecological studies of bird communities of Silent Valley and neighbouring forests. *Ph.D.* Calicut Univ., pp. 99.

Pearson, D.L. 1975. The relation of foliage complexity to ecological diversity of three Amazonian bird communities. *The Condor*, 77: 453–466.

Rai, N.D. 1991. A study of heterospecific flocking and non-breeding bird community structure of Rajaji National Park. *M.Sc.* (Wildlife Science) dissertation. Saurashtra University, Rajkot.

Recher, H.F. 1969. Bird species diversity and habitat diversity in Australia and North America. *American Naturalist*, 103: 75–80.

Sale and Berkmuller. 1988. *Manual of Wildlife Techniques for India*. Food and Agriculture Organization of the United Nations, Dehradun.

Srinivasulu, B., Srinivasulu, C., Rao, V.V., Koteshwarulu, C. and Nagulu, V. 1997. Avian use of paddy agro-ecosystem. *Pavo*, 35(1 and 2): 75–84.

Whistler. 1928. *Popular Handbook of Indian Birds*. (Eds.) Gourney and Jackson, London.

Wiens, J.A. 1989. *The Ecology of Bird Communities*. Vol. I: Foundations and Patterns. Cambridge University Press.

Woodcock, M.W. 1980. *Collins Hand Guide to the Birds of the Indian Subcontinent*. William Collins Sons and Co. Ltd., London.

Chapter 13

Studies on the Daily Activities of Rhesus Monkeys, *Macaca mulatta* (Zimmermann) in Saraswati and Veer Sonti Forests of Haryana

S.K. Gahlawat[1], R.K. Gupta[1], R.C. Sihag[1] and R.C. Gupta[2]*

[1] Department of Zoology and Aquaculture, CCS Haryana Agricultural University, Hisar – 125 004

[2] Department of Zoology, Kurukshetra University, Kurukshetra – 136 119

ABSTRACT

The present study was carried out on six troops of rhesus monkeys (*Mucaca mulatta*) in Saraswati forests (29°58′N latitude and 76°2′E longitude) and Veer Sonti forests (30°0′N latitude and 77°1′E longitude) of district Kurukshetra (Haryana). Three troops of rhesus monkeys in each forests were identified and observations were recorded on becoming active in the morning and evening feeding bouts, departure from roosting sites to central core activity center and roosting in the night. The correlation between sunrise with time of becoming active and sunset with roosting time of the monkeys was found significant

* E-mail: gahlawatsk@rediffmail.com

at $P \leq 0.01$ level. The rhesus monkeys preferred long and branched trees like *Dalbergia sissoo, Acacia nilotica* and *Albizzia lebbek* for roosting in the night.

Keywords: *Daily activities, Macaca mulatta, Saraswati forest, Sonti forest.*

Introduction

The distribution of free ranging rhesus monkey *Macaca mulatta* (Zimmermann) ranged from 70°E longitude in Afghanistan to 120°E coast of China and extends northward upto 27°N latitude (Lindburg, 1971). In India, these monkeys confined to mixed deciduous forests of northern part only (Seth and Seth, 1983). The population of rhesus monkeys has progressively declining over the years due to fast demolition of its popular habitats and ever increasing demand in laboratories for experiments. Many factors such as seasonal fluctuations, availability of food, orientation of alphamale and home range influence the daily activities singly or in different combinations (Chance, 1956; Altmann, 1962; Mukherjee, 1969; Pirta *et al.*, 1977; Singh, 1981 and Gupta and Kumar, 1992, 1993). Keeping in mind the habitat and different behaviours of rhesus monkeys in relation to environmental conditions, the present study was carried out on six troops residing in Saraswati and Veer Sonti forests of Haryana. The troop size was ranging from 25 to 92 and the mean troop size was 59. The sex ratio in respect of adult male to adult female was 3 : 1 and the mean age-sex wise percentage in respect of adult males, adult females, juveniles and infants was 6.53, 19.54, 54.39 and 19.54 per cent, respectively (Gupta and Kumar, 1993).

Materials and Methods

Saraswati forest is located on Kurukshetra-Pihowa-Cheeka road, around 35 km west of Kurukshetra university and located at 29°58′N latitude and 76°32′N longitude. Its total area is 11231 acres. The major vegetation of species like *Acacia nilotica, Dalbergia sissoo, Eucalyptus* sp., *Ficus bengalensis, F. glomerata, Morus alba, Prosopis juliflora* and *Syzygium cumini.* Veer Sonti forest is located on Kurukshetra-Saharanpur road, around 20 km east of Kurukshetra university campus and located at 30°0′N latitude and 77°1′E longitude. Its total area is 474.50 acres. Almost similar type of vegetation is present in this forest. Three troops of rhesus monkeys in each forests were identified for the present study. About twenty four observations were recorded for a period of one year at successive intervals between 5.00 A.M. to 7.30 P.M. in each day. Each phase of activity merges with the successive one and all activities can be distinguished from one another. The observations were recorded on the following activities *i.e.*, time of becoming active, departure from roosting site, time taken for depart, morning feeding bout, departure to the central activity place, arrival at the central activity place, time taken to reach the central activity place, departure from central activity place, time spent at central activity place, arrival at feeding/ roosting site, evening feeding bout and time of retirement. The following formula used for statistical analysis of correlation coefficient between time of becoming active and sun rise; time of roosting and sun set:

$$r = \frac{\Sigma xy}{N\,'x\,'y}$$

where,

$x = (X_1 - X_2); y = (Y_1 - Y_2)$

$'x$ = Standard deviation of series x

$'y$ = Standard deviation of series y

N = No. of pairs of observations

r = Correlation coefficient

X_1 = Time of becoming active

Y_1 = Time of sun rise

X_2 = Time of roosting

Y_2 = Time of sun set

Results and Discussion

Observations of all the six troops on daily activities were recorded along with time of sunrise and sunset as well as maximum and minimum temperatures. In order to know the covariance between sunrise with becoming active and sunset with roosting time, four graphs have been plotted (Figures 13.1–13.4). The correlation coefficient is found to be 0.9322 and 0.9509 for sun rise with becoming active and 0.8857 and 0.9332 for sunset with roosting at Saraswati and Veer Sonti forests respectively which are significant at $P \leq 0.01$

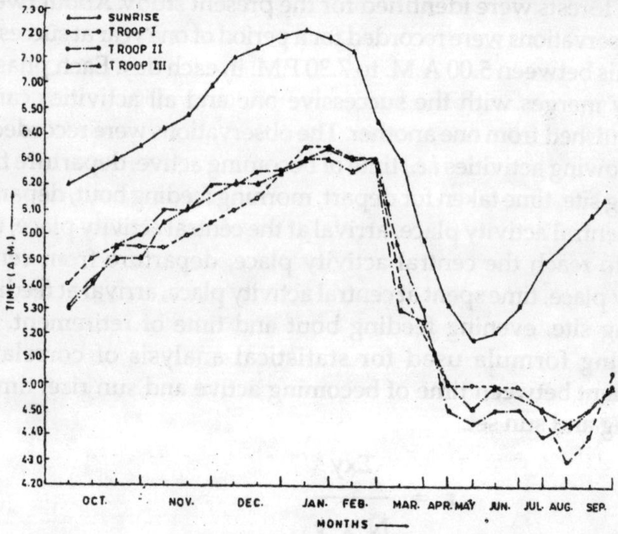

Figure 13.1: Graphical Representation of Covariance of Sunrise Time with the Corresponding Time of Becoming Active of Troops No. I, II, and III (Saraswati Forests)

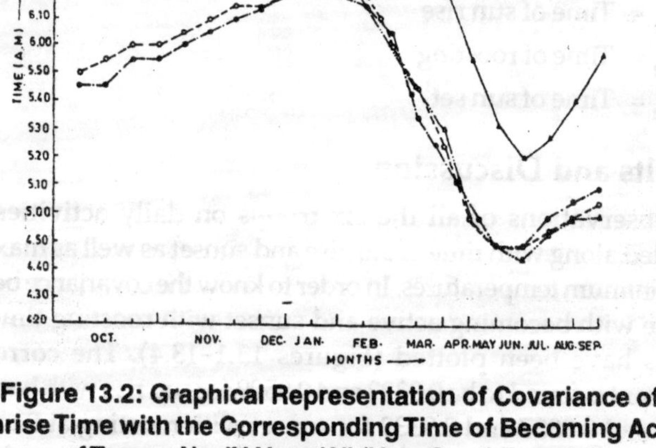

Figure 13.2: Graphical Representation of Covariance of Sunrise Time with the Corresponding Time of Becoming Active of Troops No. IV, V, and VI (Veer Sonti Forests)

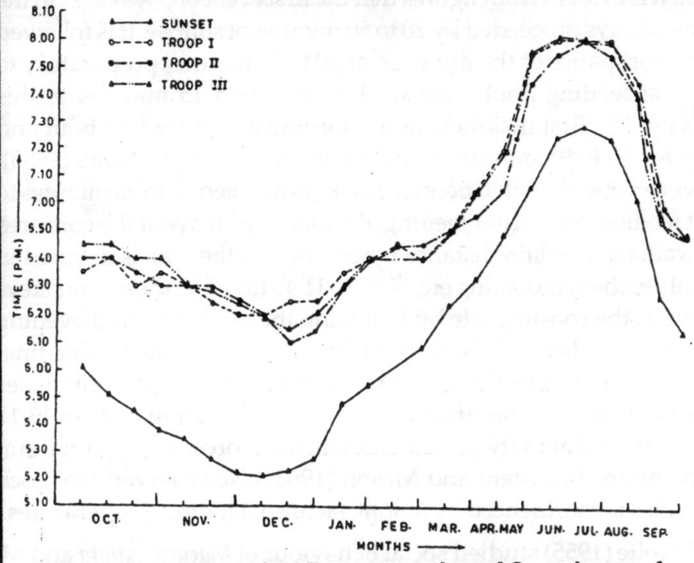

Figure 13.3: Graphical Representation of Covariance of Roosting Time with the Corresponding Sunset Time of Troops No. I, II, and III (Saraswati Forests)

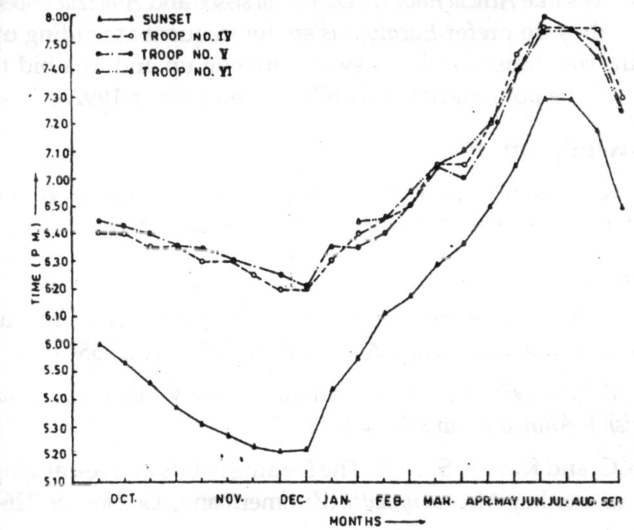

Figure 13.4: Graphical Representation of Covariance of Roosting Time with the Corresponding Sunset Time of Troops No. IV, V, and VI (Veer Sonti Forests)

level. It is evident from figures that the first event of '"waking" in the day is always proceeded by 20 to 60 minutes of sunrise. It is followed by a short pause of the duration of 10 to 30 minutes preparatory to the first feeding bout. Nolte (1955) observed 15 minutes to this activity. The first major event in morning is first feeding bout and lasts for 50 to 95 minutes (Gupta and Kumar, 1992). Nolte (1955), however, observed that bonnet monkeys devoted 60 to 90 minutes to first feeding bout. After feeding, the monkeys arrive at the core area by walking a certain distance where most of the time is devoted in aerial aerobics, grooming etc. (7.50 to 11.45 hrs). The rhesus monkeys arrive at the roosting site by following the same path and evening feeding bout lasts for 50 to 105 minutes prior to sunset. The final event relates to positioning at roosting site which generally takes place 20 to 60 minutes after sunset. It is evident from Figures 13.1–13.4 that total activity period exceeds the cooresponding total sun shine hours. Bernstein and Mason (1963) also observed that such activities are influenced by abiotic factors including temperatures.

Nolte (1955) studied social behaviour of *Macaca radiata* and *M. mulatta* and observed that daily routine began with mutual grooming, defecation and urination etc. The rhesus monkeys prefer branched and long trees like *Acacia nilotica, Dalbergia sissoo* and *Albizzia lebbek.* However, they do prefer *Eucalyptus* sp. for frequent ascending up during the day time to enjoy a swift climbing up and in a bid to escape from the potential dangers (Gupta and Kumar, 1992a).

Acknowledgement

The authors are thankful to Dr. Rajbir Singh Hooda for varied help and forest department, Haryana to work in the forests.

References

Altmann, S.A. 1962. A field study of the sociobiology of rhesus monkeys, *Macaca mulatta. Ann. N.Y. Acad. Sci.*, 102: 338–435.

Chance, M.R.A. 1956. Social structure of a colony *Macaca mulatta. British J. Animal Behaviour*, 4: 1–13.

Gupta, R.C. and Kumar, S. 1992. The feeding habits of free-ranging rhesus monkey, *Macaca mulatta* (Zimmermann). *Geobios*, 19: 226–230.

Gupta, R.C. and Kumar, S. 1992a. On the home range of free-ranging rhesus monkeys, *Macaca mulatta* (Zimmermann) in Saraswati and Veer Sonti forests in Haryana. *J. Natcon*, 4(1): 25–30.

Gupta, R. and Kumar, S. 1993. On troop size and composition of rhesus monkey, *Macaca mulatta* (Zimmermann) in Haryana. *J. Natcon*, 5(2): 111–114.

Lindburg, D.G. 1971. The rhesus monkey in north India: An ecological and behavioural study. In: *Primate Behaviour*, (Ed.) L.A. Rosenblum. Academic Press, New York, 2: 1–106.

Mukherjee, R.P. 1969. A field study on the behavior of two roadside groups of rhesus monkeys *Macaca mulatta* (Zimmermann) in northern U.P. *J. Bomb. Nat. Hist. Soc.*, 66(1): 47–56.

Nolte, A. 1955. Field observations on the daily routine and social behaviour of common Indian monkeys with special reference to the bonnet monkey (*Macaca mulatta* Geoffroy). *J. Bomb. Nat. Hist. Soc.*, 53: 177 –184.

Pirta, R.S., Parkash, P. and Singh, M. 1977. A population study of two species of non-human primates: *M. mulatta* and *M. radiata*. *J. Bomb. Nat. Hist. Soc.*, 77: 429–434.

Seth, P.K. and Seth, S. 1983. Population dynamics of free-ranging rhesus monkeys in different ecological conditions in India. *Am. J. of Primatology*, 4: 61–67.

Singh, R. 1981. Study of behavioural patterns of age-sex classes during transitional periods in free-ranging rhesus monkey (*Macaca mulatta*). *Ph.D. Thesis*, University of Rajasthan, Jaipur.

Chapter 14

Food and Feeding Habits of Free-ranging Rhesus Monkey in Jammu Region (J&K State)

Deep N. Sahi and Shubhra Sharma

P.G. Department of Zoology, University of Jammu, Jammu – 180 006

Introduction

Studies regarding feeding of rhesus monkey (*Macaca mulatta*) have been undertaken by various workers like Gunn (1969), Lindburg (1971, 1976), Southwick *et al.* (1976), Makwana (1978), Rehman and Parthasarthy (1978), and Malik and Southwick (1985) in various places in India but no such study has ever been done in and around Jammu, the southern part of the Jammu and Kashmir State which connects the undulating plains and foot hills of Shivalik range. The present work is exploratory in nature and is expected to provide comprehensive backdrop for future studies on food and feeding habits of *Macaca mulatta*.

Study Area

The study was conducted on 10 different troops residing at different ecological habits *viz.*, temple (Bahu Fort), city (Panjthirthi), forest (Mahamaya and Patnitop) and roadside (Nandini) for nearly four years (August 1996–May 2000).

All these ecologically distinct habits fall in district of Jammu except Patnitop which lies in the district Doda of the J&K State and harbour sizeable population of rhesus monkey. The vegetation of these localities varies from sparse to dense and comprises mixture of herbs, shrubs and trees.

Temple Area (Bahu Fort)

Bahu Fort is an old structure built by Maharaja of Jammu. It lies in the southeast of river Tawi on a small hillock at an altitude of 325 m and houses a temple within its premises where large no. of devotees visit every day. The area harbours a sizeable population of rhesus monkey. This troop was numbered as Troop no. 1. The vicinity of fort has human settlements.

City Area (Panjhtirthi)

It is thickly populated urban area in the east of Jammu city on bank of river Tawi. This locality is peculiar in having old secretariat building, built by Maharaja of Jammu. Presently it houses several government offices. Several troops of monkeys inhabit this congested human dwelling area. Troops studied here were numbered as troop no. II and III.

Forest Area (Mahamaya and Patnitop)

Mahamaya is about 3 km north of Bahu fort and is presently known as city forest. On the top of hill and within forest, there is a small temple, which is often visited by devotees mostly on Tuesday and Sunday. The vegetation cover found here is fairly dense and comprises semievergeen forests with poor undergrowth of herbs and shrubs. Troops of this area were numbered as IV and V.

Patnitop is a hill resort in district Doda at an elevation of 2050 mts and is about 110 km in the north of Jammu on Jammu-Srinagar National Highway. Coniferous forests of *Cedrus deodara* and *Pinus wallichiana* with poor undergrowth of shrubs and herbs dominate this area. Two troops residing here were numbered as VI and VII.

Roadside (Nandini)

A small village Nandini lies 26 km north of Jammu on Jammu-Srinagar National Highway at an elevation of 410 mts. Both sides of highway have thick semi-evergreen forest of *Pinus roxburghii*. Some broad leaved trees such as *Mollotus phillipensis, Dalbergia sissu, Butea monosperma* etc. are also common in this area. Troops studied here were numbered as VIII, IX and X.

Methodology

Data on feeding habit has been recorded throughout the year by making visits to different ecological regions at least twice in a month. Observations on each visit were made from dawn to dusk at different time intervals. Samples of plants consumed were collected and pressed for identification. Initially local names were used to identify them until their final identification.

Observations

Feeding plays most important role in determining the daily activities. On an average, 15–33.3 per cent of the daily activity of rhesus monkey centres on feeding. Studies conducted at all study sites showed that although monkeys are basically omnivorous but herbivorous feeding habit predominates. The main three food resources are: natural vegetation, agricultural crops and cooked or raw food provided by human beings.

The food provided by humans was found to be fairly similar, almost ritualistic in all the study areas except Patnitop where humans occasionally fed them. At Bahu Fort and Mahamaya, the level of provisionalization was much higher in comparison to other sites. In both these habitats, it has been observed that monkeys prefer to receive ready made food, which accounts for 63.75–76.25 per cent of the total feeding period in a day.

Natural vegetation consumed by rhesus was found to differ seasonally (Table 14.1–14.5). Throughout the year rhesus exploited the leaves of a wide variety of plants but during fruiting, season, fruits of some plants (*Morus alba, Zizyphus* sps.) etc. formed their choicest food. Tender stem of congress grass (*Parthenium hysterophoms*) was also occasionally seen to be consumed by juveniles of troop nos. II and III. Consumption of clay by members of all troops throughout the year was also noticed.

Feeding pattern not only revealed seasonal fluctuations but also shows variation during different times of the day. During hot summer days, monkeys were observed invading feeding sites in early morning hours (5.30 a.m.–6 a.m.) but during cold and cloudy days, their feeding activity was noticed to start as late as 9 a.m. During monsoon

Table 14.1: List of the Plants Consumed Month-wise by Troop No. I (Bahu Fort)

Species	Oct.	Nov.	Dec.	Jan.	Feb.	Mar.	Apr.	May	Jun.	Jul.	Aug.	Sep.
Acacia modesta	–	–	L	–	–	–	L,TL	L,TL	L	L,Po	Po	–
Citus limon	L	L	–	–	L	TL	TL	L	L	L	L	–
Cynodon dactylon	L	L	L	L	L	–	L,R	L,R	L,R	L	L,R	–
Dedonea viscosa	St	–	–	–	–	–	–	–	–	–	–	–
Ervatemia divaricata	–	–	–	–	–	–	TL	Bd	–	–	–	–
Ficus religiosa	–	–	L	–	–	TL	TL	–	L	L	L	L,F
Hibiscus sp.	St	St	–	–	–	–	–	–	–	–	–	–
Lantana camara	–	–	Bd	–	–	FL	FL	–	–	–	–	–
Laucenai leucipinta	–	–	–	–	–	L	L	–	–	–	–	–
Malvastrum sp.	–	–	–	–	–	L	–	–	–	–	–	–
Medotheca sp.	–	–	–	–	L	L	–	–	–	–	–	–
Morus alba	–	–	–	–	–	–	F	F	–	–	–	–
Poa amma	–	–	–	–	L	L,R,S	–	L	L	L	–	–
Rosa indica	–	–	L,B	L	–	–	TL	B,Bd	–	Bd	–	–
Solanum nigrum	–	–	–	–	L	L	L	–	–	–	–	–
Sonchus arvensis	–	–	–	–	R	R	–	–	–	–	–	–
Stelaria media	–	–	S	–	S	L	–	–	–	–	–	–
Vicia faba	–	–	–	–	L,R	L,R	L,S	–	–	–	–	–

L: Leaves; TL: Tender leaves; Po: Pods, R: Roots; Bd: Buds; Fl: Flower; F: Fruit; S: Seeds; St: Stem; B: Bark; Infl: Inflorescence; P: Petiole; UF: Unripe fruit; N: Nectar.

Table 14.2: List of the Plants Consumed Month-wise by Troop No. II and III (City Area)

Species	Oct.	Nov.	Dec.	Jan.	Feb.	Mar.	Apr.	May	Jun.	Jul.	Aug.	Sep.
Acacia modesta	L	–	–	L	–	TL	TL	L,FL	L,TL	L,FL,P	L	L
Achryanthus aspera	–	Infl	–	–	–	–	–	–	–	–	–	–
Adhatoda vesica	–	–	N	N	–	–	N	–	–	–	–	–
Albizzia lebbek	L,P	L,P	P,L	P	–	L,TL	TL	L	L	L	L,P	L,P
Citrus limoa	L	L	L	–	–	TL	TL	L	L	L	L	L
Citrus medica	L	L	L	L	–	TL	TL	T	L	L	–	L
Citrus sp.	L	L	L	–	–	L	TL	L	L	L	L	L
Cyanodon dactylon	L	L	L	L,R	L,R	L	L	L	L,R	L,R	–	–
Dalbergia sisoo	L,P	L	L	L	L	TL	TL,P	TL,P	L,P	L,P	L,P	L
Ervetamia sp.	–	–	–	–	–	P,F,Bed	P,F,Bd	F,B	F,B	–	–	–
Ficus benghalensis	L	L	L	L	–	–	L	L	L	L	L	L
Ficus palmata	–	–	–	–	–	–	–	–	L	L	–	L
Ficus religiosa	L	L,F	L	–	–	TL	TL	L	L	L	L	L
Jasminum sp.	–	–	–	–	–	Bd,L	Bd,TL	Bd,TL	–	–	–	–

Contd...

Table 14.2–Contd...

Species	*Oct.*	*Nov.*	*Dec.*	*Jan.*	*Feb.*	*Mar.*	*Apr.*	*May*	*Jun.*	*Jul.*	*Aug.*	*Sep.*
Lantana camara	–	–	–	–	–	FL	FL	FL	–	–	–	–
Mangifera indica	–	L	L	–	L	TL	TL	UF	F	F	–	–
Morus alba	–	–	–	–	–	L	UF	F	F	L	–	–
Occimum sanctum	L	–	–	L	L	L	–	–	–	S	S	Infl
Parthenium hysterophorus	–	–	–	–	–	–	TS	TS	TS	–	–	–
Psidium guava	–	–	–	–	–	–	–	–	–	F	F	–
Rosa indica	L,B	L,B	L	–	L	Bd,L	–	Bd,B	Bd	TL	–	–
Syzygium cuminii	–	–	–	–	–	P,L	P,L	UF,L	F	F	–	–
Thevetia peruviana	–	–	–	–	–	–	–	–	N	N	N	–
Trichosanthes dioca	–	–	–	–	–	–	–	–	L	L	–	–
Viccia hirsuta	–	–	–	–	–	–	Po	Po	Po	–	–	–
Withania sommifera	–	F	–	–	–	–	–	–	–	–	–	–
Zizyphus sp.	L,F	L,F	F	–	–	–	–	–	–	–	–	L

L: Leaves; TL: Tender leaves; Po: Pods, R: Roots; Bd: Buds; Fl: Flower; F: Fruit; S: Seeds; St: Stem; B: Bark; Infl: Inflorescence; P: Petiole; UF: Unripe fruit; N: Nectar.

Table 14.3: List of the Plants Consumed Month-wise by Troop No. IV and V (Mahamaya)

Species	*Oct.*	*Nov.*	*Dec.*	*Jan.*	*Feb.*	*Mar.*	*Apr.*	*May*	*Jun.*	*Jul.*	*Aug.*	*Sep.*
Acacia sp.	–	–	L,B	–	–	–	–	–	B	–	–	–
Adhotoda vasica	N	–	–	N	N	–	–	–	–	–	–	–
Bauhinia sp.	–	–	S	S	–	–	–	–	–	–	–	–
Bombex ceiba	–	–	–	–	–	–	–	–	B	–	–	–
Bougainvillea sp.	Bd	–	–	–	–	–	–	–	–	–	–	–
Butea monosperma	L	L	L,F	L,F	–	–	–	–	–	–	–	L
Carrissa opaca	–	–	F,L	F	F	–	–	–	–	–	–	–
Cassia fistula	–	–	–	–	–	–	–	–	F,L	–	–	–
Chenopodium album	–	L	L	L	–	–	–	–	–	–	–	–
Cyanadon sp.	L	L	L	–	–	L	L,R	L,R	L,R	L,R	–	–
Dalbergia sisoo	P	L	P	–	–	–	L,TL	L,TL	–	Po	Po	L,Po
Dodonia viscosa	–	–	–	–	–	L	L,TS	TS,L	–	–	–	–
Ehretia laevis	L	–	–	–	–	–	–	F	F	–	–	L
Flacourtia indica	–	–	–	–	–	–	–	–	F,L	–	–	–

Contd...

Table 14.2–Contd...

Species	Oct.	Nov.	Dec.	Jan.	Feb.	Mar.	Apr.	May	Jun.	Jul.	Aug.	Sep.
Grevillea robusta	–	F	F	F	–	TL	TL	–	L	–	–	–
Lannea coromandelica	–	–	–	–	–	–	–	–	L	–	L	L
Lantana camara	–	–	Bd	–	–	L,FL	FL	FL	–	–	–	–
Leucenea loucocephala	–	–	–	–	–	L	–	–	–	–	–	–
Mallotus phillipinensis	S	S	–	–	–	–	–	–	–	–	–	–
Meha azadirach	B	–	–	–	B	–	–	–	–	–	–	–
Poa amma	–	–	–	–	–	–	–	–	L	L,Infl	L,Infl	L,Infl
Pupalia sp.	F	–	–	–	–	–	–	–	–	–	–	–
Rosa indica	–	L,B	L	L	L	–	TL	B,Bd	Bd	–	–	–
Thevetia peruviana	–	–	–	–	–	–	–	–	N	N	N	N
Viccia hirsuta	–	–	–	–	–	–	Po	Po	Po	–	–	–
Zizypus sp.	L,F	L,F	L,F	L,F	L	–	–	–	–	–	–	–

L: Leaves; TL: Tender leaves; Po: Pods, R: Roots; Bd: Buds; Fl: Flower; F: Fruit; S: Seeds; St: Stem; B: Bark; Infl: Inflorescence; P: Petiole
UF: Unripe fruit; N: Nectar.

Table 14.4: List of the Plants Consumed Month-wise by Troop No. VI and VII (Patnitop)

Species	*Oct.*	*Nov.*	*Dec.*	*Jan.*	*Feb.*	*Mar.*	*Apr.*	*May*	*Jun.*	*Jul.*	*Aug.*	*Sep.*
Aliaria officinalis sp.	–	–	–	–	–	L	–	L	L	L	–	L
Berberis lycium	–	L	–	–	–	–	–	–	F	F	F	–
Brassica campestris	–	–	–	–	F	F	–	–	–	–	–	–
Caltha palustris	–	–	–	–	–	–	F	F	–	–	–	–
Cedrus deodara	–	R	R	R	–	–	–	–	–	–	R	R
Cyanodon dactylon	L	L	L	L	–	L	–	L,R	L,R	–	–	–
Dedonea viscosa	St	–	–	–	–	–	L	L	L	–	–	–
Duchnesia indica	–	L	L	–	–	F	F	F	F	–	–	–
Ficus palmata	–	–	–	–	–	–	–	–	–	–	L	L
Fragaria vesica	L	–	–	–	–	–	F	F	F	–	–	L
Hedera nepalensis	–	S	S	S	–	–	–	–	–	–	–	–

Contd...

Table 14.4–Contd...

Species	*Oct.*	*Nov.*	*Dec.*	*Jan.*	*Feb.*	*Mar.*	*Apr.*	*May*	*Jun.*	*Jul.*	*Aug.*	*Sep.*
Oxalis corniculatus	–	–	–	–	L	L	–	–	–	–	–	–
Pinus sp.	S	S	–	S	–	S	B,S	–	–	–	–	Se
Princepia utilis	–	–	–	–	–	–	L	–	–	–	–	–
Punica gramutum	–	–	–	–	–	–	–	–	–	F	F	F
Pyrus communis	–	–	–	–	–	–	–	–	F	F	F	–
Pyrus mallus	–	–	L	L	–	–	–	–	F	F	F	F
Rosa indica	B	B	B	–	B	–	–	–	L	L	L	–
Rubus ellipticus	–	–	–	–	–	–	–	F	F	F	–	–
Sacococca pruniformus	–	–	–	–	–	–	–	F	F	–	–	–
Solanum pseudocapsicum	–	–	–	–	–	–	–	F	F	–	–	–
Viburnum grandiflorum	L	L	–	–	–	–	–	F	F	F	–	–
Viola canescens	–	–	–	–	F	F	–	–	–	–	–	–

L: Leaves; TL: Tender leaves; Po: Pods, R: Roots; Bd: Buds; Fl: Flower; F: Fruit; S: Seeds; St: Stem; B: Bark; Infl: Inflorescence; P: Petiole; UF: Unripe fruit; N: Nectar.

Table 14.5: List of the Plants Consumed Month-wise by Troop No. X and XI (Nandini)

Species	Oct.	Nov.	Dec.	Jan.	Feb.	Mar.	Apr.	May	Jun.	Jul.	Aug.	Sep.
Acacia modesta	–	L	L	–	–	–	L,TL	F,Fl,TL	L,FL	L,FL,Po	L	–
Butea monosperma	L	L	L,F	–	F	–	–	–	–	–	–	–
Calandula arvensis	–	–	–	–	–	Bd	Bd	–	–	–	–	–
Carrissa opaca	–	–	F	F,L	F,L	F	–	–	TL	–	–	–
Dalbergia sisoo	P	–	–	–	–	–	L,TL,P	L,TL,Po	L,Po	Po	–	Po
Emblica officianalis	–	L	L	–	F	F	F	–	–	–	–	–
Eriophorum comosum	–	–	–	–	–	L	L	–	–	–	–	–
Ficus palmata	–	–	–	–	–	–	–	–	F	L	L	–
Ficus religiosa	L	L,F	L	–	–	TL	TL	L	–	–	L	L
Flacoutia indica	–	–	–	–	–	F	F	–	–	–	–	–
Mangnifera indica	–	L	–	L	L	–	–	UF	F	–	–	–
Morus alba	–	–	–	–	L	–	F	F	F	–	–	–
Paspalidium flavidum	–	–	–	–	–	–	–	–	Infl	Infl	Infl	–
Pinus roxburgii	–	S	S	S	–	S	–	–	–	–	–	–
Punica granatum	–	–	–	–	–	–	–	–	–	–	F	F
Terminalia arjuna	B	B	–	–	–	–	–	–	–	–	B	B
Woodfordia fructicosa	–	–	–	–	–	N,L	–	N	N	–	–	–
Ziziphus maurmana	L,F	F	F	–	–	–	–	–	–	–	L	I

L: Leaves; TL: Tender leaves; Po: Pods, R: Roots; Bd: Buds; Fl: Flower; F: Fruit; S: Seeds; St: Stem; B: Bark; Infl: Inflorescence; P: Petiole; UF: Unripe fruit; N: Nectar.

period, depending upon rains, they were seen to start their feeding activity at irregular times. Generally two principle feeding bouts in a day were noticed, one in the morning and other in the evening.

From the food provided by humans, monkeys were seem to relish bananas more in comparison to grains, peanuts and wheat chappati. It has been found that level of aggression depends upon the type and quality of food. Smaller food items such as rice, peanuts and grams did not produce as much aggressive interaction as larger food articles such as bananas. During feeding, large food items would lead to more threatening and chasing behaviour by the dominant members of the troop. The sub-ordinate members were observed to approach food items only after dominant had his pick.

Rhesus groups occasionally modified their feeding behaviour depending upon specific habitat. At times, they were found to forage upon agricultural crops and even upon waste material thrown by humans.

The monkeys use their cheek pouches as temporary storage organs mostly when food is in abundance or is of choice or when the individuals were in hurry. Before consumption, food picked from ground was seen to be dusted. Inedible parts of fruits and nuts were seen to be removed before their consumption.

Feeding bouts were occasionally interrupted by environmental exploration. In case of any possible danger and on any threat perception, monkeys gave high pitched calls to warn other members of the troops.

Troop no. 1 residing at temple area fulfill its food requirement by feeding on the offering of pilgrims on account of some religious sentiments. The monkeys of temple area even received highest direct hand out from people. However, they do supplement their diet with vegetative matter and insects. Diets of troop no. II and III consisted of food offered by people, natural vegetation and even after raiding the houses and shops. Troop no. IV and V fulfilled their requirement of food from natural vegetation and also through devotes visiting the temple. However, they were also observed raiding agricultural fields particularly during maize crop season. Troop no. VI and VII were observed making use of natural vegetation as their primary food resource. However, they were also seen feeding occasionally on food provided by visitors to the picnic spots on National Highway. Monkeys residing on roadside *viz.* VIII, IX and X also have become semi-commens, because they mostly depend for food on bus passengers and they also consume natural vegetations.

Significant variation exists in the amount of time spent in feeding by different sexes and age classes in different months of the year with maximum feeding done by females. Adult males on the other hand were seen spending least amount of time in feeding.

Discussion

There appears to be great controversy regarding feeding habit of rhesus monkey (*Macaca mulatta*). In the study on the feeding behaviour of free ranging rhesus monkey of Tughlaquabad, Malik and Southwick (1985) reported their herbivores nature of feeding. Gupta and Kumar (1992) also found them vegetarian. During present investigations, rhesus monkey inhabiting distinct ecological habitats in Jammu region of J&K state were invariably found to consume grass hoppers, wasps, bees and ants besides large amount of vegetative matter. This finding is in accordance with the observations made by Mukherjee (1969) and Lindburg (1977). During the present study spanning over a period of four years, rhesus monkeys were found to consume insects only during hot months of the year (May to August) against Lindburg (op. cit.) finding of consumption of insects in small quantities throughout the year.

Time spent in feeding varies in different habitats and ranges from 15 per cent to 33.3 per cent of their total activity pattern. Monkeys living around a temple depend heavily on direct handout food *i.e.*, 63.75 per cent to 76.25 per cent than natural resources. Siddiqui and Southwick (1979) while studying rural monkeys from Chhatari, North India have noticed that monkeys consume 83 per cent of their food from people, 10 per cent from agricultural crops and the remaining 7 per cent from natural vegetation.

In the present study, members of all sexes and age groups were seen consuming clay and was not found to be limited to any particular month or season but was observed at different times of the year. However, Malik and Southwick (1988) have reported only juveniles consuming clay. On the contrary, Blanford (1888), Roonwal (1956), Mukherjee and Gupta (1965), Lindurg (1971) and Puget (1974) have reported rhesus monkey consuming clays is a phenomenon which cannot be easily explained. Some nutritional need must be the reason for its consumption.

Two principle feeding bouts, one in the morning and other in the evening as also reported by Siddiqui and Southwick (1979), Malik *et al.* (1985) and Gupta and Kumar (1992) were also recorded during present survey.

Existence of relation between type of food and aggression has been observed in the present study where dominant monkey was found to be quite aggressive while feeding on large edible items like apples, bananas than on smaller items (ground nuts, bengal grams) etc. Chalmer (1978) and Southwick and Siddiqui (1976) have also reported same type of behaviour in rhesus.

Rhesus groups were seen modifying their feeding behaviour depending upon specific habitat and environmental conditions as is evident during the present study that rhesus residing in different areas consume different food plants available there.

References

Blanford, W.T. 1888. *The Fauna of British India including Ceylon and Burma*. Mammalia (1). Taylor and Francis, London.

Chalmer, N.R. 1978. The social behaviour of free living mangabeys in Uganda. *Folia Primatol*, 8: 263–231.

Gunn, C.R. 1969. *Abrus precatorius*: pretty but poisonous. *Science*, 164: 245–246.

Gupta, R.C. and Kumar, S. 1992. The feeding habits of free ranging rhesus monkey (*Macaca mulatta*). *Geobios*, 19: 226–230.

Lindburg, D.G. 1971. The rhesus monkey in North India: An ecological and behavioural study. In: *Primate Behavioural*, (Ed.) Rosenblum. Academic Press, New York, p. 1–106.

Lindburg, D.G. 1976. Dietary habits of rhesus monkey (*Macaca mulatta*) in Indian forests. *J. Bom. Nat. His. Soc.*, 73(2): 261–279.

Lindburg, D.G. 1977. Feeding behaviour and diet of rhesus monkey (*Macaca mulatta*) in Shivalik forests in North India. In: *Primate Ecology*, (Ed.) Clutton Brock. Academic Press, New York, p. 223–249.

Makwana, S.C. 1978. Observation on some dietary components of Hanuman Langur (*Presbytis entellus*) in nature. *Z. angew. Zool.*, 64(4): 369–373.

Malik, I. and Southwick, C.H. 1985. Feeding behaviour of free ranging rhesus monkey of Tughlaquabad. *J. Bom. Nat. His. Soc.*, 54: 171.

Malik, I. and Southwick, C.H. 1988. Feeding behaviour and activity pattern of rhesus monkeys (*Macaca mulatta*) at Tughlaquabad. In: *Ecology and Behaviour of Food Enhanced Primate Groups*. Allan R. Liss, Inc., Indis, p. 95–111.

Mukherjee, A.K. and Gupta, S. 1965. Habits of rhesus macaque in Sunderbans, 24 Paraganas, West Bengal. *J. Bom. Nat. His. Soc.*, 62: 145–146.

Mukherjee, R.P. 1969. A field study on behaviour of two roadside groups of rhesus monkey in Northern U.P. *J. Bom. Nat. His. Soc.*, 66(1): 47–56.

Puget, A. 1974. Observation on rhesus macaque, *Macaca mulatta* in Afghanistan. *Mammalia*, 35(2): 199–203.

Rahaman, H. and Parthasarthy, M.D. 1978. Behaviorual variants of bonnet macaque (*Macaca radiata*) inhabiting cultivated gardens. *Ibid*, 75: 406–425.

Roonwal, M.L. 1956. Macaque monkey eating mushrooms. *J. Bom. Nat. His. Soc.*, 54: 171.

Southwick, C.H. and Siddiqui, M.F. 1976. Demographic characteristics of semi-protected rhesus groups in India. *Yearb. Phys. Anthropol.*, 20: 245–252.

Siddiqui, M.F. and Southwick, C.H. 1979. Feeding behaviour of rhesus monkey (*Macaca mulatta*) in northern India plains. *Proc. Zool. Soc.*, Kolkata, 31: 53–57.

Part II

Animal Reproduction

Chapter 15

Space, Prey Feed Modulation, Reduviid Predator and its Prey Densities on the Reproductive Potential of *Acanthaspis pedestris* Stal (Insecta: Heteroptera: Reduviidae)

K. Sahayaraj, G. Raju and J. Francis Borgio*

Crop Protection Research Centre, St. Xavier's College (Autonomous), Palayamkottai – 627 002, Tamil Nadu, India

ABSTRACT

Laboratory study was carried out to understand the effect of prey density, space, predator density and prey feed modulation on the biology and reproductive potential of the reduviid predator *Acanthaspis pedestris* Stal (Insecta: Heteroptera: Reduviidae) *Corcyra cephalonica* Stainton as a prey. Nymphal instars adults of *A. pedestris* were fed with caterpillers of *Corcyra cephalonica* Stainton in three prey densities such as one, two and three caterpillars/day/predator. Among the three densities tested, three caterpillars/day/predrator was chosen (50.0 days and 1.57 eggs/female/day) and proceeded with space availability (2 litre, 1 litre, 500 ml and 250 ml containers)

* E-mail: ttn_ksraj@sancharnet.in

experiments. One litre container was selected based on the development and reproduction. This category was further subjected with four predator densities (5, 10 and 20 predators/ one litre container/3 preys). Maximum egg production and moderate developmental period were observed in 10 predator density. *C. cephalonica* larvae was reared separately on wheat, rice and jowar. Developmental rate, mortality rate, preoviposition period, ovipostion period, post oviposition period, fecundity rate/ day, hatchability and total animal length of treated organism in each treatment were recorded in prey density, predator densities, space availability and prey feed modulated categories and discussed.

Keywords: *Prey density, Space, Predator density, Prey feed modulation, Reproductive potential,* Acanthaspis pedestris.

Introduction

Reduviid predator are continuously regarded as important natural enemies of pest in agriculture systems (Ambrose, 1999; Schaffer, 1988; James, 1994; Sahayaraj, 1998, 2000, 2004) but their actual effect on pest population in field situation were poorly documented, largely because of the difficulties inherent in quantifying predatory activity. Reduviids constitute a prominent group of predators due to their amenability to mass production (Sahayaraj, 1998) and potential for use in pest management (Sahayaraj, 1999). One of the potential factors that need to be taken into account to understand field situation is the effect of prey scarcity or rarity. Some results indicate that prey density or prey scarcity positively affect the predator development, growth rate and fecundity (Dunbar and Bacon, 1972; Bozer *et al.*, 1996: Drummond *et al.*, 1984; Kumaraswami, 1991). Prey scarcity caused predatory shokbugs to compromise reproductive output for metabolic maintenance until sufficient prey was available to provide the, necessary energy for reproduction (Widenmann and O'Neil, 1990). Reduviids individually required more food and consumed more preys than other predators (Evans, 1962; Schaffer, 1988: Ambrose and Kumaraswami, 1993, Ambrose and Claver, 1997; Sahayaraj, 2004).

Pest suppression efficacy of reduviids on various crop pests is extensively studied throughout the world (Schaffer, 1988; Cohen, 1990: James, 1994; Sahayaraj and Ambrose, 1997; Ambrose, 1999;

Sahayaraj, 1995, 2002, 2004). Different prey densities on the laboratory rearing of *N. therasii* (Sahayaraj and Ambrose, 1994) and *Rhynocoris marginatus* Fab. (Sahayaraj, 2002) suggested that predators development and fecundity were enhanced by *C. cephalonica* larvae and the need for further studies on the effect of prey density on the development and fecundity of reduviid biocontrol agent.

Acanthaspis pedestris Stal (Insecta: Heteroptera: Reduviidae) a generalist reduviid predator is predominantly found in agro-ecosystems such as cotton, soybean, tomato and groundnut crops, scrub jungles, semi-arid zones and tropical rain forests of Tamil Nadu, India. It is considered to be an effective biological control agent against nearly 20 insect pests of various cultivated crops (Sahayaraj and Ambrose, 1994; Sahayaraj, 1995; Sahayaraj, 1998; Ambrose and Sahayaraj, 1996). Biological control by augmentation of predators and parasites that have been mass produced in insectaries provides a means of insuring the presence of numbers that are adequate to suppress pest populations to a desired level. Production of parasites and predators for augmentation has generated considerable interest in workers in biological control since the beginning of the twentieth century. The chief obstacle to the wide spread application of the augmentation method of pest cor.trol is the difficulty in mass producing or mass collecting entamophagous arthropods at economical costs (Beirne, 1974).

Utilization of commercially produced predators is interwined intimately with a detailed knowledge of the predator capabilities and preferences, and life histories of predators in relation to various biotic and abiotic factors. The concept of commercial production in promoting biological control in India was first conceived by many authors. Ambrose (1995) and Schaefer (1988) expressed the urgent need for evolving strategies to mass rear the predators for their large scale release into the pest infested agro-ecosystem and to assess their biological control potential. Selection of a host is very important in the mass culture programme of any natural enemy including the reduviid predators. Finney and Fisher (1964) reviewed the use of natural and unnatural hosts for rearing entomophagous arthropods. They defined a natural host as one that is usually attacked in nature by parasites or predators. An unnatural host is not normally attacked, because of some isolating mechanism, but it serves as a suitable host for rearing a biological control agent in the insectary. The advantages of using an unnatural host are reduced production costs, convenience and easy for handling. The eggs and larvae of rice moth, *Corcyra cephalonica* Stainton is serving as unnatural hosts for many natural enemies including the reduviid predators. Joseph (1959) used *Corcyra* larvae as food for the reduviids by releasing in plastic containers. These larvae were also used by other reduviid biologists (Lakkundi, 1989 and Sahayaraj, 1991, 1998, 2002) for rearing reduviids. One of the basic requisites for a successful biological control agent is the availability of a sound, low cost rearing and suitable development technique, which also was not found to be available for *Acanthaspis pedestris.* Predator density, space and prey density are the important limiting factors in the development of *Acanthaspis pedestris.*

Materials and Methods

Irrespective of the experimental category, newly emerged (> 1 hr) unmated adult females were individually placed in plastic vials. Then adult males (> 1 hr) were introduced into the vials. Each treatment was replicated 10 times. Predator longevity and daily numbers of eggs were separately maintained till the hatching. Unhatched eggs were noted. From this observation the hatchability percentage was calculated for all the categories.

Rearing of *Corcyra caphalonica*

C. cephalonica eggs were purchased from Tamil Nadu Agricultural University (TNAU), Coimbatore and the larvae were reared on the diet containing, wheat flour (1.5 kg), rice flour (0.5 kg) and jowar flour (0.5 kg), bran (500 g), crushed groundnut kernel (50 g), yeast granules (50 g), multivitamin tablet (500 mg) and streptomycin powder (1.0 g) in plastic troughs (3 L). Newly emerged fourth and fifth instar larvae of *C. cepharlonica* from the laboratory stock culture were chosen as prey for this study. Since *C. cephalonica* larvae spun webs and remained hidden under them, the reduviids predator often failed to attack and often got entangled in the webs and died. So the reduviids bugs were supplied with *C. cephalonica* larvae, whose heads were crushed gently to avoid webbing. Kopec (1922) observed life necessity of brain for the inception of insect metamorphosis by making ligatures around the body of the caterpillars. Care was taken to crush the head of larvae, in order to avoid the hemolymph oozing. The head crushed larvae were found to be alive for about 30 days without feeding and it remained fresh in appearance. The larvae fed upon by reduviids turned brownish black that facilitated to record the prey eaten by the predator.

Reduviid Predator

Adults and nymphal instars of *A. pedestris* were collected from Sivathipatti scrub jungle bordering agro-ecosystem (70° 47'E and 8° 30'N), Tirunelveli district, Tamil Nadu, India. They were maintained under laboratory conditions (30–32°C; 11–13h photoperiod and 75–80 per cent relative humidity) in plastic vials (6 × 6 cm) on larvae of rice moth *Corcyra cephalonica* Stainton (Lepidoptera: Collericdae). Sahayaraj (1998) recommended this prey for the laboratory rearing of reduviids. First nymphal instars hatched from the laboratory laid eggs were chosen for the experimental studies. They were reared on three different prey densities. Best category was selected for the selection of suitable space for rearing the reduviid. Then the reduviid was subjected to three different predator density such as 5, 10 and 20 predators. Best category was selected to study the suitability of the media for rearing the prey *C. cephalonica*. Sixty replications were maintained for each category separately during the nymphal stage of the reduviid.

Prey Density

The present study was carried out to understand the effect of different prey densities on the biology of *A. pedestris* in 250 ml plastic containers. The predator was maintained on three prey density such as one prey (category 1), two preys (category 2) and three preys (category 3) (caterpillars/day/predator) density. In each category 60 replications were maintained upto the nymphal development then 10 pairs of male and females were maintained in the adult stage. By using the total nymphal developmental period, and reproductive capacity of the predators, best category was selected and proceeds the next experiment. Minimum developmental periods or moderate developmental period, egg laying rate (number of eggs laid/female/day) were considered for the selection of the suitable category for the mass production of this reduviid.

Space Availability

Three preys per container was selected for this experiment. Reduviids were reared in 500 ml (category 4), 1 litre (category 5) and 2 litres (category 6) containers separately and each category was provided with three uniform size prey and the biological parameters of *A. pedestris* were recorded.

Predator Density

From the above experiment one litre container was selected and used for this experiment. Five (category 7), ten (category 8) and twenty (category 9) predators were taken in a container with three preys. Best category was selected and proceeds the next experiment.

Prey Feed Modulation

A laboratory experiment was conducted to study the suitability of different cereals for rearing the prey *C. cephalonica*. Initial culture of the test insect (adult) was obtained from the District Agricultural Office, Palayamkottai, Tamil Nadu. Three cereals, *viz.*, rice (category 10), wheat (category 11) and jawar (category 12) were used in this study. These cereals were grounded into small pieces of uniform size. One kilogram of each of the crushed cereal was taken separately in the plastic troughs having 3 litres capacity. It was fortified with crushed groundnut (200 g), yeast (5 g), streptomycin (15 mg) and multivitamin (500 mg) (Glaxo). Two hundred first instar (12 hours old) larvae of *C. cephalonica* were released into the food stuffs and covered with muslin cloth firmly tightened with rubber bands. All sets were replicated thrice. Fourth and fifth instar nymphs emerged from the 3 diets separately were supplied to *A. pedestris* and their biology was recorded.

Biological Parameters Observed

In all the above mentioned studies the total nymphal developmental duration and survival rate of *A. pedestris* were observed. The laboratory emerged adults from each category was grouped as one pair separately in plastic containers and totally 10 replications were maintained in each category. Pre-oviposition period, oviposition period, post-oviposition period, total number of eggs laid by a female, fecundity rate (number of eggs laid/oviposition period/female), hatchability (number of nymphs hatched/ number of eggs laid × 100) were recorded. After the death of the female, they were preserved in 70 per cent alcohol and their total animal length was measured by using stage and ocular micrometer.

Statistical Analysis

Student 't' test was performed to determine the significance between the categories and the significance was expressed at 5 per cent level.

Results

Total nymphal developmental period, nymphal mortality rate, pre-oviposition period, oviposition period, post-oviposition period, total egg laid by a female, egg laying rate, hatchability and animal total length are presented in Tables 15.1–15.4.

Table 15.1: Impact of Prey Density on the Developmental Period (in days), Reproduction and Animal Size (mm) of *Acanthaspis pedestris*

Parameters	*Prey Density*		
	1	*2*	*3*
Developmental period (in days)	55.0	52.4	50.0
Mortality rate (%)	20.8	9.2	11.5
Pre-oviposition period (in days)	14.2	17.1	14.9
Oviposition period (in days)	73.5	68.2	52.8
Postoviposition period (in days)	15.3	14.3	8.8
Total egg laid by a female	55.125	83.204	82.896
Egg laying rate (egg/day)	0.075	1.22	1.57
Hatchability (%)	79.3	85.8	86.0
Animal total length (mm)	1.798	2.327	2.152

Table 15.2: Impact of Predator Density on the Developmental Period (in days), Reproduction and Animal Size (mm) of *Acanthaspis pedestris*

Parameters	*Predator Density*		
	5	*10*	*20*
Developmental period (in days)	68.5	65.0	63.2
Mortality rate (%)	2.0	3.7	3.8
Pre-oviposition period (in days)	31.7	14.0	29.6
Oviposition period (in days)	71.6	70.2	65.3
Post-oviposition period (in days)	15.5	14.0	14.0
Total egg laid by a female	259.192	126.36	164.556
Egg laying rate (egg/day)	3.62	1.8	2.52
Hatchability (%)	73.7	82.9	78.5
Animal total length (mm)	2.01	2.23	2.32

Table 15.3: Impact of Prey Feed Modulation on the Developmental Period (in days), Reproduction and Animal Size (mm) of *Acanthaspis pedestris*

Parameters	*Prey Feed Modulation*		
	Rice	*Wheat*	*Jawar*
Developmental period (in days)	67.0	61.2	50.4
Mortality rate (%)	16.6	16.6	47.4
Pre-oviposition period (in days)	57	66	58
Oviposition period (in days)	86	95	79
Post-oviposition period (in days)	32.8	31.8	18.5
Total egg laid by a female	301	351.5	292.3
Egg laying rate (egg/day)	3.505	3.7	3.7
Hatchability (%)	74.3	89.0	85.6
Animal total length (mm)	1.78	2.167	2.35

Table 15.4: Impact of Space Availability on the Developmental Period (in days), Reproduction and Animal Size (mm) of *Acanthaspis pedestris*

Parameters	*Space Availability*		
	2L	*1L*	*500 ml*
Developmental period (in days)	56.2	50.7	46.2
Mortality rate (%)	13.2	11.2	8.3
Pre-oviposition period (in days)	27.9	23.9	19.5
Oviposition period (in days)	91.6	89.7	77
Post-oviposition period (in days)	4.6	8.5	17
Total egg laid by a female	200.604	191.061	128.59
Egg laying rate (egg/day)	2.19	2.13	1.67
Hatchability (%)	81.8	84.4	88.0
Animal total length (mm)	2.473	2.626	2.236

Developmental Period

The developmental period was higher in the case of category 1 (55 days) followed by category 2 (52.4 days) and category 3 (50 days). Statistical analysis between 1 prey (category 1) with 2 preys (category

2) and also to the 3 preys (category 3) categories were significant at 5 per cent. But category 1 and category 2 were not significant at 5 per cent level by student 't' test. Nymphal developmental period was further reduced to 46.2 days, when *A. pedestris* reared in 500 ml container (category 4). However, the nymphal developmental period was higher (56.2 days), if the reduviid reared at 2 litre container (category 6) and they were significant statistically ($P < 0.05$). In 1 litre container (category 5) the developmental period was 50.7 days it was statistically insignificant when compared to the category 2. In prey feed modulation studies, *C. cephalonica* reared in rice (category 10) medium fed on *A. pedestris* shows developmental period (67.0 days) followed by wheat (category 11) (61.2 days) and jowar (category 12) (50.4 days). Statistical analysis showed that they were significant at 5 per cent level. The developmental period was higher (68.5 days) in category 7 followed by category 8 (65.0 days) and minimum in category 9 (63.2 days).

Mortality Rate

It was observed that the mortality rate was higher in category 1 (20.8 per cent), followed by category 10 and category 12 (16.6 per cent) reared *C. cephalonica* fed *A. pedestris*. In space availability the mortality rate was higher in category 6 (13.2 per cent) followed by category 5 (11.2 per cent) and category 4 (8.3 per cent). At the same time the mortality rate was minimum in the case of predator density (2.0, 3.7 and 3.8 per cent for 7, 8, and 9 categories respectively). However, the mortality rate was higher in the case of prey density categories. For instant in category 1, it was 20.8, and 11.5 and 9.2 per cent for 3 and 2 categories, respectively.

Reproduction

Pre-ovipositon Period

The pre-oviposition period was notably longer (66 days) in *A. pedestris* fed with wheat (category 11) grown *C. cephalonica*, followed by jawar (category 12) (58 days) and shorter in rice (category 10) grown (57 days) prey fed predator. In prey density categories higher pre-oviposition period was observed in category 2 (17.1 days) followed by category 3 (14.9 days) and lower in category 1 (14.2 days). The pre-oviposition period was higher in five-predator (category 7) density (31.7 days) followed by 20-predator (category 9) density (29.6 days) and minimum in 10-predator (category 8) density

(14 days). In space availability test, the pre-oviposition period was higher in category 6 (27.9 days) followed by category 5 (23.9 days) and lower in category 4 (19.5 days).

Oviposition Period

The oviposition period was recorded longer (95 days) in *A. pedestris* fed with wheat (category 11) grown *C. cephalonica*, followed by category 10 (86 days) and shorter (79 days) in category 12. In prey density experiment higher oviposition period was recorded in category 1 (73.5 days) followed by category 2 (68.2 days) and lower in category 3 (52.8 days). The oviposition period was higher (71.6 days) in category 7 followed by category 8 (70.2 days) and minimum in category 9 (65.3 days). In space availability test the oviposition period was higher in the category 6 (91.6 days) followed by category 5 (89.7 days) and the category 4 (77 days).

Post-oviposition Period

Post-oviposition period was higher in category 10 (32.8 days) followed by category 11 (31.8 days) and minimum in category 12 (18.5 days). In space availability experiment the post-oviposition period was higher in category 4 (17 days) followed by category 5 (8.5 days) and lower in category 6 (4.6 days). At the same time the post-oviposition period was minimum in the case of predator density categories (15.5, 14 and 14 days) for 7, 8 and 9 categories, respectively. The post-oviposition period was higher in category 1 (15.3 days) followed by category 2 (14.3 days) and category 3 (8.8 days).

Fecundity

When we reared the predator with different densities of preys, it was recorded that the fecundity was higher in category 2 (83.204) followed by category 3 (82.89) and category 1 (55.125). It was in category 6 (200.04) followed by category 5 (191.061) and category 4 (128.59). In prey feed modulation category, *C. cephalonica* reared in wheat (category 11) medium fed with *A. pedestris* showed higher fecundity rate (351.5) followed by category 10 (301.4) and category 12 (292.3). However, the fecundity rate was decreased in category 7 (259.192) followed by category 9 (164.556) and minimum in category 8 (126.36). Among all the categories, the fecundity rate was higher when *A. pedestris* was fed with jowar reared *C. cephalonica* (3.7 egg/day). It was minimum in category 1 (0.075 egg/day)

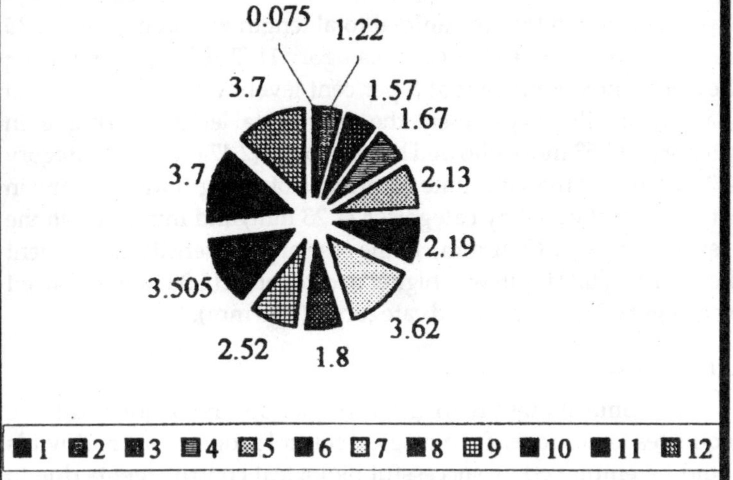

Figure 15.1: Impact of Prey Density (categories 1, 2 and 3), Space Availability (categories 4, 5 and 6), Predator Density (categories 7, 8 and 9) and Prey Feed Modulation (categories 10, 11 and 12) on the Egg Laying Rate per day of *Acanthaspis pedestris*

(Figure 15.1). The statistical analyses between these two categories were significant an 5 per cent level.

Hatchability

The hatchability was higher in the case of category 3 (86 per cent) followed by category 2 (85.8 per cent) and category 1 (79.3 per cent). In space availability experiment, the hatchability was higher in category 4 (88 per cent) followed category 5 (84.4) and category 6 (81.8 per cent). In prey feed modulation studies, *C. cephalonica* reared in wheat (category 11) medium fed with *A. pedestris* showed higher hatchability (89 per cent) followed by category 12 and category 10 (74.3 per cent). The hatchability was higher (82.9 per cent) in category 8 followed by category 9 (78.5 per cent) and minimum in category 7 (73.7 per cent). Among all the categories higher hatching was obtained in wheat-reared (category 11) *C. cephalonica* fed *A. pedestris* (89 per cent) followed by category 4 (88 per cent). The comparison between wheat and jowar categories was insignificant at 5 per cent level in student 't' test.

Animal Total Length

It was noted that the animal total length was maximum (2.35 mm) in category 12 followed by category 11 (2.167 mm). However, they were not significant at 5 per cent level by student 't' test. In space availability experiment, the animal total length was higher in category 5 (2.63 mm) followed by category 6 (2.47 mm) and category 4 (2.24 mm). At the same time the animal total length was 2.32 mm in category 9 followed by category 8 (2.23 mm) and minimum in the case of category 7 (2.01 mm). In the case of prey density experiment the animal total length was higher in category 2 (2.33 mm) followed by category 3 (2.15 mm) and category 1 (1.80 mm).

Discussion

Economical mass rearing techniques are the major and basic requisites for a successful biological control programme. Reduviids could be employed as successful biological control agents due to their greater predatory potential. Literature available on the nutritional influence on post-embryonic development, growth and reproduction of reduviids is meagre. Many natural and fictitious preys have been tested to reduviids. Among them *C. cephalonica* was suggested by many authors (Sahayaraj and Amborse, 1994; Sahayaraj, 2001). One-prey category (category 1) *A. pedestris* survived 103 days and produced very minimum eggs (55.13). Hence, during the food scarcity, *A. pedestris* tap their energy reserves to maintain survival. In these circumstances maintaining the cost of the normal metabolic processes of *A. pedestris*. One way to reduce the costs is to lower fecundity. By this manner, energy that would normally be used for egg production can be diverted to energy consumption process such as searching for food or avoiding predators to maintain survival. This homeostatic response is an appropriate adaptation by organism to maximize their fitness under varying food conditions is the common phenomena of any animal. Evans (1982) correlated body size (length and weight) of *Podisus maculiveatris* (Heteroptera: Pentatomidae) with predator rates. He found that body size was largest early in the season when preys were most abundant. At midseason, as prey numbers declined, mean body size of the predator declined. Such a relationship has not been observed in prey ration experiment. In general the ratio of prey, predator favours more frequent encounters with preys. Even when predators do encounters each other; there are significant costs that may restrain attacks on siblings. Cannibalism is associated with the vulnerability of the two individuals involved or a shortage of prey. Furthermore, sibling cannibalism is more likely to arise as a result of low prey availability than as a consequence of physical disparities between individuals. Because in our study we have not observed any disparity between the siblings reared either in relation to ration or with predator density.

The shortest stadial period was observed in the nymphal instars of *N. therasii* fed with *C. cephalonica* (67.59 days) followed by *Dysdercus cingulatus* Fab. (69.33 days) and *Odontotermes obesus* Rambur (101.30 days) (Sahayaraj and Ambrose, 1994) while provided the prey at *ad libitum*. Eveleigh and Chant (1981) and Ambrose and Claver (1997) suggested that low prey density minimizes the stadial period with regard to the higher prey density. In contrast in one prey category the nymphal developmental period was 55 days and it was reduced to 50 days in 3-prey density category. Generally predators that attack prey more frequently develop faster than predators, which attack the prey less frequently (Baumgaertner *et a1.*, 1981; Lenski, 1984). Bozer *et al.* (1996) pointed out that when the prey was scarce, allelochemicals in the diet of prey had a greater negative impact on the sting bug growth rate than when the prey was plenty. Lakkundi (1989), Vennison (1989), Sahayaraj (1995) and Ambrose (1997) reported that the fifth and the second stadia were the longest and the shortest respectively in most of the reduviids studied.

There was a positive relationship between the percentage of *A. pedestris* survival and prey density. Only few predators died prior to the moult and the time of moulting on different prey density feeding experiments (Wiedenmann and O'Neil, 1990). However, Eveleigh and Chat (1981) stated that there was additional mortality from the low prey density category due to cannibalism and other factors. Since we have reared the predators in enough special containers, cannibalism had no role in mortality.

Low prey inputs didn't increase the longevity in other predators such as pentatomid (Mukerjee and Le Roux, 1969), chrysopid (Ganey, 1977), mantis (Matsura and Morooka, 1983), oib-weaving spider (Wise, 1975), carabid beetles (Baars and Vanduk, 1984), predatory stink bug (Lagaspis and O'Neil, 1993), green lacewing (Sarode and Sonalkar, 1998) and mired (Torreno, 1994) and also in other reduviid (Ambrose and Claver, 1997).

The pre-oviposition period was greatly reduced in *Rhynocoris fuscipes* F. when the prey density was higher. However, the predators fed with excess prey showed delayed pre-oviposition in *Anthocoris confusus* (Rcuter) (Evans, 1976), and *Podisus maculiventris* (Say) (Wiedenmann and O'Neil, 1990). The number of eggs laid per female was in direct correlation with prey density and female adult longevity. Crawley (1975) stated that predators that attack more prey lay more eggs than the predators attacking less prey. A higher level of survival coupled with decreased egg production suggests a trade off between metabolic demands for these two processes (Quiring and McNeil, 1984). Slansky (1980) attributed that the number of eggs laid by an insect shows a close relationship with the amount of food consumed and the adult female may cause a delay in the onset of egg-production due to lack of food. Similarly, the total number of eggs produced by a female decreased with low prey input in *Cyrtorhinus lividipennis* Reuter (Chua and Mikil, 1989), *P. maculiventris* (Wiedenmann and O'Neil, 1990), *Delphastus pusillus Le* Conte (Heinz *et al*., 1994), *Cyrtopeltis tenuis* (Torreno, 1994), *R. fuscipes* (Ambrose and Claver, 1997). Predatory beetle *Calosoma sycophanta* L. (Weseloh, 1993) and pentatomide *P. maculiventris* (Legaspi and O'Neil, 1993) and reduviid *N. therasii* (Sahayaraj, 2001) reproduced only when the prey number was high. Ambrose and Claver (1997) stated that, if the food becomes more available, the amount of energy that could be used for reproduction would be increased resulting in greater egg production.

As observed in fecundity, the number of nymphs hatched and hatching percentage were also high at three-prey density level, and, under limited prey conditions, predators oviposited infertile eggs *e.g.* other reduviids *viz., Rhynocris margiatus* Fab. (Ambrose *et al*., 1990) and *Rhynocoris kumari* Ambrose and Livingstone (Ambrose and Sahayaraj, 1991). The hatchability was high when the predator had the lowest (80 per cent) and the highest (100 per cent) frequency of hatching. Further more, Ambrose and Claver (1997) stated that percentage of the viability of egg decreased when the prey density decreased.

Both for the nymphal instars and for adults, the total food consumption increased as the prey supply was increased. In general, the number of prey consumed at the beginning of each stage was always higher than at the end. Baumhaertner *et al*. (1981) stated that the weight gain or number of prey consumed by the predator could be described as a function of age. Predatory rate or rate of feeding per predator per day was lower in the immature stages than in the adults of reduviids. An opposite trend was observed by Lakkundi (1989), Evans (1962), Ambrose and Kumaraswami (1993) and Ambrose and Claver (1997) in other reduviids. Evans (1962) considered the predatory value as a general guide to measure the potential value of the predator. Among the nymphal instars, fifth nymphal instar consumed/killed maximum number of *Corcyra* larvae, and hence nymphal instars can be employed for quicker control of pest under field condition than the adult of *A. pedestris.* Reduviids either consumed or killed more prey than need to satiate themselves (Ambrose, 1999). However, similar kinds of studies are necessary on field crop pests with this predator. The data clearly indicate that the number of prey eaten by the predator was a function of the number of prey provided. These findings differ from those of Turnbull (1962), who observed that spiders consume virtually the same amount of food irrespective of food supply.

The lower the developmental period, and the pre-oviposition period, the higher adult longevity and survival rate, fecundity and hatchability were obtained in the treatment with three larvae/ predator/day, it can be considered as the optimum prey density for rearing *A. pedestris* in the laboratory for small scale production. The developmental period of this predator increased as the predator density decreased. However, statistical significances were not observed between the predator categories. The developmental rate of individual predator was significantly affected by predator density. Growth, development and reproduction of any insect vary according to the dietary composition of the food supplied to the insect. For instance jowar enhanced the development and reproduction of *C. cephalonica* (Sahayaraj *et al.*, 2001). Similarly, the jowar (category 12) reared *C. cephalonica* fed *A. pedestris* developed quickly and reproduce highly. Ambrose *et al.* (1990) reported that the total nymphal period of *R. marginatus* was 84.70 ± 1.01 days when it was reared in solitary condition on *Odontotermes obesus* Rambur. However, the developmental period was very short (38.82 ± 1.64 days) when it was reared on *S. litura* in isolation (Sahayaraj, 1991). It was reported that rearing in groups decreased the total nymphal period from 70.47 days to 61.03 days in *A. pedestris* and 78.84 days to 61.03 days in

Acanthaspis quinquespinosa Fab. as compared to rearing under solitary condition (Sahayaraj, 1991). The present study revealed that the developmental period decreased from 68.5 days to 46.2 days in *A. pedestris* when it was reared in category 7 to category 4.

Nymphal mortality increased as the predator density increased. In the 20 predator (category 9) group, the total nymphal mortality was 3.8 per cent. With increasing predator density, the survival rates of the immature instars decreased. This decrease in survival riles at high predator densities occurred because not all of the prey were equally available to each predator and underfed predators lend to become cannibalistic (Eveleigh, 1982). It has been observed that the oviposition period was longer than pre-oviposition in all the categories ($P < 0.05$), which is a desirable feature from the viewpoint of biological control. The result agrees with the result of Ambrose *et al.* (1988).

From the experimental results, it is very clear that 10 (category 8) *A. pedestris* provided with three *C. cephalonica* that reared on jowar (category 12) density in one-litre (category 5) capacity rearing container is suitable for the mass production. This category has minimum nymphal developmental period (50.4 days) and nymphal mortality (3.4 per cent) and maximum egg laying rate (3.7 eggs/ female/oviposition period). This method is suggested for the mass production of this reduviid. However, further studies on the selected category on *A. pedestris* with environmental factors like temperature and humidity shall provide fruitful results in the mass multiplication of this predator.

Acknowledgements

The senior author is very grateful to the Council of Scientific and Industrial research (CSIR), Govt. of India for the financial assistance. We are grateful to Rev. Dr. A. Antonysamy, S.J. Principal and Prof. M. Thomas Punithan, Head, Department of Zoology for the institutional facilities.

References

Ambrose, D.P. and Claver, M.A. 1997. Functional and numerical responses of the reduviid predator *Rhynocoris fuscipes* (F.) (Het., Reduviidae) to control leafworm *Spodoptera litura* (F.) (Lep., Noctuidae). *J. Appl. Ent,.* 121: 331–336.

Ambrose, D.P. 1997. Assassin bugs in biological control–Success and strategies: A review. In: *Biological and Cultural Control of Insect Pests: An Indian Scenario,* (Ed.) Ambrose, D.P. Adelme Publishers, India, p. 262–284.

Ambrose, D.P. and Sahayaraj, K. 1996. Long-term functional response of the reduviid predator, *Acanthaspis pedestris* Stal (Heteroptera: Reduviidae). In relation to its prey *Pectinophora gossypiella* Saunders (Lepidoptera: Noctuidae) density. *Hexapoda,* 8(2): 77–84.

Ambrose, D.P., Sahayarani, M.R. and Vennison, S.J. 1988. Effect of crowding on the camouflaging behaviour, size, development, oviposition pattern and hatchability of offspring of *Acanthaspis pedestris* Stal. A potential predator of insect pests (Heteroptera: Reduviidae). *Monitore Zool. Ital. (N. S),* 22: 111–1120.

Ambrose, D.P., Saju, T. and Sahayaraj, K. 1990. Prey influence on the development, reproduction and size of assassin bug, *Rhynocoris marginatus. Env. & Ecol.,* 8(1): 280–287.

Ambrose, D.P. 1995. Reduviids as predators: Their role in biological control. In: *Biological Control of Social Forests and Plantation Crops Insects,* (Ed.) T.N. Ananthakrishnan. Oxford and IBH Publishing Co. Ltd., New Delhi, p. 153–170.

Ambrose, D.P. 1999. *Assassin Bugs.* Science Publishers, New Hampshire, U.S.A. pp. 337.

Ambrose, D.P. and Kumaraswami, N.S. 1993. Food requirement of *Rhynocoris fuscipes* Fabr. (Heteroptera: Reduviidae). *J Biol. Control,* 7: 102–104.

Ambrose, D.P. and Sahayaraj, K. 1991. Camouflaging behaviour and the effect of decamouflaging on behaviour and development of *Acanthaspis pedestris* Stal in solitary and crowded conditions (Insecta: Heteroptera: Reduviidae). *Mitt. Zoo. Mus. Berl.,* 67: 325–337.

Baumgaertner, J.U., Gutrierrez, A.P., Summers, C.G. 1981. The influence of aphid prey consumption on searching behaviour, weight increase, developmental time and mortality of *Chrysopa* carnea (Neuroptera: Chrysopidae) and *Hippodamia convergns* (Coleoptera: Coccinellidae) larvae. *Canadian Entomology,* 113: 1007–1014.

Bears, M.A. Vanduk, T.S. 1984. Population dynamics of two carabid beetles at a dutch heathland II. Egg production and survival in relation to density. *J. Anim. Ecol.*, 53: 389–400.

Beirne, B.P. 1974. Status of biological control procedures that involve parasites and predators. In: *Proceedings of the Summer Institute on Biological Control of Plant Insects and Diseases,* (Eds.) F.G. Maxwell and F.A. Harris. University Press of Mississippi, Jackson, p. 69–76.

Bozer, S.F., Traugott, M.S., Stamp, N.E. 1996. Combined effects of allelochemical fed and scarce prey on the generalist insect predator *Podisus maculiventris. Ecol. & Ent.*, 21: 328–334.

Chua, T.H., Mikil, E. 1989. Effect of prey number and stage on the biology of *Cyrtorhinus lividipennis* Reuter (Hemiptera: Miridae): A predator of *Nilaparvata lugons. Environmental Entomology,* 18(2): 251–255.

Cohen, A.C., 1990. Feeding adaptations of some predaceous Hemiptera. *Annals of the Entomological Society of America,* 93: 1215–1223.

Crawley, M.J. 1975. The numerical responses of insect predators to changes in prey density. *J. Anim. Ecol.*, 44: 877–892.

Drummond, F.A., James, R.L., Casagrande, R.A., Faubert, H. 1984. Development and survival of *Podisus nuiculiventris* (Herriptora: Pentatomidae), a predator of the Colorado potato beetle (Comeoptera: Chrysomelidae). *Environmental Entomology,* 13: 1283–1286.

Dunbar, D., Bacon, O.G. 1972. Feeding, development and reproduction of Geocoris punctipes (Heteroptera: Lygaeidac) on eight diets. *Annals of the Entomological Society of America,* 65: 892–895.

Evans, D.E. 1962. The food requirements of *Phonoctonus* nigrofasciatus Stal (Hemiptera). *Entomological Experiment and Application,* 5: 33–39.

Evans, H.F. 1976. The effect of prey density and host plant characteristics on oviposition and fertility in *Anthocoris confusus* (Reuter). *Ecol. Ent.*, 1: 157–161.

Eveleigh, E.S. and Chant, D.A. 1982. Experimental studies on acarine predator-prey interactions: The numerical response of immature and adult predator (Acarina: Phytoseiidae). *Can. J. Zool.*, 60: 630–638.

Eveleigh, E.S. and Chant, D.A. 1982. Experimental studies on acarine prdator-prey interactions: The numerical response of immature and adult predators (Acarina: Phytoseiidae). *Canadian Journal of Zoology*, 59: 1407–1418.

Finney, G.L. and Fisher, T.W. 1964. Culture of entomophagous insects and their hosts. In: *Biological Control of Insect Pests and Weeds*, (Ed.) P. De Bach. Chapman and Hall Ltd., London, p. 329–355.

Ganev, R.E., 1977. Laboratory rearing of the common chrysopid. *Rastiteina zaschita*, 25(11): 41–44.

James, D.G. 1994. Prey consumption by *Pristhesancus plagipennis* Walker (Hemiptera: Reduviidae) during development. *Aust. Ent.*, 21(2): 43–47.

Joseph, M.T. 1959. Biology bionomics and economic importance of some reduviids collected from Delhi. *Indian J. Ent.*, 21: 46–58.

Kopec, S. 1922. Studies on the necessity of the brain for the inception of insect metamorphosis. *Biological Bull.*, 42: 323–342.

Kumaraswami, N.S. 1991. Bioecology and ethology of chosen predatory bugs and their potential in biological control. *Ph.D. Thesis*, Madurai Kamaraj University, Madurai, India.

Lakkundi, N.S. 1989. Assessment of reduviids for their predation and possibilities of their utilization in biological control. *Ph.D. Thesis*, Indian Agricultural Research Institute, New Delhi.

Legaspi. J.C., O'Neil, R.J. 1993. Life history of *Podisus muculiventris* given low numbers of *Epilwhna varivestis* as prey. *Environment Entomology*, 22: 1912–1920.

Lenski, R.E. 1984. Food limitation and competition. A field experiment with two *Carabus* species. *Journal of Animal Ecology*, 53: 203–216.

Matsura, T., Morooka, K. 1983. Influence of prey density on fecundity in a mantis, *Poratenodera angastipennis* (S.). *Oecologia*, 56: 306–312.

Mukerji, M.K., L.E Roux, F.J. 1969. A quantitative study of food consumption and growth of *Podidus maculiventris*. *Canadian Entomology*, 101: 387–403.

Quiring, D.T., and McNeil, J.N. 1984. Influence of intraspecific larval competition and mating on the longevity and reproductive performance of females of the leaf miner *Agromyza frontella* (Rondani) (Diptera: Agromyzidae). *Canadian Journal of Zoology*, 62: 2197–2200.

Sahayaraj, K. 1991. Bioecology, Ecophysiology and Ethology of chosen predatory hemipterans and the potential in Biological control (Insecta: Heteroptera: Reduviidae). *Ph.D. Thesis*, Madurai Kamaraj University, Madurai, India.

Sahayaraj, K. 1995. Developmental stages and biocontrol potential of a reduviid predator, *Acanthaspis pedestris* Stal against termites on groundnut. *Int. Ara. News.*, 15: 57–59.

Sahayaraj, K. 1998. Laboratory rearing of predaceous bugs with special reference to reduviids (Insecta: Hemiptera: Reduviidae). *Zoo's Print*, 13(5): 17–18.

Sahayaraj, K. 2000. Evaluation of Biological control potential of *Rhynocoris marginatus* on four groundnut pests under laboratory conditions. *Int. Arach. Newslet.*, 20: 72–74.

Sahayaraj, K. 2001. A qualitative study of food consumption, growth and fecundity of a reduviid predator in relation to prey density. *Entomologica Croatica*, 5(1 and 2): 19–30.

Sahayaraj, K. 2002. Small-scale laboratory rearing of reduviid predator *Rhinocoris marginatus* (Fab.) (Hemiptera: Reduviidae) on *Corcyra cephalonica* Stainton larvae by larval card method. *Journal of Central European Agriculture*, 3(2): 137–148.

Sahayaraj, K. 2004. *Indian Insect Predators in Biological Control*. Daya Publication House, Delhi, pp. 326.

Sahayaraj, K. and Ambrose, D.P. 1994. Stage and host preference and functional response of a reduviid predator *Acanthaspis pedestris* Stal four cotton pests. *J. Biol. Control*, 8(1): 23–26.

Sahayaraj, K. and Ambrose, D.P. 1994a. Prey influence on laboratory mass rearing of *Neohaematorrhophus therassi* Ambrose and Livingstone, a potential biocontrol agent (Insecta: Heteroptera: Reduviidae). *Bio-science Research Bull.*, 10(1): 35–40.

Sahayaraj, K. and Ambrose, D.P. 1997. Biocontrol potential of *Acanthaspis pedestris* Stal (Insecta: Heteroptera: Reduviidae) to *Helicoverpa armigera* Hubner of bhendi. *Madras Agricultural Journal*, 84(5): 294–295.

Sahayaraj. K. 1999. Effect of prey and their age on the feeding preference of *Rhinocoris marginatus* Fab. *International Arachis Newsletter*, 19: 39–40.

Sahayaraj, K., Sathyamoorthi, P. and Rajamanickam, K. 2001. Rearing media on *Corcyra cephalonica* Stainton biology and life table. *Shashpa*. 8(2): 125–130.

Schaefer, C. W. 1988. Reduviidae (Hemiptera: Heteroptera) as agents of biological control; In *Bicovas*, (Eds.) K.S. Ananthasubramanian, P. Venkatesan and S. Sivaraman. Loyola College, Chennai, 1: 27–33.

Slanksy, F. J.R. 1980, Effect of food limitation on food consumption and reproductive allocation by adult milkweed bugs. *Oncopeltus fasciatus*. *Journal of Insect Physiology*, 26: 79–84.

Srode, S.V., Sonalkar, U.V. 1998. Effect of *Corcyra cephatonica* (Stainton) density on the develepment of *Chrysoperia carnea* (Stephens). *Journal of Biol. Control*, 12(2): 153–156.

Torreno, H.S. 1994. Predation behavior and efficiency of the bug, *Cyrtopeltis tenuis* (Hemiptera: Miridue) against the cut worm *Spodoptera litura* (F.) *Philippines Entomology*, 9: 426–434.

Turnbull, A. L., 1962. Effect of prey abundance on the development of the spider *Agelenopsis potteri* (Black wall). *Canadian Entomology*, 97: 141–147.

Vennison, S. J., 1989. Bioecology, Ecophysiology and ethology of chosen assassin bugs. *Ph.D. Thesis*. Madurai Ka,araj, University, Madurai, 175 p.

Waseloh, R.M. 1993. Adult feeding affects fecundity of the predator, *Calosoma sycophanta* (Coleoptera: Carabidae). *Entomophaga*, 338: 4 5–439.

Wiedenmann, R.N., O'Neil, R.J. 1990. Effects of low rates of predation on selected life-history characteristics of *Podisus maculiventris* (Say) (Heteroptera: Pentatomidae). *Canadian Entomology*, 122: 271–283.

Wise, D.H. 1975. Food limitation of the spider *Lroyphia marginata*: Experimental field studies. *Ecology*, 56: 637–646.

Chapter 16

Effects of Mercury on Reproductive Physiology in Freshwater Teleost *Danio equipinatus*

Kaushalendra Kumar, Binay Kumar Singh, Md. Shamsul Haque and Prerna Azad

Department of Zoology, Government P.G. College, Shahdol – 484 001, M.P., India

ABSTRACT

In freshwater teleost the phenomenon of reproductive physiology runs under the direct influence of environmental factors. Any alterations in the aquatic medium due to contamination can cause irregularities in the physiology. Studies report that industrial effluents containing heavy metal salts like mercury cause severe hazardous effect on the reproductive physiology in the experimental fish, *Danio equipinatus*. The observations on gonosomatic index state clearly that decreased functional capacity of gonads with lower index value in the freshwater fish might be due to toxic effect of mercury present in the surrounding water.

Keywords: *Oogenesis, Pycknosis, Gonosomatic index, Oocytes.*

Introduction

With the advent of industrial development, the problem of intoxication of freshwater organisms increased simultaneously. The industries release their effluents containing heavy metals into the nearby river. A large number of workers (Affleck, 1952; Carpenter, 1930; Spehar, 1976; Saxena, 1980 and Singh,1991) have explained the impact of heavy metal salts on the reproductive physiology of fish species but much more new facts in this field have to be explored.

Present report deals with the toxic effects of mercury on the reproductive physiology of freshwater teleost, *Danio equipinatus* inhabitant of river Sone. As the water of river Sone gets intoxicated by mercury salts through the industries like Orient Paper Mills, Amlai and Hukum Chand Jute Mills above the acceptable limits of WHO.

Materials and Methods

Female freshwater teleost, *Danio equipinatus* were collected separately from river Sone and from other freshwater body (Mudna river) free from mercury intoxication for a complete year. Fishes of both the groups were brought to laboratory and sacrificed. Their ovaries were dissected out and gonosomatic index was calculated. Experiment was conducted for a complete year by collecting sample fishes in every month.

Results and Discussion

In fishes from other freshwater body the ovaries were packed with all stages of oocytes including matured stage also during spawning period. In fishes from river Sone it was found that the ovary was loosely packed with the oocytes of earlier stages *i.e.*, chromatin nucleolus stage and perinucleolus stage with corpora atretica. Indistinct ovigerous lamella and marked pycknotic changes have also been observed.

Quantitative analysis of reproductive physiology have been represented by gonosomatic index given in Table 16.1. During the course of experiment it has been noticed that the fishes collected from river Sone exhibiting decreased value of gonosomatic index as compared to the fishes from other water body in every phases of reproductive cycle *i.e.*, pre-spawning, spawning and post-spawning phase.

Table 16.1: Showing the Gonosomatic Index in Female *Danio equipinatus*

Month	*Genosomatic Index of Fishes Collected from River Sone*	*Gonosomatic Index of Fishes Collected from Other Locality(Mudna River)*
January	1.1382	1.4568
February	1.5169	2.0134
March	3.0012	3.4439
April	3.8034	4.5458
May	4.1279	5.4290
June	4.2865	5.8012
July	5.9657	11.1189
August	6.1036	11.5276
September	2.9153	5.6357
October	2.7649	3.7392
November	0.5173	0.9001
December	0.6387	1.1179

Probably the mercury-intoxicated water of river Sone may cause alterations in the overall reproductive physiology of freshwater teleost fish. Decreased values of gonosomatic index in the fishes collected from river Sone supports this opinion. Report submitted by CSE (2004) claims that water of river Sone contains mercury between 0.028–0.054 ppm.

Observations of Carpenter (1930), Affleck (1952) and Brown (1974) justify that water intoxicated by Zinc, Copper and other heavy metal salts altering the process of reproductive physiology by causing deformities in gonads with decreased rate of reproduction in a variety of fish species.

Acknowledgement

The senior author is indebted to dedicate this article to his Late Father, Binay Kumar Singh is also grateful to UGC, CRO, Bhopal for providing the financial assistance vide sanction letter No. 4–001 (03) 2004 (MRP/CRO), Dated 27–03–2004.

References

Affleck, R.J. 1952. Zinc poisoning in a trout hatchery. *Aust. J. Mar. Freshwater Res.*, 3: 142–169.

Brown, Y.M., T.L. Shaw and D.G. Shurben. 1974. Aspect of water quality and toxicity of copper to Rainbow trout. *Water Res.*, 8: 797–803.

Carpenter, K.E. 1930. Further researches on the action of metallic salts on fishes. *J. Exp. Zool.*, 56: 407–422.

Spehar, R.L. 1976. Cadmium and zinc toxicity to Flagfish, *Jordanella floridae*. *J. Fish, Res. Board, Canada*, 33: 1939–1945.

Saxena, S.K. 1980. Studies on the hypothalamo neurohypophyseal system in a freshwater fish, *Nandus nandus* (Ham) in relation with reproductive cycle. *Ph.D. Thesis*, University of Sagar.

Singh, B.K. 1991. Studies on hypothalamo neurohypophyseal system in a freshwater fish, *Danio equipinatus* (Ham) and neural control on reproduction. *Ph.D. Thesis*, A. P. S. University, Rewa (M.P.).

Mashi, S. 2000. Studies on the hypothalamo neurohypophyseal neurosecretory system in a freshwater teleost in relation with reproductive cycle. *Ph.D. Thesis*, A. P. S. University, Rewa (M.P.).

Chapter 17

The Annual Female Reproductive Cycle in Brown River Turtle, *Kachuga smithi*

V.K. Gupta*

Division of Pharmacology, Regional Research Laboratory, (CSIR) Canal Road, Jammu – 180 001

ABSTRACT

The pattern of reproductive cycle in *Kachuga smithi* studied from Jammu is essentially similar and confirms to the annual type of reptilian female reproductive cycle. That in the exhibition of this type of reproductive cycle, *Kachuga smithi* fall in line with the cycle typical of temperate reptiles. The major events of the cycle in this turtle may be summed up as initiating with (*i*) enlargement of follicles or vitellogenic phase, followed by (*ii*) ovulation and terminated by (*iii*) egg laying, where upon, the ovary and the ducts enter a short period of (*iv*) quiescence, which represents the brumating samples. This cyclic sequence is paralleled by the rise and fall in G.S.I. and the growth and activity of the oviducts.

The eggs are elliptical in shape, pinkish white in colour, having hard but fragile shells and are of on an average 45.18 mm ± 3.9 in length, 24.43 mm ± 2.5 wide and a weight of 14.800

* E-mail: vgupta_rrl@yahoo.com

± 2 g. The clutch size average 6.1 ± 2.1 within a range of 3–11; the size of the clutch bearing a negative correlation with the size of the layer (r = 0.17; y = 4.65 ± 0.03 x). Likewise the reproductive potential was found to be 16.65 ± 4.54 and also bore a negative correlation with the size of the turtle (r = 0.16; y = 4.65 + 0.06 x). Incubation time averages 72 days and colour pattern on the shell of the hatchling was the copy of its adult colour pattern.

Keywords: *Female, Reproductive cycle, Ovary, Oviduct, Emydidae,* Kachuga smithi.

Introduction

Emydidae is the largest, most diverse turtle family and includes approximately 95 species in 33 genera. Its members are distributed throughout North America, northern South America, Europe, northwestern Africa, and Asia. In India, the family is represented by 17 species in 7 genera, of which 6 species belong to the genus *Kachuga*. Despite the largest family of the living turtles, not much is known about its reproduction.

There are numerous studies (Altland, 1951; Legler, 1960; Gibbons, 1968; Powell, 1967; Ernst, 1971; Moll and Legler, 1971; Iverson, 1977; Plummer, 1977; Cox and Marion, 1978) available on the reproductive cycles of freshwater turtles, but relatively very little attention has been so far paid to the reproductive aspects and breeding ecology of the Indian turtles (Duda and Gupta, 1978, 1981; Gupta, 1979, 2002, 2004; Gupta and Duda, 1998; Singh, 1974; Verma, 1992; Verma *et al.*, 1999). Since virtually nothing about reproduction except its clutch size, is known in *Kachuga smithi* (Minton, 1966), the present study was, thus, undertaken to add detailed information dealing with the seasonal changes that occur during the female reproductive cycle in this turtle from Jammu waters, which is geographically a warm temperate (= sub-tropical) area of J&K state, India.

The Study Area

The specimens of emydid turtle, *Kachuga smithi* studied during the present investigation were collected from a slow running irrigation by-canal, New Ghomanasan Khul, situated about 15 km South-west of Jammu city. The canal has a running distance of about

23,000 ft. and it receives water supply from a major irrigation canal, the Ranbir canal.

The vegetation of the canal is almost similar in all its segments, *Vallisneria spiralis, Hydrilla verticellata, Potamogeton crispus, P. pectinatus, Marsillia quadrifolia* and filamentous algae (Zygnemacea) appear to be its common floral elements. The stream also supports a good variety of aquatic animals, the common vertebrate faunal elements being: among fishes, *Puntius sophore, P. sarana, P. conchinious, P. ticto, Mystus vitatus, M. seenghala, Channa punctatus, C. gachuga, Wallago attu, Hetropneustus fossilis, Rasbora rasbora, Danio rario, D. devario, Notopterus notopterus, Xenantodon cancila;* among amphibians, *Rana tigrina, R. cyanophlyctis, Bufo viridis,* and among reptiles, *Natrix piscator, Kachuga tectum tectum, K. smithi, Lissemys punctata punctata* and *Trionyx gangaticus.*

Materials and Methods

Collecting of turtles was done every fortnight by hand picking of partially immersed or basking turtles. Net collecting was also resorted to. Measurements and weighings were done in the laboratory from the live animals. After preliminary weighings and measurements, the turtles were anaesthetized before sacrifice for examination. Gross morphology of ovaries and oviducts was done after dissections. The ovarian weight, the number and size of ovarian follicles and corpora lutea; the number, size and weight of shelled oviducal eggs; and weight of oviducts, were noted separately for right and left side. After the system proposed by Legler (1960) the ovarian follicles were grouped according to diameter as Class I the large, greater than 15 mm diameter; Class II the medium, 11–15 mm in diameter; and Class III small, 6–10 mm diameter. Follicles smaller than 6 mm were innumerable and hence not considered.

Gonadosomatic index (GSI) referred in the text was calculated by the formula:

$$\text{GSI} = \frac{\text{wt. of gonad}}{\text{wt. of turtle}} \times 100$$

An approximate date of oviposition as used in the present study was determined when a specimen contained large corpora lutea in its ovaries but no eggs in its oviducts or had more corpora lutea in the ovary than the number of eggs in its oviducts.

Reproductive potential was taken as the number of eggs a female is capable of producing in a year. Ova of > 15 mm, plus the eggs actually laid and the shelled oviducal eggs (or the corpora lutea representing the last 2 categories) provides a total count of eggs produced in a year by a female. Number of clutches laid per year was estimated by the technique proposed by Christiansen and Dunham (1972) which was useful particularly in absence of any data. The technique involved counts of distinct sets of corpora lutea and determination of clutch frequency.

Observations

Ovarian and Oviducal Cycle (Tables 17.1 and 17.2)

The ovarian cycle in *Kachuga smithi* begins in December on the conclusion of its laying season. In this month, ovary (GSI = 1.63) and oviduct (2.30 per cent body weight) shows a decrease in their weight from the November data (Figure 17.1). The mean number of eggs (6 mm and above) during the month was observed to be 2.35 per ovary (Figure 17.2). The average number of class I follicles per individual was one, that of class II follicles 1.60 and of class III 2.1 (Figure 17.3). Class I follicles were observed only in six out of ten individuals.

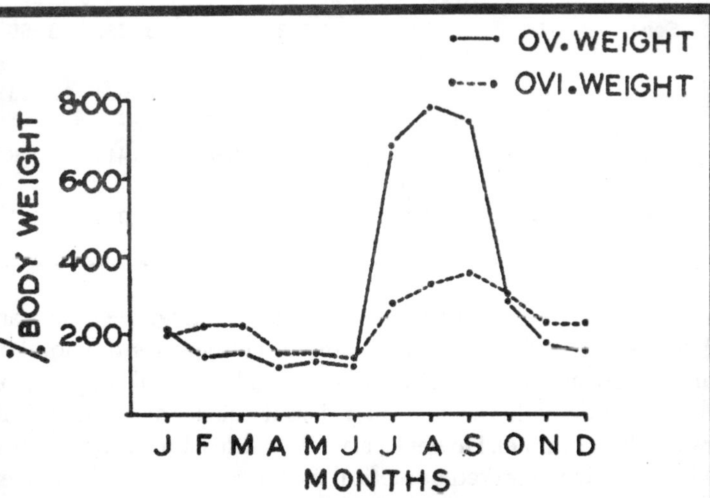

Figure 17.1: Ovarian (G.S.I.) and Oviducal Weight Per cent Body Weight in *Kachuga smithi*

Table 17.1: Monthly Records of Ovarian and Oviducal Characteristics of *Kachuga smithi*

Months	Number of Females Studied	Body weight (g)	Ovarian		Oviducal	
			Weight (g)	% Body Weight (GSI)	Weight (g)	% Body Weight
Jan	4	994 (880–1090)	21.155	2.128	20.251	2.037
Feb	7	809 (633–1000)	11.568	1.429	17.885	2.210
Mar	5	883 (795–980)	13.920	1.576	19.682	2.228
Apr	6	843 (613–1160)	9.762	1.158	12.971	1.538
May	3	1025 (980–1080)	14.165	1.381	15.567	1.518
Jun	6	1000 (820–1282)	11.516	1.151	13.981	1.398
Jul	5	1003 (775–1280)	68.196	6.799	28.004	2.792
Aug	13	1037 (900–1230)	80.816	7.793	34.254	3.302
Sep	7	882 (780–1000)	65.863	7.467	31.367	3.556
Oct	4	886 (730–960)	25.074	2.830	27.324	3.083
Nov	9	1002 (713–1298)	18.076	1.803	23.411	2.336
Dec	10	908 (700–1327)	14.804	1.630	20.915	2.303

Animals sacrificed during the months of January, February and March representing brumating sample of the species, did not show any significant change in their ovarian and oviducal weights from the December values. Class I and class II follicles also do not differ radically in number. However, class III follicles during the month of March showed marked numerical increase (Figure 17.3) and were recorded to be 6 per individual. The average number of follicles of 6 mm diameter and above in the month of March showed a increase in their number and were found to be 4 per ovary.

Table 17.2: Monthly Records of Three Size Classes of Ovarian Follicles in *Kachuga smithi*

Months	*Number of Females Studied*	*Ovarian Follicles (Average Number)*		
		Class I (> 15 mm)	*Class II (11–15 mm)*	*Class III (6–10 mm)*
Jan	4	1.25	0.50	1.50
Feb	7	0.28	0.28	2.85
Mar	5	0.60	1.60	6.00
Apr	6	1.16	2.16	5.33
May	3	0.00	1.66	12.66
Jun	6	0.00	3.16	14.16
Jul	5	12.00	8.20	5.60
Aug	13	15.07	3.92	5.46
Sep	7	12.42	1.71	2.57
Oct	4	6.50	1.50	1.50
Nov	9	6.50	0.55	1.44
Dec	10	1.00	1.60	2.10

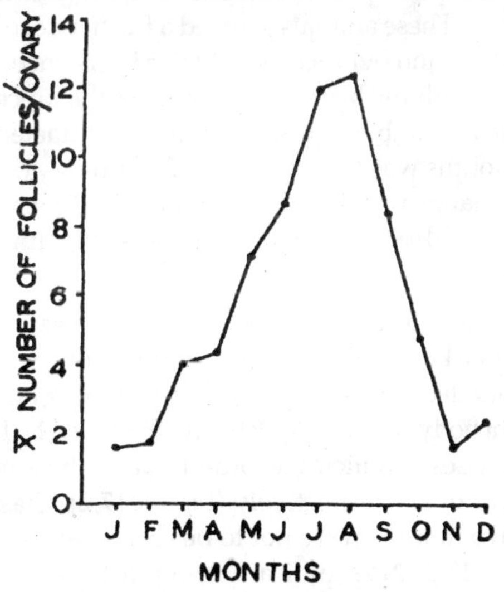

Figure 17.2: Mean Number of Follicles of 6 mm Diameter and Above per Ovary in *Kachuga smithi*

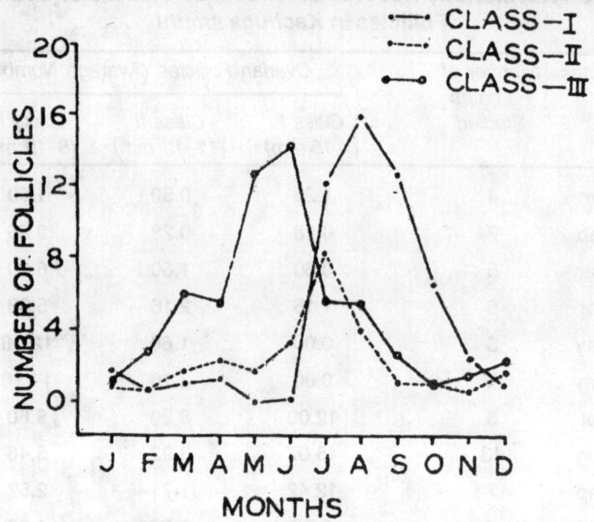

Figure 17.3: Mean Number of Class I, Class II, and Class III Follicles per Individual in *Kachuga smithi*

The April sample represented animals freshly emerged from winter brumation. These animals showed a further decrease in their ovarian (GSI = 1.15) and oviducal weight (1.53 per cent body weight) when compared with the brumating turtles of the preceding three months. The mean number of follicles of 6 mm diameter and above during these months was found to be 4.33 per ovary (Figure 17.3). No significant change was, however, found in the average number of follicles in any of three classes, when compared with the data for the month of March.

In the months of May and June while the ovarian weights did not show any marked change from the April values, the oviduct showed a further decrease in their weight, weighing now only 1.51 and 1.39 per cent body weight respectively (Figure 17.1). The average number of all classes of follicles recorded during these months was 7.16 and 8.66 per ovary respectively (Figure 17.2). Class I follicles during these two months were not to be seen in any of the turtles dissected (Figure 17.3). No significant change in the average number of Class II follicles was noticed, although Class III follicles showed a marked increase in their number and reached 12.66 in May and 14.60 in June per individual (Figure 17.3).

All turtles autopsied during the months of July and August contained pre-ovulatory follicles (= Class I follicles). The ovaries of the pre-ovulatory July turtles, showed a significant increase in their weight, and weighed 6.79 per cent body weight (Figure 17.1). The minimum weight of ovaries in a turtle in this month turned out to be 15.500 g and the maximum 55.00 g. The average number of follicles 6 mm diameter or greater was 11.9 per ovary (Figure 17.2). Class III follicles, showed a marked decrease in their number and fell to 5.6 per individual from 14.66 that in the previous month (Figure 17.3). On the other hand a marked increase in the number of class II and class I follicles was registered and were found to be 8.2 and 12 per individual respectively. In the pre-ovulatory August samples, a G.S.I. 7.79 was the highest recorded for the year (Figure 17.1). The weight of the ovary varied within a range of 19.010–84.100 g in the August sample. The mean number of ovarian follicles of 6 mm diameter and above per ovary during the month was 12.33 (Figure 17.2). Numerically class I follicles were the most numerous during this month. On an average, the number of class I follicles per individual was 15.07, while in the classes II and III, follicles decrease (Figure 17.3) numerically from that in the July sample to 3.92 and 5.46 per individual respectively. The oviduct during these months markedly increased in their weight (Figure 17.1). Percentage of oviduct in relation to body weight recorded during these months was 2.79 and 3.30 respectively.

Animals autopsied in the month of September showed a little decrease in their G.S.I. (7.46) in comparison to the August animals (Figure 17.1). During this month the oviducts were found to show the highest development, accounting for 3.55 per cent of the total body weight of the turtle. Eggs of 6 mm diameter and above were recorded on an average to be 8.35 per ovary. There was an overall shedding of numbers in all three classes of follicles in this month (Figure 17.3). The mean number of follicles of class I was 12.42, class II I. 71 and class III 2.57 per individual.

In both October and November samples, a significant decrease in their ovarian weight (Figure 17.1) was registered. In the month of October the minimum weight of the ovary was found to be 4.600 g and the maximum 18.500 g, while in November the minimum was 3.01 g and the maximum 18.300 g. Ovarian follicles of all the three classes per ovary for the two months averaged 4.75 and 2.22

respectively. The number of class I follicles showed a decrease from previous two months in numerical strength and was now only 6.5 (Figure 17.2) per individual respectively. However, class II follicles per individual during these two months were roughly equal to that of September samples. Class III follicles, however, showed a little decrease from the September turtles and was now 1.5 in October and 1.44 in November per individual (Figure 17.3). The oviduct showed a gradual decrease in its weight during these two months and weighed 3.08 and 2.33 per cent body weight respectively.

Courtship

In captivity *Kachuga smithi* showed the first signs of courtship and mating in the month of September. A sexually inclined pair (1 male and 1 female) was isolated and kept under observation in a separate aquarium. The temperature of the water was around 25°C ± 2°C, the pair was watched from 8 a.m onwards. At 1.00 p.m. when the water temperature was about 26°C, the male started pursuing the female partner. During the chase which lasted for about 10 minutes, the male repeatedly bites the female hindquarters especially at thighs. Thereafter, the male moved to her side, came ahead of her and turned back to lie face to face with the partner. In this position also the male was aggressive and resumed biting her now at fore legs and head. After lapse of another 10–15 minutes the male pulled in its head slightly, extend its forelegs and with palms outward, began titillating the sides of the female face and neck. After several series of titillation which lasted for 5 minutes, the male moved around her side back to rear of the female and attempted mounting her, with his hind feet on the posterior edge of her carapace, the fore limbs rested on or around the middle of her carapace. In 3–4 minutes it brought its tail under the tail of the female to effects coitum. The male made several more unsuccessful attempts to copulation, but finally broke contact at 1.45 p.m. The female remained passive throughout the event, only occasionally changing places. Thereafter no indication of sexual intention was exhibited by the pair for two days at an end, when the observation was withdrawn.

Ovulation and Egg Laying

As evidenced by the presence of shelled eggs in oviducts in 53.84 per cent of turtles collected after the second week of August, the first ovulation occurs during mid-August. Most females had already ovulated by the middle of November. No female collected after middle of November had any egg in their oviducts.

All specimens autopsied in September and October (Table 17.3) and none of late November and December sample possessed oviducal eggs, and that fresh corpora lutea were observed in all the specimens studied in the month of November, would tune oviposition around the months of late October and November.

Table 17.3: Monthly Incidence of Oviducal Eggs and Corpora Lutea in *Kachuga smithi*

% Sample with	*Jan*	*Feb*	*Mar*	*Apr*	*May*	*Jun*	*Jul*	*Aug*	*Sep*	*Oct*	*Nov*	*Dec*
Oviducal Eggs	–	–	–	–	–	–	–	53.83	100	100	11.11	–
Corpora Lutea	–	–	–	–	–	–	–	53.83	100	100	100	90

Eggs (Shape, Size and Weight)

The eggs of *Kachuga smithi* are variable in size, elliptical in shape, pinkish-white in colour and have a hard but fragile shells to the outside. A sample of 81 eggs ranged from 39–51 mm in length, with a mode of 44 and averaged 45.18 ± 3.9 mm in length, 23 to 29 mm in width, with a mode of 25 and averaged 24.43 ± 2.5 wide (Figures 17.4 and 17.5) and ranged from 11.900 to 16.908 g in weight at an average of 14.800 ± 2 (Table 17.4).

Table 17.4: Egg Characteristics in *Kachuga smithi*

Egg	*Minimum*	*Maximum*	*Mode*	*Average*	*SD ±*
Length (mm)	39	51	44	45.18	3.4
Width (mm)	23	29	25	24.43	2.5
Weight (g)	11.900	16.908	–	14.800	2.6

An abnormally small, yolkless egg, the runt, along with 5 normal eggs was recovered from a female dissected on 4 October. The runt was pure white with a very hard thick calcareous shell. It measured 25 mm long and 15 mm vide and weighed 3.900 g.

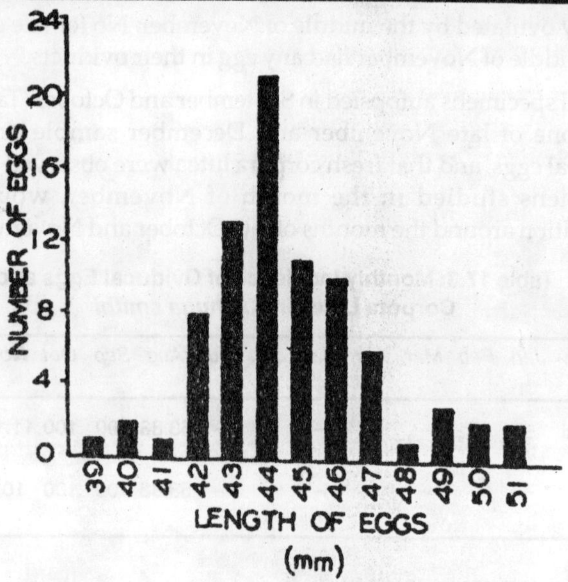

Figure 17.4: Frequency Distribution of Egg Length in *Kachuga smithi*

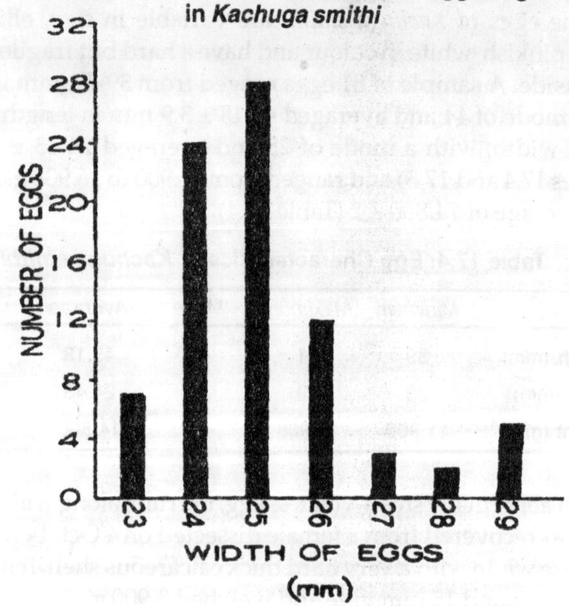

Figure 17.5: Frequency Distribution of Egg Width in *Kachuga smithi*

Clutch Size and Reproductive Potential

The clutch size of the turtle, *Kachuga smithi* is based on 17 counts of oviducal eggs. The number of eggs in each clutch varied from 3–11, the most frequent number, the mode was 6 (Table 17.5; Figure 17.6) average being 6.77 ± 2.1. More than 40 per cent of the females contained 6 oviducal eggs per individual. Clutch size showed no correlation with the size of the layer (Figure 17.7; r = 0.17; y = – 0.02 + 0.03 x).

Table 17.5: Minimum, Maximum, Mode and Average Clutch Size of *Kachuga smithi*

Number of Individuals	*Number of Eggs Obtained*	*Number*		*Mode*	*Average*	*SD ±*
		Minimum	*Maximum*			
17	93	3	11	6	6.17	2.1

Reproductive potential in 23 turtles showed a range of 9 to 26 (Table 17.6) at an average of 16.65 ± 4.54. Reproductive potential did not show positive correlation with size of the laying female (Figure 17.8; r = 0.16; y = 4.65 + 0.06 x)

Hatchling

After the incubation for 72 days at 23–32°C temperature only four out of 40 eggs hatched successfully in the laboratory. These eggs pertained to the group of eggs incubated at a temperature of 30 ± 2°C.

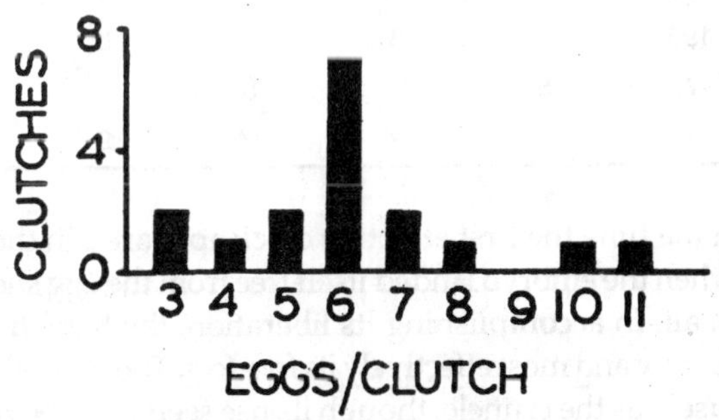

Figure 17.6: Frequency Distribution of the Number of Eggs per Clutch in *Kachuga smithi*

Table 17.6: Maximum Reproductive Potential of *Kachuga smithi*

Sl.No.	*Plastron Length (mm)*	*Number of Oviducal Eggs (No)*	*Number of Corpora Lutea (NCL)*	*Number of Eggs Laid (NCL-No)*	*Number of Large Follicles (> 15 mm)*	*Total Reproductive Potential*
1	173	3	3	–	12	15
2	176	6	12	6	11	23
3	191	6	6	–	11	17
4	160	8	8	–	7	15
5	210	–	10	10	6	16
6	194	5	5	–	15	20
7	189	–	–	–	12	12
8	190	–	–	–	12	12
9	190	–	–	–	9	9
10	197	7	7	–	18	25
11	190	7	7	–	14	21
12	192	6	6	–	11	17
13	195	6	6	–	10	16
14	199	10	10	–	16	26
15	195	6	6	–	16	22
16	173	4	4	–	11	15
17	176	–	–	–	14	14
18	180	6	6	–	12	18
19	192	6	6	–	12	18
20	187	11	11	–	4	15
21	193	3	3	–	5	8
22	177	5	10	5	1	11
23	215	–	14	14	2	16

From the time the first effective crack appeared in the shell to the time when the embryo landed itself free from the egg shell it took 6 hours in all. In accomplishing its liberation, the hatchlings used most frequently and most effectively its fore feet. The only other organ found in use was the cruncle, though its use seemed to be only very little. The hatchling carried with it a yolk sac, which measured 5 mm in diameter. The yolk was completely absorbed after 7 days of

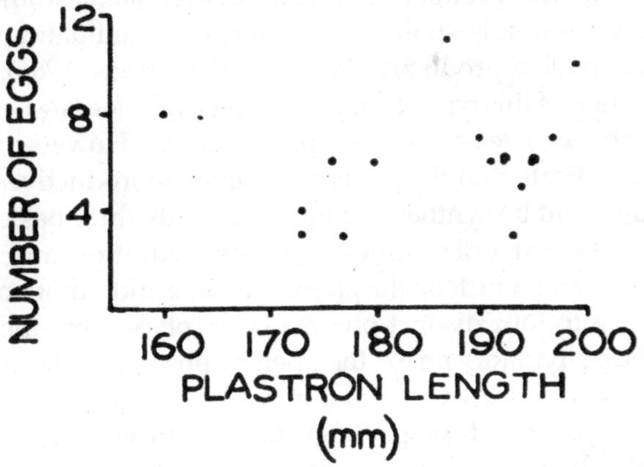

Figure 17.7: Relationship Between Plastron Length (Abscissa) and Number of Eggs per clutch (Ordinate) in *Kachuga smithi* (r = 0.17; y = – 0.02 + 0.3x)

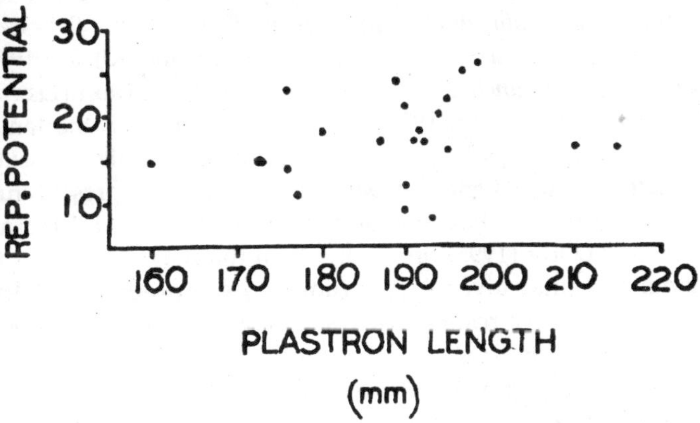

Figure 17.8: Relationship Between Plastron Length (Abscissa) and Maximum Reproductive Potential (Ordinate) in *Kachuga smithi* (r = 0.16; y = 4.6 + 0.06x)

its hatching. The cruncle disappeared after 11 days after its emergence from the egg. Colour pattern on the shell of the hatchling was a copy of its adult colour pattern.

Discussion

The female reproductive cycle in *Kachuga smithi* studied from Jammu, is essentially similar and confirms the annual type of the reptilian female reproductive cycle of Saint Girons (1966). That in the exhibition of this type of a reproductive cycle, *Kachuga smithi* fall in line with the cycle typical of temperate reptiles. However, the data obtained indicate that the pattern of female reproductive cycle in this turtle could be synthesized into essentially three basic events; (1) the follicular enlargement, (2) the ovulation and (3) the oviposition, and which on the premise of time and duration, can be separated into four distinct phases. These phases are: the resting phase, the preparatory phase, the ripening phase, and the ovulatory phase, which follow one another in sequence and reflect changes that occur in ovaries during different times of the year, unless phases or events themselves are drastically offset by very adverse environmental conditions.

A perusal of the published literature indicates that turtles exhibit varying number of sexual phases during their annual reproductive cycle. Three phases in turtle, *Chrysemys picta* (Gibbons, 1968), four in, *Kinosternon subruburum hippocrepis, K. flavescens flavescens, Sternotherus carinatus, S. odoratus* (Mahmood and Klicka, 1972), *Lissemys punctata punctata* (Gupta and Duda, 1998) and six in *Lissemys punctata granosa* (Singh, 1977) have been reported in the female gonadal cycle. In exhibiting only four readily recognizable sexual phases, the seasonal female sexual cycle in turtle of the present study corresponds well with that of the Mahmood and Klicka (1972) and Gupta and Duda (1998), in all of which the major events of the gonadal cycle may involve (*i*) recrudescence or preparatory phase of (*ii*) a quiescent or resting ovary, succeeded by (*iii*) maturation or ripening of the follicles, which eventually (*iv*) ovulate and get extruded before the ovary relapses into quiescence.

Number of clutches laid, clutch size and morphology of eggs in turtles is well documented by Moll (1979), in "Turtle: Perspectives and Research". Morphologically, the eggs of *Kachuga smithi* are elliptical in shape and coincides with the general morphological characteristics of the eggs of other species of the same family emydidae (Moull, op. cit.).

The data on clutch size in *Kachuga smithi* presented here from Jammu compare well with that reported by Minton (1966) from

Pakistan ($x = 7.0$; range 4–10), an area of similar ecology, which only emphasizes a simple correlation between ecological similarity and reproductive strategy of a form. When compared for clutch size with other emydid turtles (Cahn, 1937; Cagle, 1950; Minton, 1966; Gibbons, 1968; Ernst and Barbour, 1972; Iverson, 1977) reveals that an average clutch size varies from 2–13 (in more than 63 per cent of species worked out) which is possibly the model size for these turtles within which the clutch size of *Kachuga smithi* also falls. However, when compared with the average clutch size in other groups of turtles especially chelydridids (Ash, 1951; White and Murphy, 1973), kinosternids (Einem,1956; Mahmood and Klicka,1972), and chelonids (Simon *et al.*, 1975), it emerges that, the clutch size in emydine forms is generally higher. What selective advantage could such a reproductive strategy offer to these over the other groups of turtles is not immediately clear.

Since the body size ultimately limits the number of eggs that can be produced, a positive correlation between body size and clutch size is not surprising. In fact the significant increase in clutch size with body size suggests that greater energy is expanded for reproduction in older females as expected by current theory (Willams, 1966; Gadgil and Bosert, 1970; Tinkle *et al.*, 1970). Whether this increase represents proportionally greater reproductive efforts is uncertain. A lack of correlation between these parameters in some turtles, *Chrysems picta* (Cagle,1954), *Pseudemys scripta elegans* (Gibbons and Tinkle, 1969), *Kinosternon subruburm hippocrepis, K. flavescens, Sternotherus odoratus,* and *S. carinatus* (Mahmood and Klicka, 1972) and in *Kachuga smithi* of the present study is meaningful.

In absence of nesting data from individual turtles, the number of clutches laid per year by the turtles in the present investigations were estimated by techniques outlined by Christiansen and Dunham (1972) and Gupta (1979). The evidences accumulated from observations on corpora lutea, oviducal eggs and enlarged follicles is greatly in favour of maximum of two clutches in *Kachuga smithi* with 6.17 ± 2.1 eggs per clutch, the annual minimum reproductive potential for these turtles is thus, $6.17 \times 2 = 12.32$. This value do not differ significantly from the maximum reproductive potential of 16.65 ($x^2 = 0.318$, $p > 0.5$) in *Kachuga smithi.* While there is no published evidence to indicate that *Kachuga smithi* exhibit a similar reproductive potential elsewhere in their distribution range, two or more clutches per year is, however, not a unique feature to this turtle alone. A similar trend has been reported for a number of turtle species (Smith, 1931; Powell, 1967; Deraniyagala, 1939; Webb, 1961, 1962; Carr, 1952; Thomas, 1972; Cagle, 1950; Tinkle, 1958; Legler, 1960; Plummer, 1977) and nearly all members of the family chelanidae and dermochelidae (Carr, 1950). Although, a higher number of clutches has been generally correlated with a warmer climate (Richmond, 1945; Christiansen and Dunham, 1972), two clutches per year appears to be a model number for most of the emydid turtles that have been so far worked out (Cagle, 1950, 1954; Legler, 1960; Powell, 1967).

Acknowledgements

The author take this opportunity of thanking Dr. P.L.Duda, former Prof. and Head, P.G. Department of Zoology, University of Jammu, Jammu for his advice and guidance during this investigation and to CSIR, New Delhi for financial assistance.

References

Altland, P.D. 1951. Observations on the structure of the reproductive organs of the box turtle. *J. Morphol.*, 89: 599–621.

Ash, R.P. 1951. A preliminary report on the size, egg number, incubation period, and hatching in the common snapping turtle, *Cheiydra serpentina. Va. Acad. Sci.*, 2: 312.

Cagle, F.R. 1950. The life history of the slider turtle, *Pseudemys scripta troostii* (Holbrook). *Ecol. Monogr.*, 20: 31–54.

Cagle, F.R. 1954. Observations on the life cycles of painted turtles (genus Chrysemys). *Am. Midl. Nat.*, 52: 225–235.

Cahn, A.R. 1937. The turtles of Illinois. *Illionis Biol. Monogr.*, 35: 1–218.

Carr, A.F. 1952. *Handbook of Turtles.* Comstock Publ. Assoc. Ithacas, N.Y., pp. 542.

Christiansen, J.L. and Dunham, A.E. 1972. Reproduction of the yellow mud turtle (*Kinosternon flavescence flavescence*) in new Mexico. *Herpetologica*, 28: 130–137.

Cox, A.W. and Marion, K.R. 1978. Observations on the female reproductive cycle and associated phenomena in spring dwelling populations of *Sternotherus minor* in North Florida (Reptilia: Testudines). *Herpotologica*, 34(1): 20–30.

Deraniyagala, P.E.P. 1939. *The Tetrapod Reptiles of Ceylone, Vol. 1: Testudinates and Crocodilians*. Dulan. Co. Ltd., London, 32: 412.

Duda, P.L. and Gupta, V.K. 1978. A case of gonadal atrophy in *Lissemys punctata punctata* (Reptilia: Testudines). *Herp. Review*, 9(2): 46.

Duda, P.L. and Gupta, V.K. 1981. Courtship and mating behaviour of the Indian soft shell turtle, *Lissemys punctata punctata*. In: *Proc. Indian Acad. Sci. Anim. Sci.*, 99(4): 453–461.

Einem, G.E. 1956. Certain aspects of the natural history of the mud turtle, *Kinosternon bauri*. *Copeia*, 1956: 186–188.

Ernest, C.H. 1971. Sexual cycles and maturity of the turtle, *Chrysemys picta*. *Biol. Bull.*, 140: 191–200.

Ernest, C.H. and Barbour, R.W. 1972. *Turtles of the United States*. The Univ. Press of Kentucky, pp. 347.

Gadgil, A.M. and Bosert, W.H. 1970. Life historical consequences of natural selection. *Am. Nat.*, 104: 1–24.

Gibbons, J.W. 1968. Reproductive potential, activity, and cycles in the painted turtle, *Chrysemys picta*. *Ecology*, 49(3): 399–409.

Gibbons, J.W. and Tinkle, D.W. 1969. Reproductive variation between turtle population in a single geographic area *Ecology*, 50: 340–341.

Gupta, V.K. 1979. Studies on the female reproductive organs of some chelones from Jammu. *Ph.D. Thesis*, University of Jammu, Jammu, pp. 352.

Gupta, V.K. 2002. Some aspects on the ecology of Indian soft-shelled turtle, *Lissemys punctata punctata* (Bonnaterre) from J&K State. In: *Perspectives in Animal Ecology and Reproduction*, (Eds.) V.K. Gupta, J.P. Sharma and Anil K. Verma. Daya Publishing House, New Delhi, p. 160–165.

Gupta, V.K. 2004. Histogensis and development of corpus luteum in fresh water turtles. In: *Perspectives in Animal Ecology and Reproduction*, Vol. 2, (Eds.) V.K. Gupta and Anil K. Verma. Daya Publishing House, New Delhi, p. 287–310.

Gupta, V.K. and Duda, P.L. 1998. Female reproductive cycle of the Indian freshwater turtle, Lissemys punctata punctata. In: *Proc. International Conf. Biology and Conservation of the Amphibian and Reptiles of South Asia*, p. 177–188.

Iverson, J.B. 1977. Reproduction in freshwater and terrestial turtles of northern Florida. *Herpetologica*, 33(2): 205–212.

Legler, J.M. 1960. Natural history of the ornate box turtle, *Terrapene ornate ornate* Agassiz. *Publ. Mus. Nat. Hist. Univ. Kans.*, 11: 527–669.

Mahmood, I.V. and Klickaj. 1972. Seasonal gonadal changes in Kinosternid turtles. *J. Herpetol.*, 6: 183–189.

Minton, S.A. 1966. A contribution to the herpetology of West Pakistan. *Bull. Am. Mus. Nat. Hist.*, 134: 27–184.

Moll, E.O. 1979. Reproduction and Development. In: *Turtles: Perspectives and Research*, (Eds.) Marion and Morlock. A Willey-Interscience Publication, John Willey and Sons Inc., N.Y., pp. 625.

Moll, E.O. and Legler, J.M. 1971. The life history of a neotropical slider turtle, *Pseudemys* scripta (Schoepff) in Panama. *Bull. Los. Angles Co. Mus. Nat. Hist. Sc.*, 11: 1–102.

Plummer, M.V. 1977. Reproduction and growth in the turtle, *Trionyx muticus. Copeia.*, 1977(3): 440–447.

Powell, C.B. 1967. Female sexual cycles of *Chrysemys picta* and *Clemmys insculpta* in Nova Scotia. *Can. Field Nat.*, 81(21): 134–140.

Richmond, N.D. 1945. Nesting habits of the mud turtle. *Copeia*, 1945: 217–219.

Saint-Girons, H. 1966. Le cycle sexual des serpents venimeus. *Mem. Inst. Butantan. Simp. Internac.*, 33: 105–114.

Simon, M.H., Ulrich, G.F. and Parker, A.S. 1975. The green sea turtle (*Chelonia mydas*): mating, nesting and hatching on a foam. *J. Zool. Lond.*, 177: 411–423.

Singh, D.P. 1974. Analysis of environmental factors regulating the gonadal cycle in tropical pond turtle, *Lissemys punctata granosa* (*Schoepff*). *Experientia*, 30: 967–968.

Singh, D.P. 1977. Annual sexual rhythm in relation to environmental factors in a tropical pond turtle, *Lissemys punctata granosa*. *Heroetologica*, 33(20): 190–194.

Smith, C.L. 1931. *The Fauna of British India, including Ceylone and Burma*. Taylor and Francis, London, pp. 188.

Thomas, K. 1972. The annual cycle of reproduction of the emydid turtle, *Pseudemys floridana floridana* (Testudinata: Testudinae) with observation on its ecology. *M.Sc. Thesis*, Auburn University Albama, USA, pp. 57.

Tinkle, D.W. 1958. The systematics and ecology of the *Sternotherus carinatus* complex. *Tulane Stud. Zool.*, 6: 3–56.

Tinkle, D.W., Wilbour, H. and Tilley, S. 1970. Evolutionary strategies in lizard reproduction. *Evolution*, 24: 55–74.

White, J.B. and Murphy, G.G. 1973. The reproductive cycle and sexual dimorphism of the common snapping turtle, *Chelydra serpentina serpentina*. *Herpetologica*, 29: 240–246.

Verma, A.K. 1992. Studies on the male reproductive cycles of some emydid turtles of Jammu. *Ph.D. Thesis*, University of Jammu, Jammu, pp. 169.

Verma, A.K., Sahi, D.N. and Gupta, V.K. 1999. Observations on the nesting of the Ganges soft-shelled turtle *Aspideretes gangeticus* cuvier (Chelonia: Trionychidae) at lake Mansar in Jammu Shiwaliks, J&K State. *J. Ecobiol.*, 11(2): 137–143.

Webb, R.G. 1961. Observations on the life histories of turtles (genus *Pseudemys* and *Graplemys*) in lake Texoma, Oklahoma. *Amer. Midl. Nat. Hist.*, 13: 29–128.

Webb, R.G. 1962. North American recent soft shelled turtles (family: Trionychidae). *Univ. Kansas Publ. Mus. Nat. Hist.*, 13: 429–611.

Williams, G.C. 1966. *Adaptation and Natural Selection*. Princeton, N.J. Univ. Press, Princeton.

Chapter 18

Synopsis of Reproductive Characteristics of Freshwater Male Emydid Turtles, *Kachuga tectum* and *Kachuga smithi* from Jammu, J&K State

Anil K. Verma

Head, Department of Zoology, Government P.G. College, Udhampur, J&K

The present work was undertaken in a bid to assess the architecture, qualitative and quantitative testicular changes in respect of testis size, weight, histology and seasonal changes in testes and epididymis in relation to the environmental factors during different months of the year. The data on these parameter was taken for the four consecutive years (1989–1991) and collated to present as a combined data for different months of one year only.

The turtles studied (*Kachuga tectum* and *Kachuga smithi*) are the two freshwater emydid, hard shelled turtles which are found inhabiting the freshwater, lotic water bodies in Jammu, J&K State (Map). These turtles are strict sympatriates (Gupta, 2002, 2004; Verma, 1993, 2002a and 2002b) and undergo a winter hibernation (December to February in *K. tectum* and November to February in *K. smithi*) every year. Their occurrence during the remaining months of the year does not show any drastic variation.

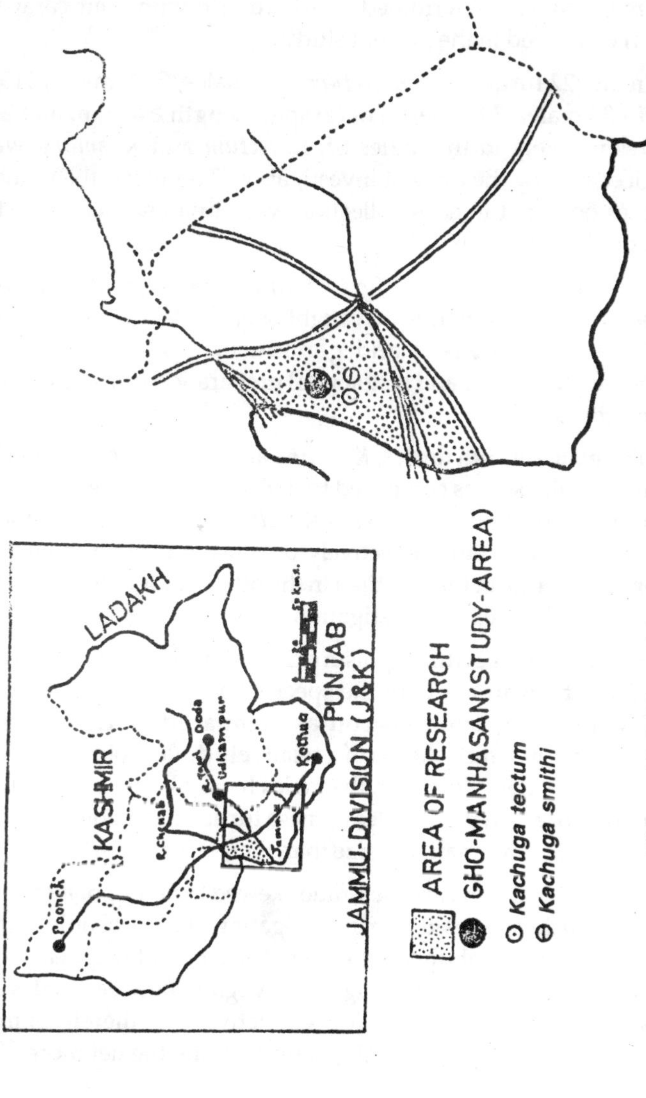

Figure 18.1: Map of Jammu Region Showing Study Area

In view of restriction on these reptiles because of conservation laws, Wildlife Act (Protection) – Shedule I, only a maximum not more than 3 or 4 per month and not less than 2 permonth (only mature specimens determined in accordance with their carapace length) were used in the present study.

In all 222 turtles (109 *K. tectum* – 36 males, 73 females; 113 *K. smithi* – 34 males, 79 females) of carapace length ≥ 60 mm and ≥ 56 mm respectively in the males of, *K. tectum* and *K. smithi,* were sacrificed during the present investigation. Therefore, the number of males obtained in each collection was obviously, less than the expected 1 : 1 sex-ratio.

The data on macroscopic (gravimetric) parameters of the gonads (testes) was taken from freshly, caught open turtles of either species, after which the gonads and gonoducts were processed for subsequent sectioning and histological preparations through routine microtechniques.

The males in both turtles, *K. tectum* and *K. smithi,* on an average attain a smaller size as compared to their female counterparts. The minimum maturity size in males of *K. tectum* and *K. smithi* was found to be 60 mm and 56 mm respectively, possibly indicating an intrinsic lower growth rate in this sex than in the other; under the conditions obtaining in the area of investigation.

A distinct sexual dimorphism exists in the turtles *K. tectum* and *K. smithi.* The males in both the species have a brightly coloured shell with a concave plastron as compared to their female counterparts, in which shell is of a relatively duller colouration and the plastron is flat. Moreover, the males have a larger tail length as compared to that in the females of these turtles. Additionally, males in these turtles are smaller in size than their female members.

Present study reveals a sex-ratio skewed in favour of females of both the turtle species under investigation. The probable factors leading to a greater abundance of females over males appears to be a difference of activity rate or agility or vagility between males and females; the females *per se* being less active and, therefore, more vulnerable to netting than males, which escape the net more often than not.

The testes in both the species in present study reflect a general plan of their architecture, shape and topography like in other turtles.

The testes are cream yellow, oval or elliptical bodies with a reddish tinge during the active phase of testicular cycle. The annual variation in the histomorphological changes concur to a type of testicular cycle, where spermatogenesis occurs in summer, attains a peak in late-summer. Spermiogenesis takes place in mid-summer and mating is a late-summer or fall event.

In their general makeup, the testes of both *K. tectum* and *K. smithi* are similar not only to each other and the reptiles in general, but also are similar to the mammalian pattern. Convoluted seminiferous tubules are embedded in an interstitial stroma, which shows mammalian type of leydig cell distribution. These turtles are remarkable in exhibiting a mantle of glandular tissue during regressed phase round the tubules like in urodeles and certain fishes on one hand, and the subtunic leydig cells beneath the *tunica albuginia* like some lizards (*Cnemidophorus, Agama*) on the other hand.

Within the seminiferous tubules, there is present a permanent germinal epithelium constituted by spermatogonia, spermatocytes (primary and secondary), spermatids, spermatozoa and the non-germinal sertoli cells, which vary quantitatively according to the stage of seminiferous tubule cycle.

There is no overall zonal distribution of germinal stages. The individual germ cells do, however, show a graded radial axis of maturation within each tubule, so that the spermatozoa are the most centrally located (Figures 18.2–18.5) and the spermatogonia the most peripherally located germ cells.

A feather like arrangement of spermatozoa characteristic of other turtles is also displayed by the active tubules of the testes in both the turtles (Figure 18.6D). Of all the cellular elements of the seminiferous tubules, only spermatogonia are consistently found throughout the year within the tubule; while all other germinal elements are seen in active testes. Inter-tubular spaces may be filled with interstitial tissue, fibrocytes, blood vessels and blood cells.

The turtle *K. smithi* like most other turtles, exhibits a single summer-peak in testicular cycle but *K. tectum,* on the other hand, shows a departure from the generalized type in exhibiting another spermatogenic activity, though of a very lower order in fall (pre-hibernal). However, the second spermatogenic activity is abortive as no spermatozoa are produced and the activity culminates because of onset of winter hibernal period.

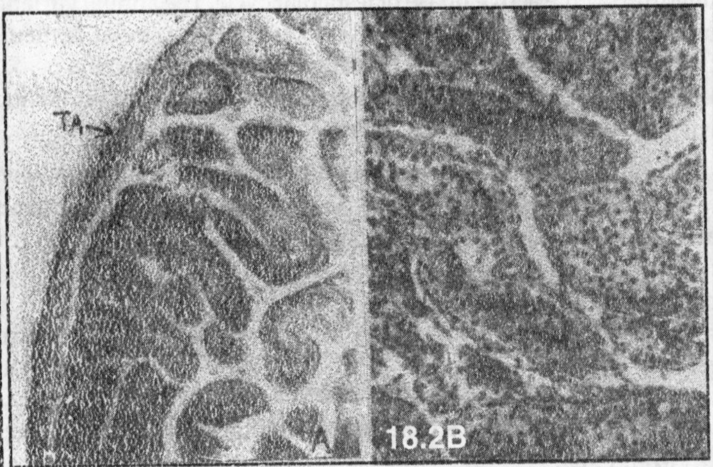

Figure 18.2A–B: Cross Section of Testis of *K. tectum* Showing *tunica albuginea* (TA) and Seminiferous Tubules (x 200)

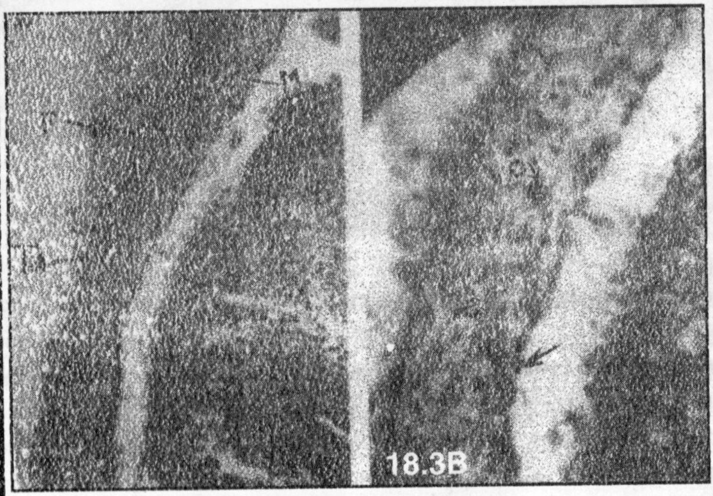

Figure 18.3A–B: Enlarged View of Testis of *K. tectum* Showing Elastic Fibres (F) and Smooth Muscle Cells (M) in (TA) *tunica albuginea* (x 600)

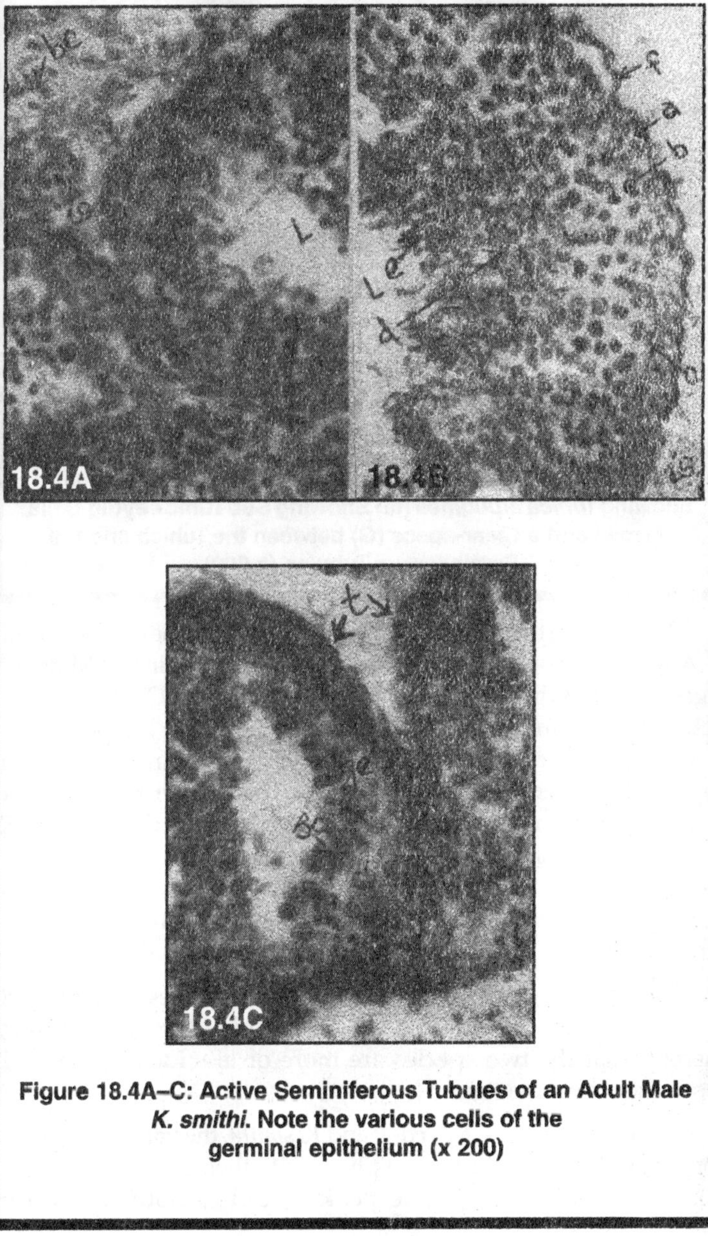

Figure 18.4A–C: Active Seminiferous Tubules of an Adult Male *K. smithi*. Note the various cells of the germinal epithelium (x 200)

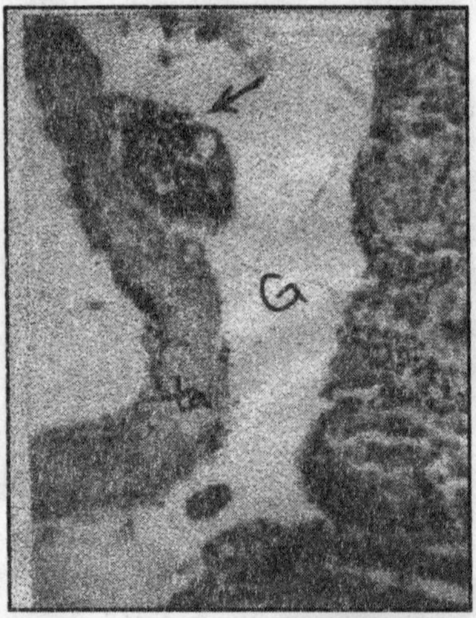

Figure 18.5: A Portion of Cross Section of Testis of *K. smithi* Showing *tunica albuginea* (ta) Showing Sub Tunic Leydig Cells (Arrow) and a Clear-space (G) between the Tunica and the Seminiferous Tubules (x 600)

The peak in testicular activity in *K. smithi* reaches in the month of August. The spermatogenesis in this turtle begins in May and regression starts in September, prior to hibernation. Contrarily, in *K. tectum,* the spermatogenesis begins, a month earlier but peak (as in *K. smithi*) is attained in August and regression begins, similarly, in September but is of a shorter duration (September-mid-October) against a longer regression phase (September–April) in case of *K. smithi.* On the basis of the present investigation, it could be, therefore, generalised that "the emydids are seasonally active and lacks an extended / or continued activity period as otherwise, reported in their kins".

The spermiogenesis, however, in both the turtles, *K. tectum* and *K. smithi,* occurs only once in a year (in July–August) indicating thereby, that the two species are more or less identical in their reproductive behaviour.

In both the turtles, *K. tectum* and *K. smithi,* the values of gonado somatic index (GSI) are well in phase with their testicular size and absolute mean weight. The peak in GSI in both the turtles

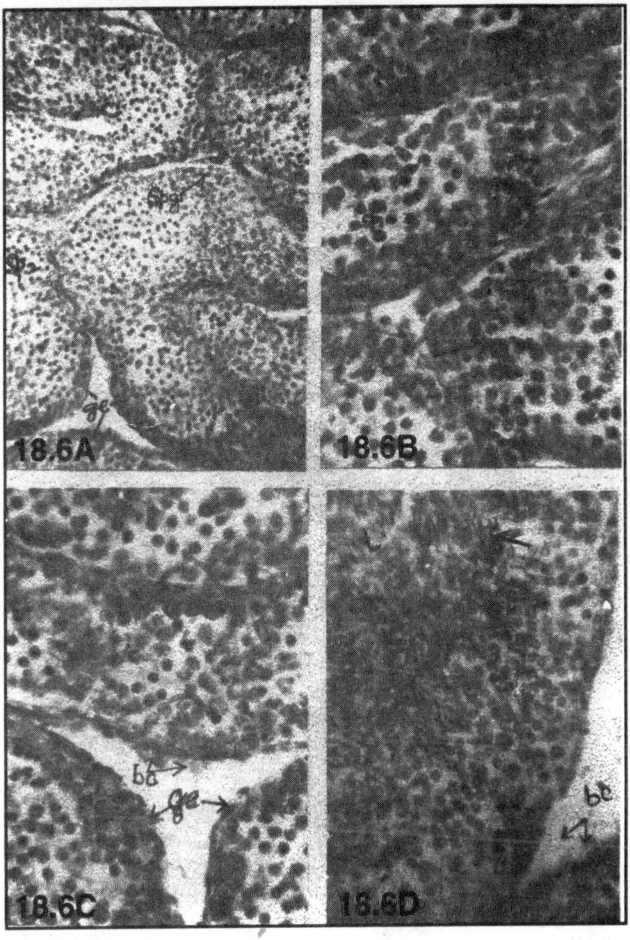

Figure 18.6A–C: Another View of the Cross-section of Testis of *K. smithi* Displaying the Active Seminiferous Tubules with a Germinal Epithelium Containing Division Products of the Spermatogonia Placed in Tiers. Note the sperms (e) arranged in bundles facing the lumen (l) (x 200)

Figure 18.6D: Cross-section Through Testis of *K. smithi* Showing Germinal Epithelium (ge) of Seminiferous Tubules, Blood Cells (bc) in the Intertubular Spaces and Feather like Arrangement of Sperms (↑) (x 450)

corresponded well to maximum average weight of the testes coinciding fairly with the period of active spermatogenesis. This functionally, unimodal late (Figure 18.7) summer peak in GSI in these two sympatric emydid turtles is not unique to these species but a more widespread phenomenon as discussed above.

Present study reveals that the Interstitial cells are very sparsely distributed within the intertubular spaces of the testes of both *K. tectum* and *K. smithi*, and are not easily detectable. Some cells are present even beneath the tunica (Figure 18.5) and other form a mantle round seminiferous tubules like those in Urodeles and some fishes. A healthy and fully developed interstitium in the month following peak spermatogenesis coincides a high secretory activity of these

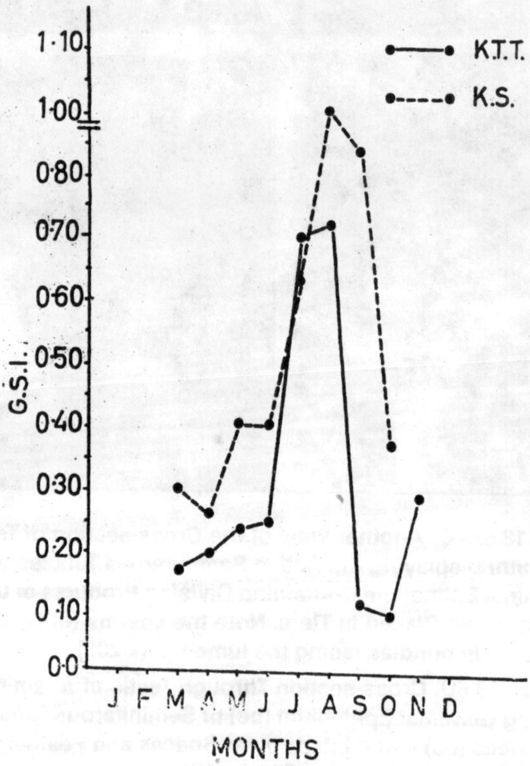

Figure 18.7

cells, producing the androgens which evokes a mating urge in these turtles during autumn or fall. This temporal dissociation between the histology and function of endocrine tissue during sexual cycle is, probably, common in reptiles, as also maintained by Licht and Pearson (1978).

The peak in spermatogenesis in both the turtles like the other turtles in general, is obtained round July/August, sperms are seen in the epididymis from August onwards (till October) and the lumen of the tubules are empty after the emergence of turtles from hibernation presumably, because of mating (Gupta, 1979), which in these turtles is an autumn or fall event.

The sexual cycle of these two turtles, therefore, conforms to a basic pattern of pre-nuptial type of cycle as reported for all the other reptiles (turtles included), studied to date from India but, however, differs from all the temperate zone turtles (Moll, 1979) which are typically post-nuptial.

Interspecific comparison of the testicular cycles between the turtles, *K. tectum* and *K. smithi* indicates concordance in some aspects and discordance in others. Although some variations are found in May samples of *K. tectum*, most males exhibits initiation of spermatogenesis in the start of this month. Initiation of spermatogenesis in *K. smithi*, however, do not occur until late May or early June, unlike that found in *K. tectum*.

Peaks in testes weight and active spermatogenesis occurs in August in both these species and follow the pattern as already reported for several other turtle populations (Ernst, 1971a, 1971b; Moll, 1973; McPherson and Marion, 1981) studied elsewhere.

Testicular regression in *K. tectum* first occurs in September and it continues to decline a month or so prior to hibernation (*i.e.*, October) such that the pre-hibernal testes are a little heavier and but a loss of a lesser magnitude occurs during hibernation so that the turtle after emergence have regressed testes. Whereas, in *K. smithi* also regression is initiated in September and testes continue to be regressed until the animal enters hibernation.

It seems, therefore, logical to presume that the male reproductive cycle in these turtles like reptiles in general (Licht, 1973) is influenced by the complex interactions of temperature, photoperiod and internal rhythms. The interspecific differences in the reproductive cycles of

K. tectum and *K. smithi*, and the differential nature of responses exhibited by these turtles to similar set of environmental cues reflects differences in the relative control of gonadal cycles by internal rhythms in concert with the environmental influences and not by the extrinsic factors alone.

Interspecific variations in the reproductive cycle of these species, probably, reflects differences in the extent to which each of the above mentioned factors is involved and the relative roles they play in controlling different phases of testicular cycle. However, some influence of other factors like rain fall, humidity, food supply and/or stored energy cannot be ruled out.

Notwithstanding the differences in their form and to some extent habits, the pattern of male reproductive cycle in the emydid turtle species *viz., Kachuga tectum* and *Kachuga smithi*, investigated during present study is essentially similar to and conforms to the annual type of male reproductive cycle of Saint-Girons (1966).

Asynchrony is, nevertheless, evident among the major events of the cycle in the two turtles. In the emydid turtle, *K. tectum* the recrudescent phase of the testicular cycle is a couple of months in advance of the similar events in the other emydid turtle (*K. smithi*), although the peak spermatogenic activity in both the species (Figure 18.8) remains temporarily in phase *i.e.,* in the month of August.

Recrudescent phase in the two turtles is, however, out of phase with one another, being pre-hibernal in *K. tectum* and post-hibernal in *K. smithi*. The onset of spermatogenic activity seems related to rising temperature, photoperiod and/or food availability.

The onset of testicular regression again seems to be an outcome of interaction between extrinsic and intrinsic factors. However, the onset of regression phase and refractoriness although temporally in phase in the two turtles, differ in their duration.

No sooner the spermiogenesis is complete in these turtle species, (*K. tectum* and *K. smithi*) spermiation occurs (in late-summer) in these turtles, and only a single crop of spermatozoa in an year, is produced.

In the seminiferous tubules of these turtles, individual germ cells show a graded radial axis of maturation so that spermatozoa are the most centrally and spermatogonia, the most peripherally located germ-cells.

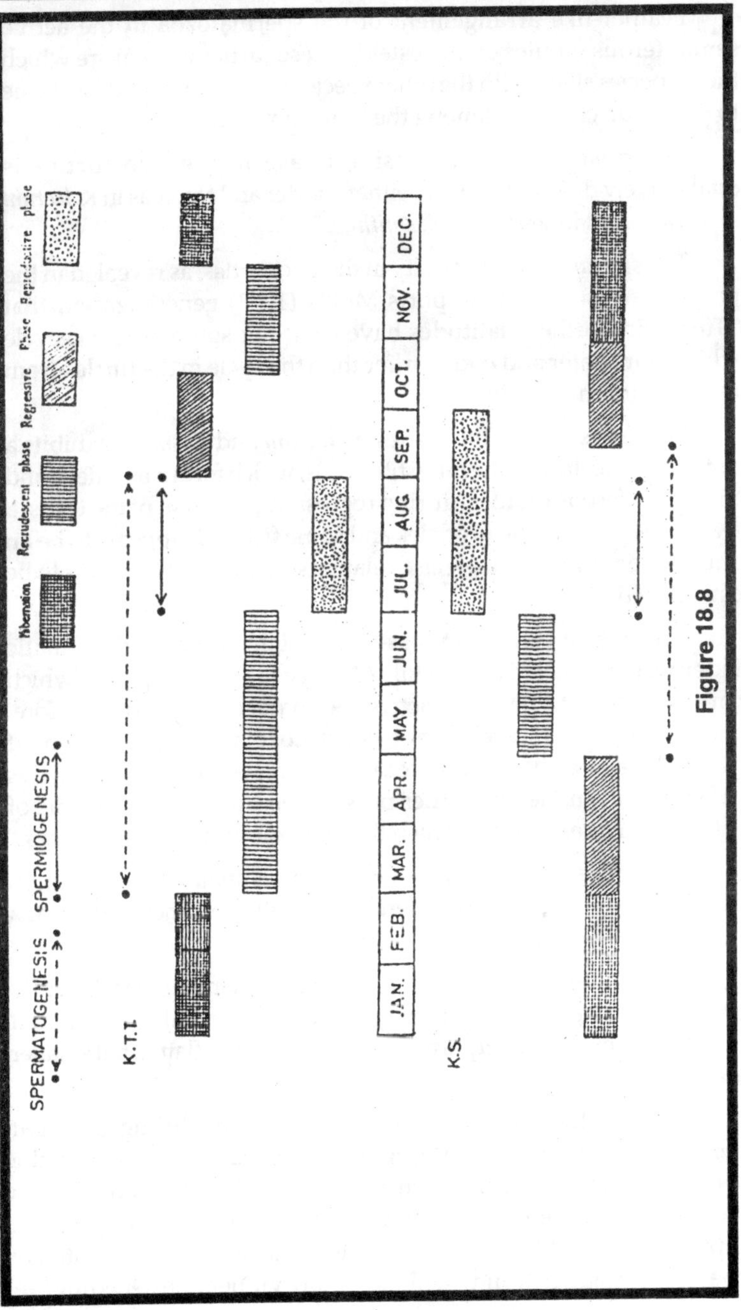
SPERMATOGENESIS
SPERMIOGENESIS
Hibernation
Recrudescent phase
Regressive Phase
Reproductive phase
K.T.T.
JAN.
FEB.
MAR.
APR.
MAY
JUN.
JUL.
AUG.
SEP.
OCT.
NOV.
DEC.
K.S.

Figure 18.8

Feather-like arrangement of the spermatozoa in the active seminiferous tubules of the testes in these turtles is a feature which these species share with the other species of the turtles and is unique to the group chelonia, among the living amniotes.

The post-nuptial regressive phase in the two turtles is qualitatively different in being either shorter and faster as in *K. tectum* or slower and longer as in *K. smithi.*

The spermatogenic activity in the two turtles, as revealed in the present investigation supports Moll's (1973) generalization that "Turtles in northern latitudes have a shorter spermatogenic cycle which starts later and ends earlier than the cycle in the turtles from more southern latitude".

The testes in both the turtles, *K. tectum* and *K. smithi* exhibits a remarkable feature of the Interstitial cells, which form a mantle round the seminiferous tubules. during regressed condition of the testis, a feature peculiar to the urodeles and some fishes. It appears to be an aquatic adaptation, the precise relationship of which remains to be established.

The highest values of testes-size, weight and Gonado Somatic Index were registered during July / August in both the species, which corresponds to the period of active spermatogenesis, and, therefore, these parameters are in perfect accord with the period of spermatogenic activity. However, the spermatogenic activity extends differentially in the two turtles possibly because of the outcome of differential interaction between the extrinsic and intrinsic factors.

In the males of *K. tectum* sexual maturity is attained at a minimum carapace length of 60 mm, whereas, *K. smithi* becomes mature at a minimum of 56 mm carapace length.

Males of turtles are invariably smaller than females, brightly coloured, have a concave plastron and a larger tail as compared with their female counterparts in which plastron is flatter, tail smaller and colour dim.

The females usually outnumber the males during different months, the sex-ratio being roughly 1 male : 2 females in both the species. The factor causing such a ratio biassed in favour of females has been attributed to greater vagility of the males who escape capture more readily than their female members which are relatively heavier and sluggish and, therefore, more vulnerable to netting.

The testicular cycle of *K. tectum and K. smithi* show minor differences in their macroscopic as well as microscopic details and also in chronology of events, triggering different phases of the cycle, despite of the fact that both the turtles are sympatric in habit and exposed to the similar environmental cues.

In final conclusion, neither environmental nor inherited cues/ factors alone seen sufficient to regulate the reproductive cycles in the male turtles and it may be the totality of extrinsic and intrinsic factors that influences the testicular cycles in these emydids in particular and turtles in general.

References

Duda, P.L. and Kaul, O.N. 1976. Subtunic leydig cells in the testes of *Agama tuberculata* (Gray). *J. Herpetol.*, 5: 539–541.

Ernst, C.H. 1971a. Sexual cycle and maturity of the turtle, *Chrysemys picta. Biol. Bull. Mar. Biol. Lab.* Woods. Hole, 140(2): 191–200.

Ernst, C.H. 1971 b. Population dynamics and activity cycles of *Chrysemys picta* in south eastern Pennsylvania. *J. Herpetol.*, 5: 151.160.

Goldberg, S.R. and Lowe, C.H. 1966. The reproductive cycle of the western whiptail lizard (*Cnemidophorus tigris*) in Southern Arizona. *J. Morphol.*, 118: 543– 548.

Gupta, V.K. 1979. Studies on the female reproductive cycle of some Chelones from Jammu. *Ph.D. Thesis,* University of Jammu, Jammu, pp. 352.

Gupta, V.K. 2002. Histomorphological studies on the oviduct of the fresh water turtles, *kachuga tectum* and *kachuga smithi*. In: *Perspectives in Animal Ecology and Reproduction,* (Eds.) V.K. Gupta, A.K. Verma and J.P. Sharma. Daya Publishing House, New Delhi, 1: 282–291.

Gupta, V.K. 2004. Histogensis and development of corpus luteum in freshwater turtles. In: *Perspectives in Animal Ecology and Reproduction,* (Eds.) V.K. Gupta and A.K. Verma. Daya Publishing House, New Delhi, 2: 287–310.

Licht, P. 1973. Thermal and photic influences on reptilian reproduction. *Excerpta. Med. Int. Cong. Series,* 273: 185–190.

Licht, P. and Pearson, A.K. 1978. Cytophysiology of the reptilian pituitary. *Int. Rev. Cytol. Supplement*, 7: 239–286.

Moll, E.O. 1979. Reproductive cycles and adaptations. In: *Turtles: Perspectives and Research*, (Eds.) M. Harlers and R.H. Morlock. John Wiley and Sons, New York, p. 305–331.

McPheroon, R.J. and Marion, K.R. 1981. Seasonal testicular cycle of the stinkpot turtle (*Sternotherus odoratus*) in Central Albama. *Herpetologica*, 37: 33–40.

Verma, A.K. 1993. Studies on the male reproductive cycles of some emydid turtles of Jammu. *Ph.D. Thesis*, University of Jammu, Jammu, pp. 169.

Verma, A.K. 2002a. Interstitial cell cycle of some freshwater hardshelled turtles (family Emydidae) from Jammu (J&K) State. In: *Perspectives in Animal Ecology and Reproduction*, (Eds.) Gupta *et al.* Daya Publishing House, New Delhi, 1: 250–258.

Verma A.K. 2002b. On the histological and cytological elements of the testes of *Kachuga tecta* (Gray) and *Kachuga smithi* (Gray) from Jammu. In: *Perspectives in Animal Ecology and Reproduction*, (Eds.) Gupta *et al.* Daya Publishing House, New Delhi, 1: 282–291.

Abbreviations Used

a: Spermatogonia.
b: Primary Spermatocytes.
c: Secondary Spermatocytes.
d: Spermatids.
e: Spermatoza.
f: Fibrocytes.
ta: Tunica albuginea.
ge: Germinal epithelium.
ve: Vasa-efferentia.
tw: Tubular wall.
s: Sertoli Cells.
i: Interstitial Cells.
C: Connective tissue.
bv: Blood Vessels.
bc: Blood Cells.
t: Seminiferous tubules.
l: Lumen.
D: Debris.
is: Interstitial space.
cc: Columnar Cells.
ss: Secretory strands.

Chapter 19

The Physiology of Hypothalamic Pituitary Gonadal Axis: The Recent Perspectives

Charanjit Kaur Dhanju
Department of Zoology and Fisheries,
Punjab Agricultural University, Ludhiana – 141 004

Introduction

Like all other physiological phenomena, mammalian reproduction is regulated by central nervous system (CNS) through coordinated control over the secretion of sex hormones and sex behaviour. The neural mechanisms involved in the regulation of gonadotrophin secretion from pituitary and sex hormones from gonads as well as sexual behaviour are located in the hypothalamus (Kaur, 1978) The gametogenic and endocrine functions of the gonads are both under the integrative control of the CNS. Gonadotrophins; follicle stimulating hormone (FSH) and luteinizing hormone (LH) secreted by the anterior pituitary are important regulators of reproductive processes in male as well as female. The hormone secretion from the anterior pituitary is controlled by the brain via chemical messengers released from the hypothalamus into hypophyseal portal vessals. The primary regulatory component of the CNS for reproductive activities is the hypothalamic peptide luteinizing hormone releasing hormone (LHRH) – a decapeptide produced by neurons located in the hypothalamic-pre-optic region. After synthesis, LHRH is transported to nerve terminals located in the median eminence, where it is released in close proximity to portal capillaries and then transported to anterior pituitary through hypophyseal portal circulation. It acts on specific membrane receptors of gonadotrophin producing cells of the anterior pituitary and stimulates both gonadotrophins, LH and FSH secretion and thus is generally called gonadotrophin releasing hormone (GnRH). These gonadotrophins in turn affect gonadal function. Now considerable evidence (Nigro Vilar *et al.*, 1982; Kalra and Kalra, 1987) indicates that central neurons play a major role in regulating the LHRH neuronal systems. Even, it is now clear that hypothalamic-pituitary axis is controlled not only by classical neuro-transmitters, but also by a large series of newly discovered peptides; amongst which specific role is played by opiods. Leukotrienes; the recently discovered novel compounds belonging to the arachidonic acid metabolites, are also known to exert effects on the gonadotrophic axis at the hypothalamic level. The release of LHRH may thus be under the control of more than one compound. It is also well established that hormonal secretions of the gonads are the primary signals that derive and sustain the intricate balance within the closed feedback loop of the hypothalamo-hypophyseal-gonadal axis. Various studies on hypothalamo-hypophyseal-gonadal axis have led to the concept that negative feedback of gonadal steroids at hypothalamic as well hypothyseal level is a major factor modifying gonadotrophin secretion.

Although the principles of GnRH secretion have been established for over two decades (Weiss *et al.*, 1989, 1991) the regulation of GnRH secretion via sex steroids and non-steroidal factors in normal reproduction (*i.e.*, menstrual cycle), different stages of life cycle (*i.e.*, gonadotrophin amplification during puberty) and pathophysiologic states (*i.e.*, polycystic ovary syndrome) is a complicated and incompletely understood phenomenon (Seminara and Crowley, 2001) Patients with single gene mutations within hypothalamic-pituitary-gonadal axis have provided unique avenues to increase our understanding of reproductive neuro-endocrinology. Genetically engineered animal models have further elucidated the roles of specific hormones in reproductive functions by confirming and extending human findings establishing species-species phenotypes. The combination of both animal and human tools have proven to be powerful, revealing paradoxes of reproductive biology and unexpected differences between species. For example, 20 years ago, a naturally occurring mouse model of hypogonadotrophic-hypogonadism (hpg mouse) was described with low levels of LH and FSH, sexual immaturity and infertility which was due to a deletional mutation in the distal half of the gene encoding GnRH and GnRH associated peptide (GAP). The paradoxes of reproductive neuro-endocrinolgy are thoroughly described and the gaps present are highlighted in this article.

Hypothalamo-Hypophyseal Axis

Harris (1955) formed the concept that hormone secretion from the anterior pituitary is controlled by the brain via chemical messengers released from the hypothalamus into hypophyseal portal vessels. At the anatomical level, mediobasal hypothalamus (MBH) was demonstrated as an important brain area in this function. It was also suggested that tuberniofundibular neurons with cell bodies in the arcuate nucleus (AN) and fibres in the median eminence (ME) could represent morphological correlate for this control. The primary regulatory component of the CNS for reproductive activities is the hypothalamic peptide, luteinizing hormone releasing hormone (LHRH), a decapaptide produced by neurons located in the hypothalamic-pre-optic region. Thus the GnRH neurons form the final central pathway regulating reproduction (Sullivan and Moenter, 2004). After synthesis, LHRH is axonally transported in nerve terminals located in the median eminence (ME) where it is released in close proximity to portal capillaries. The decapeptide is then transported to the anterior pituitary (adenohypophysis) through the hypophyseal portal circulation where acting on specific membrane receptors of its gonadotrophin producing cells, it stimulates LH and FSH secretion. These gonadotrophins in turn affect gonadal function. It is now clear from the evidences accumulated during the last few decades that LHRH plays a major modulatory role in the regulation of gonadal function by acting at different levels within the hypothalamic-pituitary gonadal axis. The secretion of LHRH seems to be under the influence of multifactorial regulatory mechanisms in which numerous extra-cellular messengers may participate including the catecholamines (Negro-Vilar *et al.*, 1979), peptides (Valenca *et al.*, 1985) and sex steroid hormones (Kalra and Kalra, 1984).

The Role of Neuroactive Substances in Hypothalamo-Hypophyseal Functions

Neurotransmitters

Considerable evidence indicates that central noradrenergic (and perhaps also adrenergic) neurons play a major role in regulating the LHRH neuronal system. Interference with normal noradrenergic neuronal activity in both rodents and subhuman primates results in an inhibition of LH release (Negro-Vilar *et al.*, 1982; Kalra and Kalra, 1983). Other monoaminergic pathways including the epinephrine (E), norepinephrine (NE), dopamine (DA), and serotonin (S) neuronal systems, as well as the cholinergic system have also been implicated in the regulation of LHRH and LH release (Negro-Vilar *et al.*, 1982; Kalra and Kalra, 1983; Morell *et al.*, 1985). The monoamine regulation of gonadotrophin secretion points towards direct excitatory effect of catecholamines on LHRH release from the hypothalamus at least during the LH surge. There are many evidences that catecholamines play an important role in the regulation of anterior pituitary function. Pharmacological agents which modify synthesis, release or degradation of brain catecholamines also affect the release of pituitary hormones (Scharrer, 1969). Moreover, NE enhances LH release by stimulating hypothalamic LHRH secretion (McCann *et al.*, 1978). It has also been suggested that catecholaminergic neurons play a role in mediating the events leading to the surge of prolactin (PRL) and LH that occurs in rats on the afternoon of pro-estrous before ovulation (Weiner and Ganong, 1978). Intraventricular injections of DA and NE are affective in inducing ovulation which infers that dopaminergic and noradrenergic mechanisms activate the release of pituitary ovulating hormones (Sato and Kimura, 1981).

Neuropeptide Y

The report that neuropeptide Y (NPY) coexisted adrenergic transmitters in some neurons in the brain stem (Hokfelt *et al.*, 1984) suggested the possibility that NPY and adrenergic transmitters may be coreleased and coact to exite LH release. The source of NPY in the hypothalamus appeared to be NPY immunopositive perikarya in the arcuate (ARC) and periventricular nucleus and cells in the brain stem (Hokfelt *et al.*, 1984; Chronwal *et al.*, 1985; Kalra *et al.*, 1991; Crowley *et al.*, 1999). These NPY-containing cells innervate the preoptic-tuberal pathway *i.e.*, in the proximity of hypophyseal portal veins in the ME, which have been previously implicated in the control of LH release and also in the projection site of LHRH (King *et al.*, 1982) and steroid concentrating neurons (Pfaff, 1980). Parallel changes in NPY and LHRH levels in the ME in association with the LH surge (Kalra *et al.*, 1987) also raised the possibility of a simultaneous discharge of LHRH and NPY into the portal circulation for an action at the level of the pituitary. McDonald *et al.* (1985) reported increased LH release when NPY was delivered to anterior pituitary cells in an *in vitro* perfusion system. However, Kerkerian *et al.* (1985) did not observe any affect of NPY on LH release in the pituitary cell incubation system. Crowley *et al.* (1987) observed that addition of NPY *in vitro* to hemipituitaries from ovariectomized rats only marginally stimulated the release of LH. However, when added with LHRH, NPY significantly potentiated the LHRH-induced LH release, but on its own failed to activate LH release from dispersed pituitary cells in culture.

Kalra *et al.* (1987) suggested that NPY may have dual sites of action, one within the hypothalamus and the other at the level of pituitary gonadotrophs and that under the physiological conditions characterized for LH release, NPY may be released in the hypothalamus to induce LHRH secretion, and into the hypophyseal portal veins, to potentiate the LHRH-induced LH response and thus a two-pronged action of NPY; stimulation of LHRH release in concert with adrenergic transmitters in the hypothalamus and stimulation of LH release in concert with LHRH at the level of the pituitary, may be important in evoking the preovulatory and ovarian-induced LH surges lasting for several hours. Horvath *et al.* (2001) have also shown the role of NPY in LH release.

Endogenous Opioid Peptides (EOPs)

It is now clear that the hypothalamic-pituitary axis is controlled not only by the classical neurotransmitters (*e.g.*, norepinephrine, dopamine, serotonin etc.), but also by a large series of newly discovered peptides. Amongst these peptides, a special role is certainly played by the opiods. However, EOPs are the best example of hypothalamic neuropeptides shown to exercise an inhibitory influence on LH release in gonadectomized and intact rats (Kalra and Leadem, 1984). Of the three classes of EOPs whose distribution in the hypothalamus has been described in detail (Watson *et al.*,

1982), β-endorphin neurons are anatomically best suited to interact with LHRH neurons (and β-endorphin readily suppressed LH release in gonadactomized rats and blocked preovulatory LH release (Leadem and Kalra, 1985). Dynorphin, another EOP abundant in the hypothalamus but with a distribution pattern distinct from that of β-endorphin (Watson *et al.*, 1982) suppressed LH release to a lesser degree (Leadem and Kalra, 1985). The enkephalins are ineffective but the methionine-enkephalin analogs that presumably undergo rapid inactivation after intracerebroventricular injections, produced a sustained decrease in plasma LH levels in ovariectomized rats and also blocked ovulation (Leadem and Kalra, 1985). It is possible that each of these EOPs may employ specific opiate receptor types and evidences have been presented that receptors may be preferentially involved in the control of LH release (Pfeiffer *et al.*, 1983). The possibility that EOP or opiate-induced decrease in LH release may primarily be a result of diminution in the LHRH output has not been experimentally evaluated. In fact morphine infusion *in vitro* failed to decrease hypothalamic LHRH release (Kalra *et al.*, 1987). However, the observations that morphine blocked the progesterone induced increase in the MBH LHRH levels (Kalra and Simpkins, 1981), the opiate receptor antagonist, naloxone, stimulated in vitro hypothalamic LHRH release (Leadem *et al.*, 1985; Kalra *et al.*, 1987) and when microinfused in the proximity of LHRH neurons in the hypothalamus, naloxone increased LH release (Kalra, 1981), suggest that EOP release in the hypothalamus suppress LHRH release. On the other hand, contrary to previous reports (Cicero *et al.*, 1977), it was later shown that morphine and β-endorphin not only decreased basal LH release from cultured pituitary cells but also attenuated the LHRH-induced LH release response (Cacicedo *et al.*, 1986; Blank *et al.*, 1986). It seems that EOP may have a dual site of action; one in the hypothalamus to inhibit LHRH release and the other at the level of the pituitary to attenuate the LHRH-induced LH release.

In spite of many experimental evidences suggesting that the EOP neurons may be hypothalamic inhibitory neural circuit, the physiological significance of EOP in the regulation of gonadotropin secretion is imprecise. EOP may participate in the dynamics of the hypothalamic neural circuitry, which controls episodic basal and preovulatory LH release in two possible ways. (1) EOP may mediate the inhibitory feedback effects of gonadal steroids on LH release

(Kalra, 1983). This view implies that gonadal steroids may monitor the inhibitory opioid tone of secretion either regulating EOP output in the vicinity of LHRH neurons or by modulating EOP receptor function. (2) EOP may exert an inhibitory tone on LH release but the stimulatory feedback action of gonadal steroids may curtail the inhibitory influence to allow preovulatory LH surge to occur on pro-estrous. Evidences have also been provided for the view that morphine inhibits the synthesis of dopamine in tuberoinfundibular dopaminergic neurons (Alper *et al.*, 1980). Reymond *et al.* (1983) have supported the view that morphine inhibits both the release and synthesis of dopamine but is more effective in inhibiting the release than synthesis of dopamine. Kalra and Hovarth (1998) revealed neuroendocrine interactions between galanin, opioids and neuropeptide Y in the control of reproduction.

Leukotrienes

Leukotrienes are newly discovered novel compounds belonging to the arachidonic acid metabolites (Samuelsson, 1983). Evidence has been obtained that leukotrienes C4 (LTC4) can be produced in slices from the rat brain under stimulation with ionophore and in the presence of arachidonic acid, with the highest concentrations in the hypothalamus (Lindgren *et al.*, 1984). Localization of LTC4 in CNS by immunoreactivity (IR) has shown LTC4-IR nerve fibres in the external layer of the median eminence (Lindgren *et al.*, 1984). Thus the distribution of LTC4-IR fibres is similar to the one described for LHRH-IR fibres in the median eminence (Barry *et al.*, 1986). Hutting *et al.* (1985) have revealed that these fibres, to a large extent are identical which suggests that LTC4-like compound is present in LHRH neurons and that the two compounds may be released together. It has also been observed that LTC4 in very low concentrations can cause LH release from dispersed pituitary cells in culture and this had a more rapid onset than LHRH induced release and was of a short duration (Hutting *et al.*, 1985). The stimulation of LHRH release at very low concentrations of leukotrienes suggested that these may also exert effects at hypothalamic level. (Gerozissis *et al.*, 1986).

Pulsatile Secretion of GnRH

GnRH neurons are the central regulators of fertility and are controlled by steroid feedback but the control mechanisms for this are poorly understood. GnRH release must occur in discrete pulses separated by periods of little to no secretion to maintain pituitary function because continuous elevation of GnRH suppresses gonadotrophin synthesis and secretion (Belchetz *et al.*, 1978) causing interfertility Modulation of the frequency of GnRH release is equally critical as GnRH pulse frequency is a primary determinant of diffrential pituitary function. Therefore pulse patterns must vary across the reproductive cycle to differentially regulate the release of the two gonadotrophins; luteinizing hormone (LH) and follicle stimulating hormone (FSH) (Wildt *et al.*, 1981; Marshall and Griffin, 1993). Low GnRH pulse frequency favours. FSH, whereas high pulse frequency stimulates LH (Wildt *et al.*, 1981). This change in relative levels of gonadotrophin is critical to proper ovarian follicular maturation. Progesterone, produced during the luteal phase of the cycle, provides negative feedback to lower GnRH/LH pulse frequency (Leipheimer *et al.*, 1984; McCartney *et al.*, 2002); Low frequency GnRH pulses continue into the early follicular phase, allowing an FSH rise that is crucial for ovarian follicular recruitment and maturation. Androgens also provide feedback to GnRH neurons. The existence of a hypothalamic pulse generator has been proposed for the pulsatile secretion of GnRH determining the pituitary gonadotrophin secretion (Hotkiss and Knobil, 1994). The observation of episodic gonadotropin releasing hormone (GnRH) secretion from immortalized murine GnRH neurons (Martinez *et al.*, 1992; Wetsel *et al.*, 1992) and the recent demonstration of pulsatile GnRH release during cultured GnRH neurons from the primate and rat embryonic olfactory placode (Terasawa *et al.*, 1999; Funabashi *et al.*, 2000) pointed to the existence of a pulse generator intrinsic to GnRH neurons. In addition in explants of the retrochiasmatic rat hypothalamus, where no immunochemically GnRH cell bodies could be observed, the secretion of GnRH was typically pulsatile *in vitro*, suggesting the existence of a pulse-generating network extrinsic to the GnRH neurons (Pumelle *et al.*, 1998). Pulsatile GnRH secretion occurred from the first postnatal day and progressively increased in frequency till the age of 25 days, before the onset of puberty (Yanamaka *et al.*, 1999; Bourguignon *et al.*, 1990, 1992). However, no pulsatile GnRH secretion could be observed with foetal explants obtained on day 20–21 of gestation (Yanamaka *et al.*, 1999).

Gamma aminobutyaic acid ($GABA_A$) receptors were found to mediate a potent inhibition of the GnRH pulse generator in the juvenile primate *in vitro* (Mitsushima *et al.*, 1994; Terasawa *et al.*, 1999) and in prepubertal rat hypothalamus *in vitro* (Bourguignon *et al.*, 1997) GABAegric inputs play both inhibitory as well as stimulatory effect on GnRH release (Hales *et al.*, 1994 Kusano *et al.*, 1995). It has also been shown that increased biosynthesis of glutamate from glutamine through glutaminase could contribute to an increase in frequency of pulsatile GnRH secretion between 15 and 25 days (Bourguignon *et al.*, 1995). Parent *et al.* (2005) studied the facilitatory and inhibitory mechanisms possibly accounting for the early postnatal occurrence of pulsatile GnRH secretion and concluded that in the foetal rat hypothalamus, GnRH inhibitory auto feedback and GABAergic inputs do not account for the absence of pulsatile GnRH secretion *in vitro*. A low rate of glutamate biosynthesis from glutamine is a possible limiting factor. Oxytocin (posterior pituitary hormone synthesized by neurons of the hypothalamus) and PGE2 can play a facilitatory role in the postnatal, occurrence of pulsatile GnRH secretion.

Effect of Gonadal Steroids on Hypothalamo-Pituitary Axis

It is now well established that hormonal secretions of the gonads are the primary signals that derive and sustain the intricate balance within the closed feedback loop of the hypothalamo-pituitary gonadal-axis. Moore and Price (1932) proposed that a tight reciprocal relationship between the gonad and the pituitary was responsible for maintaining gonadal functions in the adult rat (Figure 19.1). Intensive investigations in the later years have confirmed the broad outline of this gonad-pituitary "push-pull" operation. The secretion of LH and FSH is generally maintained to a low basal range in intact rats, and in the absence of gonadal feedback signals, as after gonadectomy, gonadotrophins are secreted at augmented rates. It is now apparent that gonadal steroids do not inhibit LH release by directly restraining the release of LH from the pituitary gonadotrophs but they monitor the response of gonadotrophs to LHRH. It is also suggested that a combined milieu of estradiol-17 β (E_2) and progesterone (P) in female rats diminish, while E_2 in the absence of P, facilitate the LHRH induced LH release. Hohlweg and Junkman (1932) proposed that the site of gonadal steroid action in the regulation of reproductive function might be the central nervous system (Figure 19.1). This concept was later supported by Harris

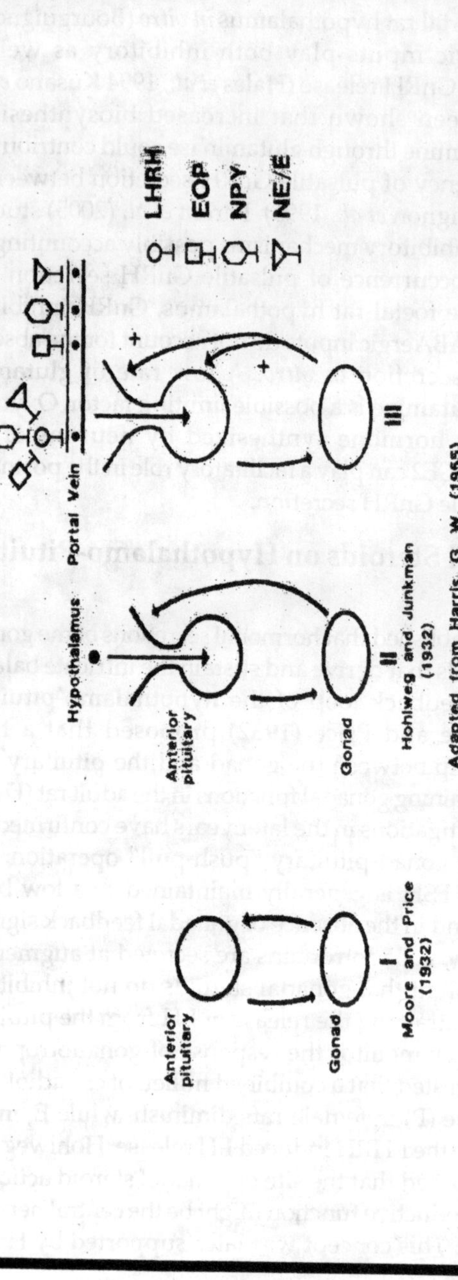

Figure 19.1: Diagrammatic Representation of the Reciprocal Relationship Between the Gonads and Anterior Pituitary as Advanced by Moore and Price (1932); Between the Gonads, Hypothalamus and Pituitary as Advanced by Hohlweg and Junkman (1932) and Supported by Harris (1955); and the Recent View of the Facilitatory (+) Nature of Gonadal Steroids on a Variety of Hypothalamic Neuropeptidergic and Classical Aminergic Neuro-transmitters in the Control of Gonadotropin Secretion. For details see text, LHRH: Luteinizing hormone releasing hormone; EOP: Endogenous opioids peptides; NPY: Neuropeptides Y; NE: Orepinephrine; E: Epinephrine (Kalra *et al.*, 1987 b).

(1955) who demonstrated that neurohormonal substances of hypothalamic origin were discharged into the hypophyseal portal circulation for passage to the pituitary where in they stimulated gonadotrophin secretion. The chemical composition of the hypothalamic neurohumoral substance was elucidated by Matsuo *et al.* (1971) and LHRH is the primary hypothalamic neural signal responsible for controlling LH release from the pituitary gonadotrophs.

Biosynthesis of gonadal steroids is regulated primarily by stimulatory action of pituitary gonadotrophins, while gonadotrophin release is subject to powerful inhibitory control by gonadal steroids and their metabolites. During sexual maturation, increased secretion of gonadotrophins stimulates the gonads to produce progressively increasing amounts of steroids without inhibiting further secretion of LH and FSH. As a result of this decrease in the sensitivity of hypothalamic pituitary system to negative steroid feedback, the concentrations of both pituitary gonadotrophins and gonadal steroids rise in concert, increasing gradually from low (prepubertal) to higher (adult) levels. Numerous studies showed that much lower doses of steroids are required to prevent post-castrational rise in peripheral LH and FSH levels in prepubertal as compared to adult individuals thus providing direct evidence for resetting of the sensitivity to steroid negative feedback during sexual maturation. Indeed, this change in the response of the hypothalamic-pituitary system to gonadal steroids is believed to represent one of the mechanisms of sexual maturation in human and animals, "the gonadostat theory" (McCann *et al.*, 1974).

Two different mechanisms have been proposed to regulate LH secretion patterns; a long loop feedback of estradiol-17β (E) from the ovaries (Brinkley, 1981) and (or) short loop for internal feedback of pituitary hormones themselves (Docke and Glaser, 1971). In some species, the increase in circulating estrogen to a critical level in late follicular phase elicits a surge release of LH by a positive feedback mechanism (Miyake *et al.*, 1977). Concannon *et al.* (1979) observed that the preovulatory surge of LH in the female dog may be a response to withdrawal of the negative feedback affect of estrogens and this withdrawal might be facilitated by a preovulatory increase in progesterone. The localization of LH in the subcellular fraction of the hypothalamus (Emanele *et al.*, 1985) led to the speculation on the role of brain based LH. It is proposed that hypothalamic LH may serve as mediator of the short loop feedback mechanism therein regulating release of LH from the anterior pituitary by intraneuronal communication between LH and GnRH containing cells (Emanele *et al.*, 1985).

Effects of Steroids on GnRH Secretion

The stimulatory and inhibitory effects of gonadal steroids upon the LHRH neuronal systems are thought to be mediated by interactions of the steroids with a variety of central neurotroansmitters systems. The best studied among these systems are the central monoaminergec neuronal system that innervate the hypothalamic-preoptic region; the main areas where LHRH neuronal cell bodies, fibres and terminals are located (Kalra and Kalra, 1984). Steroidal feedback regulates GnRH secretion but the central mechanisms for steroid feedback are poorly understood. Steroids may act directly on GnRH neurons, via activation of nuclear receptors expressed in these cells. Convincing evidence *in vivo*, however, exists only for expression of the beta isoform of estradiol receptor in rats (Hrabovszky *et al.*, 2000, 2001); in mice, this message has been reported (Herbison and Pape, 2001), but the protein not yet been detected (Hrabovszky *et al.*, 2002). Likewise neither androgen (Herbinson *et al.*, 1996) nor progesterone receptors (Skinner *et al.*, 2001) have been detected in GnRH neurons except in small subpopulation of cells in the guinea pig (King *et al.*, 1995). However, the GnRH neurons comprise the final common pathway of an estrogen-sensitive pattern generator controlling fertility.

Estradiol targets many functions of the central nervous system including cognition cell survival, behaviour and reproduction (McEwen *et al.*, 1997). The GnRH neurosecretory system provides a unique model for examining central estradiol actions because steroid feedback effects on this are well studied at the whole-animal level. GnRH neurons play a pivotal role in reproduction by forming the final common pathway for the central control of fertility. GnRH release is distinctly patterned with pulses of secretion separated by prolonged periods of little detectable hormone release (Levine *et al.* 1985; Moenter *et al.*, 1991) and GnRH initiates secretion of the gonadotropins LH and FSH (Levine *et al.*, 1991; Padmanabhan and McNeilly, 2001).

Estradiol is an important feedback regulator of GnRH release LH pulses and a volleys of hypothalamic electrical activity, two markers of GnRH secretion, and GnRH release itself are modulated by estradiol (Kaur and Mangat, 1978; Karsch *et al.*, 1987; Moenter *et al.*, 1991; Mori *et al.*, 1991; Evans *et al.*, 1994; Kato *et al.*, 1994). As there is no evidence for estrogen accumulation (Shivers *et al.*, 1983) or receptor expression (Herbison and Theodosis, 1992; Lehman and Karsch, 1993) in GnRH neurons, it suggests that estradiol feedback is conveyed to GnRH neurons via estrogen sensitive afferents (Herbison, 1998). Recent studies using more sensitive methods indicate GnRH neurons expressing estrogen receptor binding suggesting the possibility of a direct action (Skynner *et al.*, 1999; Harbovszky *et al.*, 2001). Thus both direct and transsynaptic mechanisms may cooperate in the regulation of GnRH release. Recently developed transgenic mice have permitted positive identification of living GnRH neurons through green fluorescence protein (GFP) expression (Spergel *et al.*, 1999; Suter *et al.*, 2000). Nunemaker *et al.* (2002) determined the estradiol (E) effects on GnRH neuron firing patterns in ovariectomized and E-treated adult transgenic mice and indicated that estradiol-sensitive afferents modulate long term episodic firing patterns of GnRH neurons and the estradiol feedback is enforced via GABAergic and or glutamatergic afferents.

Alternately, steroid feedback may be conveyed to GnRH neurons transsynaptically via steroid sensitive neurons. A substantial body of work supports the amino-acid neurotransmitters, gamma aminobutyric acid (GABA) and glutamate action on $GABA_A$, a subtype of GABA receptors ($GABA_AR$) as a mediator (Herbison 1998). GABAergic afferents provide a major anatomical and functional input to GnRH neurons Sim *et al.*, 2000; (Jensen *et al.*, 2003; Pompolo *et al.*, 2003; Sullivan *et al.*, 2003, Sullivan and moenter 2004) and hypothalamic GABAergic neurons are known to express progesterone (Leranth *et al.*, 1991) and estrogen (Flugge *et al.*, 1986) receptors. The role of $GABA_AR$ activation in the control of GnRH neurons is controversial. Inhibition is suggested by observations of an inverse correlation between GABA and LH levels in rats and sheep (Jarey *et al.*, 1988; Herbison *et al.*, 1991; Robinson *et al.*, 1991) and increased LH release or advanced puberty following hypothalamic application of $GABA_A$ antagonists or blockade of GABA synthetic enzymes in rats and primates (Adler *et al.*, 1986;

Lamberts *et al.*, 1983; Kimura and Jinnal, 1994; Terasawa *et al.*, 1999). In rat explants treated *in vitro*, both inhibition and stimulation of GnRH release in response to $GABA_AR$ antagonism have been noted, depending on age (Felder *et al.*, 1996) or explant boundary (Donose *et al.*, 1992). In above studies, it is important to bear in mind that in these studies the afferent neuronal circuitry controlling GnRH release is also subjected to the pharmacological treatment; thus the possibility that the inhibition or stimulation observed may be indirect cannot be eliminated. In this regard, it is of interest that inducible localized increases in GABA near GnRH nerve terminals in the median eminence in vivo stimulated the hypothalamic-pituitary ovarian-axis in rats (Bilger *et al.*, 2001).

$GABA_AR$ activation opens an intrinsic chloride ion channel; the direction of chloride flow depends on resting membrane potential and intracellular chloride levels Unlike most mature neurons, GnRH neurons maintain high intracellular chloride (De Fazio *et al.*, 2002). Direct studies on GnRH neurons have shown both depolarizing and hyperpolarizing actions of applied GABA (De Fazio *et al.*, 2002; Ran *et al.*, 2002) and inhibitory and stimulatory roles for endogenous GABA (Ran *et al.*, 2004; Moenter *et al.*, 2004). Studies of alterations in GABAergic drive to GnRH neurons induced by negative energy balance and neuromodulators involved in conveying metabolic signals to the reproductive endocrine system indicate a direct physiological correlation between increased $GABA_AR$-mediated drive to these cells and increased reproductive activity This explantation being contronersial, the advent of transgenic models in which GnRH neurons can be visualized in living brain slices (Spergel *et al.*, 1999; Suter *et al.*, 2000; Ran *et al.*, 2004) allows direct testing of GABAergic mechanisms by which steroidal feedback may be communicated to GnRH neurons. Sullivan and Moenter (2005) examined GABAergic postsynaptic currents by the whole cell patch clamp technique to record from GnRH neurons in brain slices from ovariectomized (OVX) adult female mice treated with physiological levels of steroid hormones *in vitro*. Spontaneous GABAergic postsynaptic currents (PSCs) were recorded to examine overall changes in $GABA_AR$-mediated signaling to GnRH neurons; miniature GABAergic postsynaptic currents (mPSCs), which are action potential independent, were recorded to begin to examine whether changes in spontaneous PSC properties were due to pre- and/or post-synaptic mechanisms. Their results demonstrated that

GABAergic neurons participate in integrating and conveying steroid feedback to GnRH neurons, defining potential central mechanism for steroid regulation of GnRH neurons during the reproductive cycle and providing one possible mechanisms for increased activity of these cells in hyperandrogenic females.

Effect of Steroids on LHRH in Male Rats

By the measurement of LHRH with radio immunoassays, it became clear that orchidectomy decreased hypothalamic LHRH levels (Kalra and Kalra, 1983). Evidences have also been accumulated to indicate that androgens may facilitate the neurosecretory function of LHRH neurons. The castration-induced depletion in the medial basal hypothalamus (MER) LHRH stores can be reinstated by androgens at levels that mayor may not inhibit LH release and the time course of the androgen-induced MBH LHRH and LH responses were markedly different suggesting that the increase in the MBH LHRH levels may be due to de novo synthesis rather than a mere consequence of diminution in LHRH release. Thus with respect to the effects of androgens on LHRH release, there is general consensus that as compared to that from intact rats, the basal and evoked LHRH *in vitro* from the hypothalami of castrated rats was decreased (Kalra and Kalra, 1986). Dluzen *et al.* (1983) reported enhanced LHRH release from the basal hypothalamus of intact as compared to castrated rats. On the other hand, Brar *et al.* (1985) failed to observe any change in portal plasma concentrations of LHRH after orchidectomy. Androgens thus exert a trophic effect on LHRH neurosecretion (Kalra and Kalra, 1983). In adult male rodents, androgens impose negative feedback (Gay and Midgay, 1969). In contrast, data from adult women (Pastor *et al.*, 1988; Eagleson *et al.*, 2000) and from models of prenatal androgenization in primates, sheep and rodents (Dumesic *et al.*, 1997; Robinson *et al.*, 1999; Sullivan *et al.*, 2004) suggest that, in females, excess androgen may disrupt communication of negative feedback signals from progesterone and/or independently stimulate GnRH release.

Effect of Steroids on LHRH in Female Rats

Luteinizing hormone releasing hormone (LHRH) neurons of the forebrain play a pivotal role in the central regulation of female reproduction and many aspects of LHRH neuronal activity are modulated by circulatory estrogens. Steroid hormones are known to be important modulator of LHRH and gonadotrophins (LH and FSH) secretion. Estradiol regulates the biosynthesis and secretion of LHRH. Estrogen (E) treatment induces daily LH surges in ovariectomized rats and there are dynamic changes in hypothalamic LHRH levels tightly coupled with LH surges, but it has not been possible to carefully examine the effects of ovarian steroids on LHRH secretion (Kalra and Kalra, 1983; Kalra, 1985). The only evidence to support the view that ovarian steroids may facilitate LHRH accumulation in the hypothalamus, possibly due to increased de novo synthesis, is the demonstrated increase in LHRH levels preceding the onset of the preovulatory LH surge (Kalra and Kalra, 1977). A similar response can be readily induced by P in estrogen-primed ovariectomized rats (Crowley *et al.*, 1982). Similar to the action of androgens in males, treatment of estrogen alone (Drouva *et al.*, 1984) or P in estrogen-primed rats augmented basal and evoked LHRH release from the hypothalamus *in vitro* (Kim and Remirez, 1985). In addition, *in vitro* studies show a correlation between basal LHRH release and E2 titers during the estrous cycle (Gallo, 1981).

A number of investigators have assessed the affect of estrogen on *in vivo* LHRH secretion in the ovariosctomized rat, sheep and monkey. Estrogen treatment rapidly decreased LH release but a corresponding decrease in LHRH release was not detected. The failure to observe a tight reciprocal relationship between estrogen and LHRH release was apparent when LHRH release was assessed either by the push pull cannula technique in the rat (Levine and Ramirez, 1980), monkey (Levine *et al.*, 1985) and sheep (Schillo *et al.*, 1985) or in the hypophyseal portal plasma in monkey and sheep (Clarke and Cummins, 1985). However, it is important that a decrease in LHRH secretion is not essential to suppress LH release after estrogen treatment in monkeys, because LH secretion decreased to a low basal range despite the presence of uninterrupted-LHRH signals (Knobil, 1980). Thus the *in vitro* evidence and failure to document a reciprocal relationship between ovarian steroids and LHRH release *in vivo* favour the hypothesis that estrogen may not adversely affect LHRH release but that the ensuring decrease in LH may be due to a decrease in the pituitary responsiveness to episodic LHRH signals and/or due to the release of hypothalamic factor(s) that inhibit LH release.

Influence of Prolactin on the Sensitivity to Steroid Feedback

It has been hypothesized that peripubertal and seasonal changes in plasma prolactin (PRL) levels represent one of the mechanisms responsible for concomitant alterations in the sensitivity to steroid feedback. Plasma PRL levels typically increase during sexual maturation, which parallels the increase in gonadal steroid output. Estradial is among the most effective stimuli for PRL release (Shull and Gorski, 1984) and testosterone increases peripheral PRL levels in the male rat (Nolin *et al.*, 1977). Prolactin enhances gonadal responsiveness to gonadotrophin stimulation by increasing the number of LH receptors (Purvis *et al.*, 1979).

In majority of animal species, reproductive functions are restricted to a well-defined and often very brief period of the year. Anntial transitions between the periods of gonadal activity and quiescence in seasonal breeders require corresponding changes in the sensitivity to negative feedback of gonadal steroids and there is a great deal of evidence that such changes indeed take place. In male golden hamster, reproductive activity is confined to spring and summer as length of photoperiod to less than 12.5 hrs. of light per day triggers suppression of plasma levels of LH, FSH and testosterone, cessation of spermatogenesis, involution of gonads and inhibition of sexual behaviour and prolonged period of infertility (Bartke, 1985). From many evidences, it has been concluded that in male golden hamster, there are major seasonal changes in the susceptibility of LH and FSH release to androgen feedback and PRL may be suspected of mediating these changes.

Early cycles of fertility and infertility in sheep and other seasonal breeders are thought to reflect the outcome of several endogenous neural and endocrine processes. Either these interactions impinge on GnRH neurons, a final common pathway in the control of reproduction, to declare a neuroendocrine rhythm of GnRH secretion and represent gametogenesis of approximately 1 yr in length. An important component of the annual GnRH rhythm in female is that it reflects a profound change in the brain's sensitivity to estradial-negative feedback (Karsch *et al.*, 1993).

Interaction of Gonadal Steroids with Neuroactive Substances

The stimulatory and inhibitory effects of gonadal steroids upon LHRH neuronal system are to be mediated by interactions of the steroids with a variety of central neurotransmitter systems.

Effects of Gonadal Steroids on Catecholamines (CAS)

The concentrations of neurotransmitters in brain are influenced by gonadal function as the concentration of CAS in the hypothalamus (Lofstrom, 1977) and hypophyseal portal blood fluctuates during estrous cycle (Ben-Janathan *et al.*, 1977). Dopamine is secreted into hypophyseal portal blood greatest during the day of proestrous (Porter *et al.*, 1983). Ovariectomy results in decreased turnover of norepinephrins in the hypothalamus (Meites *et al.*, 1977), probably accounting for the increased release of LH and FSH. Administration of estrogen can inhibit gonadotrophin secretion probably by exerting opposite effects on the turnover of hypothalamic catecholamines as catecholaminergic neurons may play a role in mediating the events leading to the surge of PRL and LH that occurs in rats on the afternoon of proestrous (Weiner and Ganong, 1978).) A specific correlation between the changes in concentration of PRL and LH with changes in the activities of DA and NE-containing neurons in ME have been found (Damarest *et al.*, 1981, Porter *et al.*, 1983). Similarly Kaur and Porter (1985) observed a complex relationship between dopaminergic neuronal activity of hypothalamus and PRL secretion from anterior pituitary on different days of the estrous cycle.

Effects of Gonadal Steroids on NPY Secretion

Since NPY stimulated LH release only in rats that were primed either exogenously (Kalra and Crowley, 1984) or endogenously (Allen *et al.*, 1985b) with gonadal steroids, it was suggested that there are several ways by which gonadal steroids may accomplish their facilitation on NPY-induced LH response (Kalra *et al.*, 1987). Crowley *et al.* (1985) studied the effects of sequential estradial-17β (Eβ) and P treatments of ovariectomized rats on NPY levels in several microdissected sites in the preoptic tuberal pathway and correlated them on a time sequence basis with changes in LHRH levels in these sites and with the rise in LH secretion. Administration of P to

Eβ primed rats induced a sequential rise and fall in the ME LHRH levels which preceded LH surge. Similarly P treatment also induced arise and a fall in the ME NPY concentrations with a time course similar to that of LHRH. Since these shifts in the ME LHRH levels appear to be responsible for stimulation of LH surge and since NPY can stimulate LHRH release from ME nerve terminals, these observations imply that ovarian steroids can independently facilitate NPY neurosecretion which, in turn, may be involved in the hypersecretion of LHRH. Also parallel alterations of NPY and LHRH concentrations in the ME are suggestive of synchronizing affects of this steroid regimen on the neurosecretory function of two disparate but functionally linked neuropeptidergic systems. Evidences also suggest that testicular steroids may promote some aspect of NPY neurosecretion in the male rat (Sahu *et al.*, 1987).

Crowley *et al.* (1985) observed that NPY failed to stimulate *in vitro* LHRH release from the MBH of ovariectomized, hormonally untreated rats. On the other hand, under indentical *in vitro* conditions, NPY readily stimulated LHRH release from the MBH of ovariectomized rats pre-treated with estrogen. Also, NPY evoked a dose related stimulation of LHRH from the MBH of Eβ-P-primed ovariectomized rats. Thus stimulation of LH release in steroid-primed, rats evoked by intracerebroventricular administration of NPY (Kalra and Crowley, 1984) appeared, in parts to be due to LHRH release. Kalra *et al.* (1987) suggested that the excitatory effects of NPY are exerted via an axo-axonic link between NPY and LHRH neurons in the preoptic-tuberal pathway and that the excitatory expression of the functional link is gonadal steroid dependent. Allen *et al.* (1985) showed that under the influence of gonadal steroids NPY may interact with adrenergic transmitters and that NPY and NE acted in concert in ovarian steroid primed, ovariectomized rats. Collectively these findings document that gonadal steroids facilitate the release of LHRH by adrenergic transmitters and NPY. Thus it appears that gonadal steroids coordinate the function of several aminergic and peptidergic systems in the regulation of gonadotrophin secretion in the rat.

Effects of Gonadal Steroids on EOP Secretion

A number of investigators have studied the effects of gonadal steroids on the hypothalamic levels and release of EOP. Gonadal steroids modify EOP levels in the hypothalamus and MPOA (Forman *et al.*, 1985) and in portal blood (Sarkar and Yen, 1985). Gonadectomy decreased the levels of β-endorphin in the hypophyseal portal plasma of monkey (Ferin *et al.*, 1983) and rat (Sarkar and Yen, 1985). Combined treatment of estrogen and progesterone reinstated β-endorphin levels in the portal plasma of ovariectomized monkeys to the range normally seen during the luteal phase of the menstrual cycle (Ferin *et al.*, 1983). On the other hand, a long-term steroid replacement therapy with estrogen in ovariactomized rats (Werdlaw *et al.* 1982) decrease hypothalamic β-endorphin levels. However, studies have strongly suggested that gonadal steroids possess the capability to affect some aspects of β-endorphin neurosecretion, presumably by acting within the β-endorphin producing perikarya in the arcuate nucleus (Morrell *et al.*, 1985).

EOP Mediation of the Inhibitory Feedback Effects of Steroids

EOPs and morphine suppress LH release (Leadem and Kalra, 1985), and naloxone and naltrexone, the opiate receptor antagonists which displace EOPs from receptor sites, increase LH secretion (Petraglia *et al.*, 1986; Piva *et al.*, 1986). This suggested that EOPs are involved in mediating the negative feedback actions of gonadal steroids on LH secretion. Previous studies have shown that the ability of naloxone to increase LH secretion is lost with time following ovariectomy (OVX) (Petraglia *et al.*, 1984) in rats and after menopause in humans (Casper and Alapin-Rubillovitz, 1985) until gonadal steroids are initially given and negative feedback is restored. Similarly, the LH releasing effect of naloxone was seen in sheep only when P levels were elevated (Malvan *et al.*, 1984). These reports suggest that EOP involvement in LH secretion is lost alongwith the disappearance of gonadal steroid negative feedback *i.e.*, there is no steroid independent, EOP suppression of LH release.

It has also been shown that naloxone stimulated pulsatile LH secretion but it was blocked by simultaneous infusion of morphine, demonstrating that the effect was mediated by EOP receptors. Moreover, naloxone had no effect on *in vivo* pituitary responsiveness to LHRH, demonstrating that the EOP receptors blocked by naloxone were not located at the pituitary level. Since LH pulses are preceded by LHRH pulses (Clarke and Cummins, 1982; Levine *et al.*, 1982), it suggests that naloxone increased pulsatile LHRH release, although naloxone or opiates have been reported to influence basal pituitary

LH secretion *in vitro* (Wiesner *et al.*, 1984). Moreover LHRH release is important for the stimulatory action of naloxone, since antagonists of LHRH block naloxone induced LH secretion (Cicero *et al.*, 1985). Naloxone has also been shown to increase LHRH release from superfused rat (Leadem *et al.*, 1985; Kalra *et al.*, 1987) and human (Rasmussen *et al.*, 1983) hypothalamic, and to increase LH release when implanted into the MPOA or arcuate nucleus (ARC)-ME region, but not closely adjacent areas of rat brain (Kalra, 1981). Naloxone has been shown to reverse the inhibitory affects of gonadal steroids on LH release Caster and Alapin-Rubillovitz, 1985). However, the reversal of steroid induced suppression of LH release by naloxone does not prove that EOPs mediate steroid-induced regative feedback. Thus there may well be both steroid independent and steroid dependent EOP suppressions of LH secretion.

EOP Mediation of the Stimulatory Feedback Effects of Steroids

EOP may participate in the regulation of steroid dependent induction of preovulatory LH release onproestrous. A number of studies have show that morphine and β-endorphin inhibited ovulation, primarily by blocking the preovulatory. LH and LHRH hypersecretion (Kalra, 1986). It has been shown that naloxone failed to stimulate LH release during the LH surge occurring either sponstaneously on proestrous or that induced by sequential estrogen, progesterone therapy (Piva *et al.*, 1985). Intraventricular β-endorphin administration decreased the high amplitude, high frequency LH pulses in ovariectomized rats (Leadem and Kalra, 1985). These observations proposed that sometime prior to and during the critical period, a decrease in opioid tone in the preoptic-tuberal pathway may trigger the preovulatory LH surge (Kalra, 1983). This hypothesis was tested by infusing naloxone between 10.00–12.00 h on proestrous and it was found that a sustained decrease in opioid tone induced LH hypersecretion and in majority of rats, the pattern and magnitude of LH surge resembled that normally seen on the afternoon of proestrous. Since it is well known that a neural clock triggers the preovulatory LH and LHRH hypersecretion on the afternoon of proestrous, and the ovarian steroid milieu in the interval preceding the critical period on proestrous facilitates the timing of the neural clock either directly or via other neuronal systems may transiently curtail the tonic inhibitory EOP influence locally in the preoptic tuberal pathway. The progressive decrease in the inhibitory EOP tone may gradually accelerate the frequency of episodic LHRH discharge from the circhoral to that approximating the inherent frequency of 2–3 pulses/hr as seen in ovariactomized rats. This high frequency LHRH discharge may eventually activate and sustain a prolonged LH surge from the pituitary in the afternoon of proestrous.

Conclusion

Thus it is clear that a variety of neuro-peptidergic circuits either connected to or in close apposition either with the LHRH neurons or with the hypophyseal portal vessels are potentially important in the central control of LH secretion (Figures 19.1 and 19.2). The evidences reviewed also reveal that gonadal steroids generally exert a facilitatory effect on the neuro-peptidergic systems in a complex fashion. There findings illustrate that gonadal steroids may modify neurosecretion (biosynthesis, storage and release) and adjust the postsynaptic response of the peptidergic signals in the hypothalamus. However, some neuropeptidergic circuits may also modulate the central and pituitary action of steroids. The gonadal steroids may interact with two distinct hypothalamic circuits, an excitatory and in inhibitory circuit, each composed of one or more of the three major classes of identified neurochemical signals. The three classes of signals are the classical aminergic neurotransmitters, the neuropetides and the amino acids. The circuits may be connected with one another by synapses or may be in close apposition to affect the neighbouring systems. The model assumes that to stimulate LH release, the excitatory circuits either directly activate the release of LHRH or they may potentiate the action of LHRH at the pituitary level. They may also independently stimulate LH release from the pituitary gland. Similarly, to inhibit LH release, gonadal steroids may mobilize the inhibitory neural circuits that either inhibit LHRH release, attenuate the effectiveness of LHRH at the level of the pituitary or directly inhibit LH release (Figure 19.3).

Pulsatile Luteinzinng Hormone Release During the Estrous Cycle

Luteinizing hormone (LH) secretion during the estrous cycle (Fox and Smith, 1985) occurs in a pulsatile manner. Plasma levels of estradial (E_2) and progesterone also vary form to stage to stage in

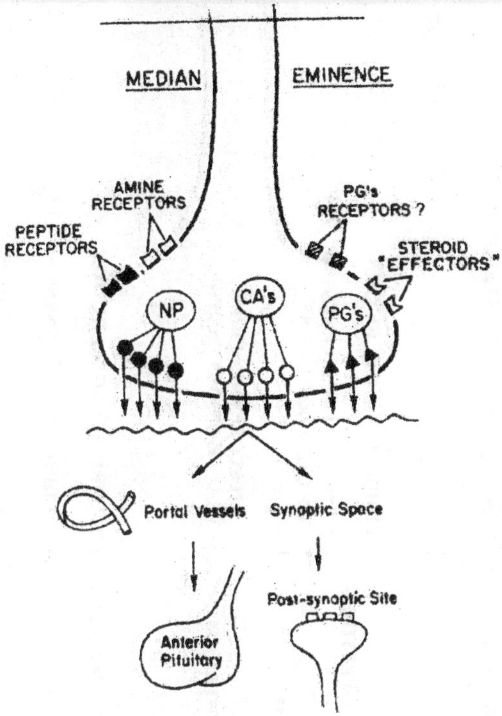

Figure 19.2: Schematic Diagram Representing an Axon Terminal from a Neuron Projecting to the Median Eminence. The secretary product of the neuron may be, for example, a neuropeptide (NP), one of the catechol amines (CA's) and/or one of the prostaglandins (PG's).The secretary products may be released and transported via the portal vessels to the anterior pituitary (as in the case of many neural peptides and dopamine) or released to act at a post-synaptic site on inrtra neuronal terminal. Negro-Vilar, *et al.*, 1982.

estrous cycle. The hypothalamic secretion of GnRH in the human triggers a cascade of events leading to gonadal sex steroid secretion, folliculogenesis and spermatogenesis. Animal studies have demonstrated a remarkable concordance between GnRH pulses in hypothalamic portal blood and LH pulses in the periphery (Clarke and Cummins, 1982) establishing that LH pulsatility is a refection of antecedent, intermittent discharge of GnRH. Not only pulsatile

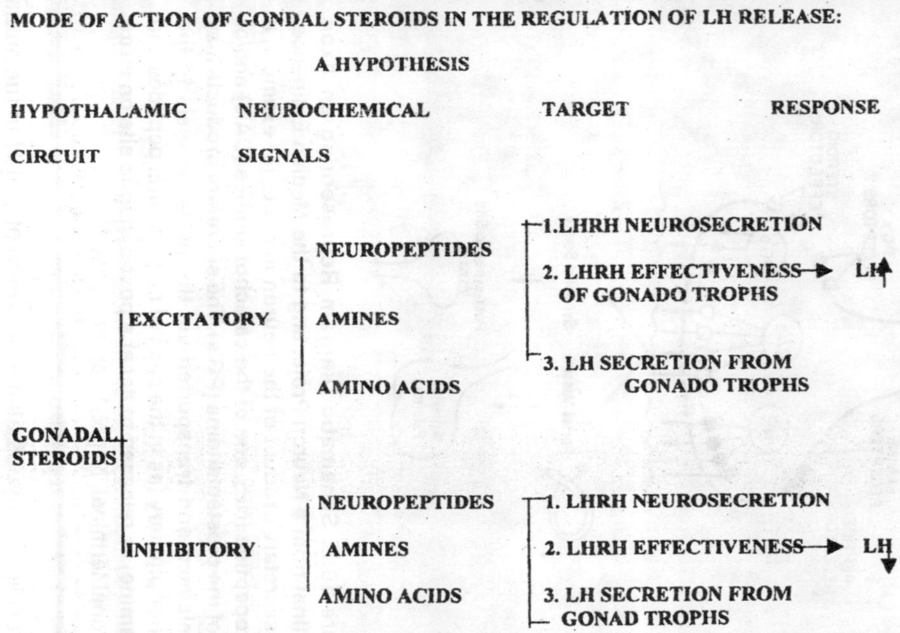

Figure 19.3: A Conceptual Model to Explain how Gonadal Steroids may Employ Excitatory and Inhibitory Hypothalamic Neural Circuits to Stimulate or Inhibit LH Release, Respectively (Kalra *et ai.*, 1987 b)

GnRH secretion necessary for maintenance of normal gonadotrophin secretion, but continuous, as opposed to pulsatile, administration of GnRH actually desensitizes gonadotrophin release (Belchetz *et al.*, 1978). In GnRH-deficient animals and humans, administration of pulsatile GnRH reestablishes normal hormonal responses and in patients with congenital hypogonadotropic hypogonadism, pulsatile GnRH reestablishes normal hormone responses and pubertal development (Hoffman and Crowley, 1982).

During the rat estrous cycle, blood LH levels are the lowest on estrous, probably due to the longest LH interpulse interval of the cycle (Gallo, 1981). Moreover, an absence of LH pulse on estrus have been reported (Fox and Smith, 1985). Between estrous and diestrous I (D_1) when plasma levels of both E_2 and P are low, there is a marked increase in LH pulse amplitude and frequency (Babu *et al.*, 1987). The amplitudes of these LH pulses are the largest of any other period of basal pulsatile LH release in the rat cycle (Gallo, 1981). Between D_1 and diestrous 2 (D_2) when the secretion of both ovarian E_2 and P is increasing, LH pulse frequency remains stable, while pulse amplitude decreases (Fox and Smith, 1985). In the next 24 h interval between the mornings of D_2 and proestrous, a second increase in E_2 secretion occurs, and the corpus luteum regresses, resulting in a decline in plasma levels and during this period, LH pulse amplitude and frequency remains stable (Leipheimer *et al.*, 1985; Babu *et al.*, 1986, 1987). Finally, in the remaining 24 h interval between the mornings of proestrous and estrous, E_2 secretion in maintained for the initial 10 h, secretion increases for about 10 h, and LH pulse frequency declines while pulse amplitude remains stable (Leipheimer *et al.*, 1986).

An analysis of ovarian steroidal regulations of the changes in amplitude and frequency characteristics of pulsatile LH secretion has demonstrated that the daily changes in secretary patterns of E_2 and P in the rat estrous cycle are decisively important in regulating the differences in basal pulsatile LH release that occurs at each stage. Moreover, there may be both steroid independent and steroid dependent EOP suppressions of pulsatile LH secretion.

Selective Modulation of FSH and LH Secretion

The regulation of the secretion of FSH, however, cannot be explained adequately with LMRH as the sole regulatory hypothalamic hormone because of a significant divergence between

FSH and LH secretion in male as well as female (Zanisi and Martini, 1975). The most direct way to explain divergent secretion of FSH and LH under selected physiological conditions would be to postulate the presence of a distinct FSH releasing factor (FSH-RF) (Igarashi *et al.*, 1973; Igarashi and McCann, 1974). But this has not been supported adequately by experimental evidences (Schally *et al.*, 1976). The concept of FSH-RF was revived by using a combination of bioassays and radioimmunoassays (Mizunuma *et al.*, 1983). Nevertheless the presence of a specific FSH-RF is still not widely accepted.

The concept of a gonadal peptide responsible for the modulation of FSH secretion was first proposed by McCullagh in 1932. Extensive work on the presence of such a peptide of gonadal origin has been carried out by several investigators (Rivier *et al.*, 1986) and a cyclic fluctuation of inhibin-like material has been reported in ovarian tissue and ovarian venous blood (Depaolo *et al.*, 1979). The ovarian peptides have been isolated and characterized. (Rivier *et al.*, 1985; Robertson *et al.*, 1986) but still extensive evaluation is needed to determine if this peptide plays a physiological role in the regulation of FSH secretion.

Neuroendocrine Regulation of FSH Secretion

It is well recognized that electrical stimulation of the preoptic suprachiasmatic nuclear region of the hypothalamus on the day of proestrus brings about a proestrus type of LH and FSH surge. Electrical stimulation of the dorsal anterior hypothalamic area (DAHA) elicits preferential FSH release (Chappel and Barraclough, 1976). Bilateral lesions of the DAHA resulted in a reduction of serum FSH levels during proestrus (Lumpkins and McCann, 1982). These observations indicate the presence of separate neural pathways involved in the control of FSH and LH and anatomical evidence for these pathways was provided by Kimura and Kawakami (1978). Further work by several investigators has shown that deafferentation of neurosecretory cells in the ARC-ME region from those in the preoptic-anterior hypothalamic area prior to 1800 h of proestrus reduced or abolished FSH release on estrus but deafferentation after 1800 h did not have any effect (Blake *et al.*, 1982). The absence of requirement for LHRH in the secretion of FSH on estrus was further confirmed by demonstrating that antibodies to LHRH administered on the day of estrus did not have any effect on FSH secretion

(Hasegawa *et al.*, 1981; Rush *et al.*, 1982). Several investigations have studied the effect of LH and FSH administration on subsequent secretion of FSH on estrus. (Ashiru *et al.*, 1981). Both hormones were able to stimulate FSH release by the anterior pituitary. Binding sites for FSH in the hypothalamus have also been reported (Davies *et al.*, 1975). The mode of administration of LHRH (Wise *et al.*, 1979; Turgeon and Waring, 1982) on the pulse intervals (Wildt *et al.*, 1981) has been shown to induce a divergence in FSH and LH secretary patterns.

Inspite of the pivotol role of FSH in reproduction, our understanding of the factors involved in the control of FSH is incomplete. The long circulating half life of FSH, the unavailability of assays to detect the variant isoforms of FSH and cross reactivity with gonadotrophin α-subunit and LH have all made it rather difficult to assess secretary patterns of FSH from peripheral measurements, Still, several advances have been made in understanding the regulation of FSH. In general, FSH appears to be secreted in two modes; a basal or constitutive modern and an episodic mode.

Basal Mode of FSH Release

This mode of FSH release is supported by the findings that FSH is continued to be released for prolonged times from (1) hypothalamic disconnected sheep (Clarke *et al.*, 1983; Padmanabhan *et al.*, 1997a); (2) hypohysectomized rats (Depaolo, 1991) and sheep, bearing pituitary transplant under the kidney capsule (Padmanabhan *et al.*, 1997a) and (3) long-term pituitary cultures (Sheridan *et al.*, 1979) This component of FSH release appears to be dictated by the availability of translatable FSH β and mRNA (Farnsworth, 1995).

Episodic Mode of FSH Release

Because pulses of FSH measured at the periphery are not that discrete as that of LH, investigators had to rely heavily on statistical approaches to define them. Monitoring FSH patterns at the secretary site appears to be the only resource by which one can establish the true nature of FSH secretion. Hypophyseal portal collection technique pioneered by Clarke and Cummins (1982) for monitoring hypophysiotropic hormones provided a tool to define FSH secretary dynamics. Because hypophyseal portal vessels are lesioned at surface of the pituitary, hypophyseal portal blood provides a resource for measuring not only hypothalamic releasing and release-inhibiting peptides but also of secretary products originating from the pituitary.Comparisons of patterns of LH in the hypophyseal portal and peripheral circulating (Midgley *et al.*, 1997) have revealed that: (1) Each (GnRH) pulse is associated with a LH pulse both at the peripheral and hypophyseal portal circulation level, (2) LH concentrations are several-fold higher in hypophyseal portal than peripheral circulation, (3) Patterns of LH are more discrete in hypophyseal portal circulation than in the periphery. Utilizing this approach, it was recently found that FSH is indeed secreted in a dual mode–a basal or constitutive mode and an episodic mode (Padmanabhan *et al.*, 1997b). Furthermore, these studies have established that each pulse of GnRH, in addition to its association with LH is a pulse of FSH and that additional non-GnRH-associated pulses of FSH exist, but how are these triggered is yet unresolved.

Neuroendocrine vs Paracrine Regulation of FSH Production/ Secretion

Believers of one releasing hormone (GnRH) hypothesis have explained the paradoxical difference in release patterns of LH and FSH as due to: (1) an alteration in an GnRH pulse frequency with slower frequencies favoring FSH release (Wildt *et al.*, 1981), (2) differential sensitivity of sub-population of gonadotrophs that secrete LH and FSH (Denef *et al.*, 1978) and consequent threshold differences in response to GnRH (amplitude modulation) with low levels favoring FSH secretion (Wise *et al.*, 1979), (3) Different response times of LH and FSH to GnRH. If GnRH is the only regulative of episodic mode of FSH release, FSH pulsatility should obliterate by blocking GnRH action. But episodic FSH secretion has been shown to persist even after elimination of endogenous GnRH activity/action (Culler and Negro Vilar, 1987; Pau *et al.*, 1991; Padmanabhan *et al.*, 1997a) thus supported the existence of additional FSH-regulatory mechanisms.

Well documented differences in metabolic clearance rates of LH and FSH (Ulloa Aguirre *et al.*, 1995), differential effects of gonadal steroids and peptides offer explanations for changes in threshold level of circulating LH and FSH but do not explain the asynchronous presence and initiation of FSH pulse especially in gonadectomized animals (Culler and Negro-Villar, 1987; Pau *et al.*, 1991; Padmanabhan *et al.*, 1997b). Two other explanations could potentially explain the existence of the non-GnRH associated pulse of FSH; (1) temporal changes in the local pituitary milieu of inhibins, activins and follistatins and/or (2) the existence of a separate neural trigger. Inhibins, activins and follistatins, are all produced at the pituitary level (Depaolo *et al.*, 1991; Mather *et al.*, 1992). In defining the local feedback loop, one needs to recognize the functional overlap that exits among them.

Inhibins and actvins have functionally opposite roles (Ying, 1988). Inhibins suppress FSH and activins stimulate FSH production (Ying, 1988). Inhibin can bind activin receptors (Mathews and Vale, 1991) and antagonize activin's action. Follistatin is a binding protein and neutralizer of activin's action (Robertson, 1992). Relative abundance of these three regulator is likely to set the local regulatory tone for FSH (Figure 19.4). Increases in activin/activin receptors or decreases in inhibin/follistatins would increase FSH production (Figure 19.4) and the converse would inhibit FSH production/secretion.

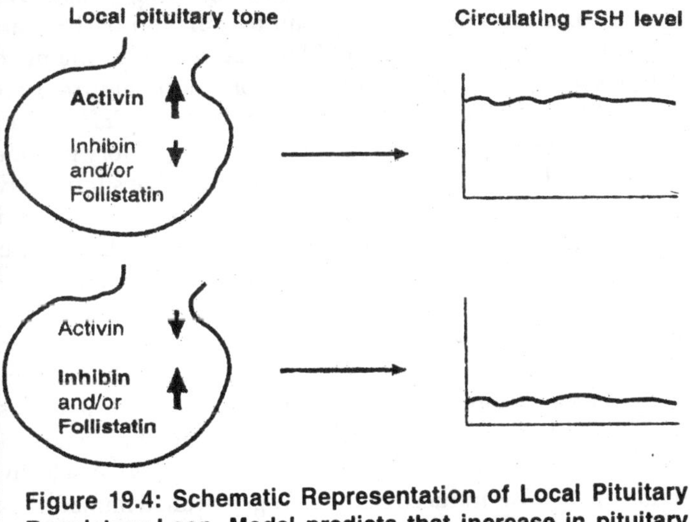

Figure 19.4: Schematic Representation of Local Pituitary Regulatory Loop. Model predicts that increase in pituitary activin or decrease in inhibin and follistatin would increase and decrease in activin or increase in follistatin/ inhibin would decrease FSH secretion (Padmanabhan, 2001).

To account for the episodic mode of FSH secretion, the regulatory tone must undergo frequent changes. Temporal changes in individual components of the FSH regulatory loop in concert with estradiol may result in "pulse" like changes in regulatory tone that lead to an episodic mode of FSH secretion. Recent studies have shown that the local regulatory loop is under neuroendocrine control and a reciprocal relationship appears to exist in the regulation of FSHβ and follistatin mRNA by GnRH High frequency GnRH pulses increase follistatin mRNA and protein (Besecke *et al.*, 1996) but decrease FSH expression (Wildt *et al.*, 1981). Low frequency GnRH stimulates only FSHβ and not follistatin mRNA expression. More recent studies have shown that GnRH may mediate its effects by changing local activin availability (Besecke *et al.*, 1996). To conclude, these findings suggest that changes in local pituitary milieu may be a means by which changes in GnRH could ultimately lead to differential secretion patterns of LH and FSH.

Regulation of the Selective Release of FSH by Steroids

The regulation of the selective release of FSH is still poorly understood. The presence of an FSH-releasing factor, gonadal FSH inhibiting. and releasing peptides, autoregulation by FSH and changes in the pulsatile patterns of LHRH secretion are among the postulates used to explain this phenomenon (Figure 19.5). Considerable evidence of differential regulation of FSH and LH secretion by steroids in the absence of gonadal regulatory peptides have been produced (Mahesh *et al.*, 1987). Extensive studies have been carried out to examine the role of progesterone in modulating gonadotrophin secretion using the low dose estradiol primed ovariectomized immature rat. When progesterone was administered in combination with estradiol, its effects were facilitative or inhibitory to gonadotrophin release depending upon the dose of progesterone used (McPherson and Mahesh, 1979). The ovary secretes a number of progesterone metabolites during the rat estrous cycle and these have been isolated from peripheral blood and ovarian venous blood in the rat on the day of proestrus (Ichikawa *et al.*, 1974). A selective secretion of FSH was induced in the estrogen primed immature rat model by the administration of progesterone metabolite 5-dihydroxy progesterone (5-DHP) while selective LH secretion was induced by 3,5-tetrahydroxy progesterone (3,5-THP). The mechanism involved

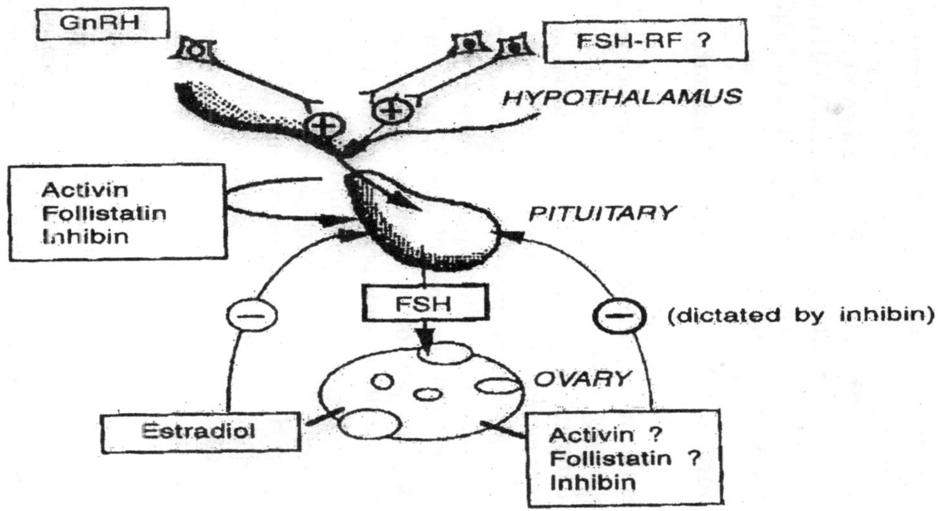

Figure 19.5: Shows the Hypothalamic, Ovarian and Pituitary Regulation of FSH Production/Secretion. At the hypothalamic level, in addition to GnRH, there is evidence supporting the existence of a separate FSH-releasing factor. At the pituitary level, locally-produced activin, inhibin and follistatin appear to dictate the threshold of FSH release. At the ovarian level estradiol and inhibin are two major negative feedback regulators of FSH (Padmanabhan, 2001).

in this selective release of FSH by 5-DHP is unclear at present. However, the selective secretion of FSH and LH induced by progesterone metabolites was confirmed in the immature female rat primed with PMSG and maintained in constant light (Murphy and Mahesh, 1985). 5-DHP was also able to induce a greater release of FSH when administered to the adult cycling rat on proestrus. The priming of the pituitary gonadotroph in secreting a high baseline level of FSH or responding to LHRH in releasing a greater amount of FSH appeared to be an important factor in selective FSH release and such priming can be brought about by 5-DHP in the absence of regulatory peptides (Murphy and Mahesh, 1985).

Evidence Supporting the Existence of a Separate FSH-Releasing Factor

This hypothesis stems from the studies, which showed selective release of FSH following ablation (Lumpkin and McCann, 1984); deafferentation (Lamperti and Hill, 1987) of the dorsal anterior hypothalamic area or electrochemical stimulation of hypothalamic regions apart from those which regulate LH secretion (Chappel and Barraclough, 1976). Lesions of anterior hypothalamic area abolished FSH but not LH pulsatility (Lumpkin *et al.*, 1989). Thus it provides strongest evidence in support of the existence of specific site of control for FSH. Selective suppression of FSH pulsatility also occurs following lesion of the caudal and mid-brain eminence (Marubayashi *et al.*, 1999). McCann's group succeeded in achieving partial separation of FSH and LH-releasing activities from extracts of porcine and ovine hypothalami (Lumpkin *et al.*, 1987). Despite the supportive neuroanatomical and physiological evidences, a physiologically relevant FSH-RH is yet to be isolated and characterized. Independent of this line of research; identification of variant forms of this GnRH has opened up anew possibilities. Since FSH is more sensitive to the negative feedback effect of estrogen than LH, factors other than changes in hypothalamic GnRH release have been assumed to account for the differential suppression of these gonadotrophins. Estradiol replacement to postmenopausal women does not completely return the FSH levels to normal, indicating that additional regulatory components, such as ovarian inhibin, may play a major role for feedback regulation (Morley *et al.*, 1992).

Receptor Mechanism to the Mode of Action of Steroids on Gonadotrophin Secretion

Synthesis and secretion of gonadotrophins from the anterior pituitary, and the end-product regulation of these processes, represents the cornerstone of reproductive physiology. The fact is greatly realized and thus resulted in the vast of experimental approaches, which have been taken to reach ever deeper into the intricacies of the complex interrelationships involved. Steroid hormones are known to be important modulations of LHRH and gonadotrophin secretion (Negro-Vilar and Oheda, 1981). In the early 1970's many workers investigated the occurrence and nature of steroid hormones receptors in the anterior pituitary (Eisenfeld, 1970; Korach and Muldoon, 1973). These studies were followed by subsequent description of the estrogen retention system in the hypothalamus (Korach and Muldoon, 1974). Interrelationships between gonadotrophin fluctuation and neural steroid hormone receptor turnover have also been examined. (Muldoon, *et al.*, 1984).

Estradiol manifests a biphasic action on gonadotrophin secretion, being actually inhibitory and chronically stimulatory. Both hypothalamic and anterior pituitary sites of estrogen action seem to be involved in this feedback mechanism (Kalra and Kalra, 1982) but the precise nature of the individual tissue responses have not been properly worked out. The secondary positive effect at the hypothalamic level may actually reflect a release of the initial negative action on the synthesis and secretion of LHRH. The observation that implants of estradiol can block the ovulatory surge of gonadotrophin (Turgeon, 1979) provides indirect evidence supporting this idea. A positive mechanism for such a self-regulation is estrogen stimulation of estrogen-2/4-hydroxylase activity in neural tissue, thus forming the metabolite, 2-hydroxyesterone which is antiestrogenic at this level (Mondschein *et al.*, 1986). Very little is known about the primary negative effect of estrogens at the anterior pituitary level. However, Aiyer *et al.* (1974) indicated a true positive enhancement of sensitivity by induction of an LHRH self-priming effect. The mechanism of this action of estrogens at LHRH receptor activity is not clearly understood, and the action may as well occur at a mechanistic site distal to the primary receptor interaction.

The effects of progesterone on gonadotrophin secretion are complicated due to its interplay with the actions of estradiol, apparently at both hypothalamic and pituitary sites of action. The effects of progesterone appear to depend upon the availability of sufficient progesterone receptors to elicit a given response. Since progesterone receptor activity is critically regulated by estrogen (Vu Hai *et al.*, 1977), this shows a dependency on the estrogen priming level of the endogenous millieu. During the rat estrous cycle, progesterone level begins to rise just as estradiol level have peaked following a long gradual rise. This represents a situation of strong estrogen priming and elevated progesterone receptor levels. Under these conditions progesterone can act immediately and has its most pronounced effects on the acute responses of the tissues to estrogen. Since estrogen is acutely inhibitory to gonadotrophin secretion, progesterone would tend to be facilitatory. In addition, progesterone independently appears to directly stimulate LHRH release, inhibit its degradation, and perhaps after pituitary sensitivity to LHRH, all of which would intensify a positive feedback effect of this hormone, On the other hand, appearance of progesterone under conditions of relative estrogen deficiency would result in limited rapid actions because of the low progesterone receptor activity. In this setting, progesterone is antagonistic to the long-term positive feedback effects of estrogen, and is thereby inhibitory to gonadotrophin secretion.

Transmembrane Signals and Intracellular Messengers Mediating LHRH Secretion

In the preceding sections, it has been concluded that a large number of neurotransmitters and blood borne chemical signals reach the LHRH neurons and interact with a specific membrane or intracellular receptors to modify the activity of these neurons, and ultimately result in LHRH secretion. It is quite obvious that an understanding of the intracellular events that mediate the action of the various neurotransmitters and hormones upon the LHRH neurons is required to unravel the complex interaction between amines, peptides and steroids that are so essential to the integrated function of the hypothalamic-pituitary-axis.

After interacting with specific membrane receptor, many hormones, neurotransmitters, growth factor and other biologically active substances generate responses by eliciting signals that are carried from the membrane into its cell interior by a variety of different routes. Production of cyclic nucleotides, calcium mobilization, protein kinase activation and inositol phospholipid breakdown are among the major intracellular events through which signals are transduced and converted into specific cellular responses. Some additional systems like arachidonic acid also play an important role in the mediating hormone secretion and details of these mechanisms is not under the perview of this topic.

Mutations of Gonadotrophin Subunits and their Receptors

Distinct models of LH and FSH deficiency occur in the inactivating mutations of the gonadotrophins subunits and their receptors. LH and FSH are dimers composed of two non-covalently associated protein subunits, the α-subunit that is shared by both, and the α-subunit that is unique and hence confers the specificity to each heterodimer. In both sexes, LH stimulates steroidogenesis, whereas FSH controls gametogenesis. In women, LH increases P 450c 17 expression in the thecal cells and catalyzes conversion of 21-C androgen precursors to androgen that are then aromatized to estrogens in granulosa cells. Hence LH is critical for achieving the full complement of ovarian steroidogenesis, inducing follicular rupture at the midcycle ovulatory surge and supporting normal coupus luteum function. In men, LH stimulates androgen production by Leydig cells, regulating sexual differentiation, androgenization sexual function and fertility (Seminara and Crowley, 2001).

Experimental evidences from mutations in gonadotrophin subunits and their receptors (Dierich *et al.*, 1998; Abel *et al.*, 2000) reveal that the relationships between gonadotrophin stimulation and sex steroid production, testicular volume, follicle development and fertility are much more complex than previously thought. Genes such as DAXI (Ikeda *et al.*, 1996: Phillip *et al.*, 1998; Lindstedt *et al.*, 1998; Seminara *et al.*, 1999) that have traditionally been implicated in secondary hypogonadism actually have direct gonadal effect in the regulation of spermatogenesis; their hypogonadism may mask their primary gonadal effects. Thus, there will be additional genes that act at all levels of the hypothalamic-pituitary-gonadal-axis and the description of new clinical disorders should continue to elucidate their presence. Moreover, gene defects that cause abnormalities in GnRH neuronal migration play important roles in axonal targeting beyond the reproductive phenotypes as parts of larger, developmental syndromes. Sophisticated knockout models and gene rescue experiments will provide additional and exciting new avenues of investigation in the coming years on this important aspect of regulation of fertility.

References

Abel, M.H., Wotton, A.N., Wilkins, V., Huhtaniemi, I., Knight, P.G., Charlton, H.M. 2000. The effect of a null mutation in the follicle stimulating hormone receptor gene on mouse reproduction. *Endocrinology*, 141: 1795–1802.

Adler, B.A., Crowley, W.R. 1986. Evidence for gamma-amino butyric acid modulation of ovarian hormonal effects on luteinizing hormone secretion and hypothalamic catecholamine activity in the female rat. *Endocrinology*, 118: 91–97.

Aiyer, M.S., Chiappa, S.A., Fink, G. 1974. A Priming effect of luteinizing hormone releasing factor on the anterior pituitary gland in the female rat. *J. Endocrinal.*, 62: 573–588.

Allen, L.G., Kalra, P.S., Crowley, W.R., Kalra, S.P. 1985. Comparison of the effect of neuropeptide Y and adrenergic transmitters on LH release and food intake in male rats. *Life Sci.*, 37: 617–623.

Alper, R.H Demarest, K.T and Moore, K.E. 1980. Morphine differentially alters synthesis and turnover of dopamine in central neuronal systems. *J. Neur. Trans.*, 48: 157–165.

Ashiru, O.A., Rush, M.E., Blake, C.A. 1981. Further studies on the effectiveness with which endogenous LH and FSH stimulate the release of endogenous FSH during rat estrous cycle *J. Endocrinol.*, 88: 103–133.

Babu, G.N., Bona-Gallo, A., Gallo, R.V. 1986. Interaction between estradiol and a non-steroid factor in procine follicular fluid in regulating LH pulse amptitude between the morning of diestrous 2 and proestrous in the rat. *Neuroendocrinology*, 44: 8–14.

Bartke, A. 1985. Male hamster reproductive endocrinology: In: *The Hamster*, (Ed.) Siegel H.I. Plenum Publ., p. 73–98.

Belchetz, P.E., Plant, T.M., Nakayi, Y., Keogh, F.J., Knobil, E. 1978. Hypophyseal responses to continuous and intermittent delivery of hypothalamic gonadotrophin – releasing hormone. *Science,* 202: 631–633.

Ben-Jonathan, N., Oliver, C., Weiner, H.J., Mical, R.S. and Porter J.C. 1977. Dopamine in hypophyseal portal plasma of rat during the estrous cycle and throughout pregnancy. *Endocrinology,* 100: 452–458.

Ben, Rafael Z., Levy T., Schoemaker, J. 1995. Pharmacokinetics of follicle stimulating hormone: clinical significance. *Fertil. Steril.,* 63: 689–670

Besecke, L.M., Guendner, M.J., Schneyer, A., Bauer-Dentoin, A.C., Jameson, J.L., Weiss, J. 1996. Gonadotrophin releasing hormone regulates follicle-stimulating hormone-beta gene expression through an activin/follistatin autocrine or paracrine loop. *Endocrinology,* 137: 3667–3673.

Bilger, M., Heger, S., Brann, D.W., Paredes, A., Oheda, S.R. 2001. A conditional tetracycline-regulated increase in gamma-amino butyric production near luteinizing hormone releasing hormone terminals disrupts estrous cyclicity in the rat. *Endocrinology,* 142: 2101–2114.

Blake, C.A., Metcalf, J.P., Handricks, S.E. 1982. Anterior pituitary gland secretion after forebrain ablation: periovulatory gonadotrophin release. *Endocrinology,* 111: 789–793.

Blank, M.S., Fabbri, A., Catt, K., Dufau, M. 1986. Inhibition of luteinizing hormone release by morphine and endogenous opiates in cultured pituitary cell. *Endocrinology,* 118: 2097–3002

Bourguignon, J.P., Gerard, A., Purnelle, G., Czajkowski, V., Yanamaka, C., Lemaitre, M., Rigo, J.M., Moonen, G., Franchi-mont, P. 1997. Duality of glutamatergic and GABAergic control of pulsatile GnRH secretion by rat hypothalamic explants. II. Reduced NR_2C and $GABA_A$ receptor mediated inhibition at initiation of sexual maturation. *J. Neuroendocrinol.,* 9: 193–199.

Bourguignon, J.P., Alvarez Gonzalez, M.I., Gerard, A., Franchimont, P. 1994. Gonadotrophin releasing human inhibitory autofeedback by subproducts antagonist at N-methyl-D-aspartate receptors: a model of autocrine regulation of peptide secretion. *Endocrinology,* 134: 1589–1592.

Bourguignon, H.P., Gerard, A., Alvarez Gonzalez, M.I., Pumelle, G., Franchimont, P. 1995. Endogenous glutamate involvement in pulsatile secretion of gonadotrophin–releasing hormone: evidence from glutamine effect and developmental changes. *Endocrinology*, 136: 911–916

Bourguignon, J.P., Gerard, A., Alvarez Gouzalez, M.I., Franchimont, P. 1992. Neuroendocrine mechanism of onset of puberty. Sequential reduction in activity of inhibitory and facilitatory N-methyl-D-aspartate receptors. *J. Clin. Invest.*, 90: 1736–1744.

Bourguignon, J.P., Gerard, A., Mathieu, J., Mathieu, A., Franchimont, P. 1990. Maturation of the hypothalamic control of pulsatile gonadotropin–releasing hormone secretion at onset of puberty: Increased activation of N-methyl-D-aspartate receptors. *Endocrinology* 127: 873–881.

Brar, A.K., McNeilly, A.S., Fink, G. 1985. Effects of hyperprolactinemic and testosterone on the release of LH-releasing hormone and gonadtropins in intact and castrated rats. *J. Endocrinol.*, 104: 35–43.

Brinkley, H.J. 1981. Endocrine signaling and female reproduction. *Biol. Reprod.*, 24: 22.

Caciedo, F., Sanchez-France, F. 1986. Direct action of epioid peptides and naoxone on gonadotrop in secretion by cultured rat anterior pituitary cells. *Life Sci.*, 38: 617–623.

Casper, R.F., Alpin-Rubillovitz, S. 1985. Progestins increase endogenous opioid peptide activity in postmenopausal women. *J. Clin. Endocrinol. Metab.*, 60: 34–36.

Chappel, S.C., Barraclough, C.A. 1976. Hypothalamic regulation of pituitary FSH secretion. *Endocrinology*, 98: 927–935.

Chronwal, B.M., Dimaggio, D.A., Massari, V.J., Vickel, V.M., Ruggier, D.A., O'Donohue, T.L. 1985. The anatomy of neuropeptidey Y containing neurons in the rat brain. *Neuroscience*, 15: 1159–1181.

Cicero, T.J., Schmoeker, P.F., Meyer, E.R., Miller, B.T. 1985. Luteinizing hormone releasing hormone mediates naloxones's effects on serum luteinizing hormone levels in normal and morphine–sensitized male rats. *Life Sci.*, 37: 467–474.

Cicero, T.J., Shaiker, B.A., Meyer, E.R. 1979. Endogenous opioids participate in the regulation of the hypothalamic–pituitary–luteinizing hormone axis and testosterone's negative feedback control of luteinizing hormone. *Endocrinology*, 104: 1286–91.

Clarke, L.J., Cummins, J.T. 1982. The temporal relationship between gonadotrophin-releasing hormone (GnRH) and luteinizing hormone (LH) secretion in ovariectomized ewes. *Endocrinology*, 111: 1737–1739.

Clarke, L.J., Cummins, J.T., de Kretser, D.M. 1983. Pituitary gland function after disconnection from hypothalamic influences. in the sheep. *Neuroendocrinology*, 36: 376–384.

Concannon, P.R., Cowan, R., Hansel, W. 1979. LH release in ovariectomized dogs in response to estrogen withdrawal and its facilitation by progesterone. *Biol. Reprod.*, 20: 523.

Crowley, W.R., Hassid, A., Kalra, S.P. 1987. Neuropeptide Y enhances the release of luteinizing hormone induced by luteinizing hormone releasing hormone. *Endocrinology*, 120: 941–945.

Crowley, W.R., Terry, L.C., Johnsson, M.D. 1982. Evidence for the involvement of central epinephrine system in the regulation of luteinizing hormone, prolactin and growth hormone release in female rats. *Endocrinology*, 110: 1102–1107.

Crowley, W.R., Tessel, R.E., O'Donohue, T.L., Adler, B.A., Kalra, B.P. 1985. Effects of ovarian hormones on the concentration of immunoreactive neuropeptide Y in discrete brain region of the female rat: correlation with serum LH and medium eminence LHRH. *Endocrinology*, 117: 1151–55.

Culler, M.D. and Negar-Vilar, A. 1987. Pulsatile follicle stimulating hormone secretion is independent of luteinizing hormone releasing hormone (LHRH): Pulsatile replacement of LHRH bioactivity in LHRH-immunoneutrilized rats. *Endocrinology*, 120: 2011–2021.

Davies, A.G., Duncan I.F., Lynch, S.S. 1975. Autoradiographic localization of 125-labeled FSH in the rat hypothalamus. *J. Endocrinology*, 66: 301–302.

De Fazio, R.A., Moenter, B.M. 2002. Estradiol feedback alters potassium currents and firing properties of gonadotropin-releasing hormone neurons. *Mol. Endocrinol.*, 16: 2225–2265.

Denef, C., Hantekeete, E., Dewaps, R. 1978. Monolayer cultures of gonadotrophs separated by velocity sedimentation: heterogeneity in response to luteinizing hormone releasing hormone. *Endocrinology*, 103: 736–747.

Depaolo, L.V. 1991. Hypersecretion of follicle stimulating hormone (FSH) after ovariectomy of hypophysectomized, pituitary-grafted rats: implications for local regulatory control of FSH. *Endocrinology*, 128: 1731–1740.

Depaolo, L.V., Shavder, D., Wise, P.M., Barraclough, C.A., Channing, C.P. 1979. Identification of inhibin-like activity in ovarian venous plasma of rats during estrous cycle. *Endocrinology*, 105: 647–654.

Dierich, A., Sairam, M.R., Monaco, L., Pimia, G.M., Gansmuller, A., Lemeur, M., Sassone-Couri, P. 1998. Impairing follicle stimulating hormone (FSH) signaling *in vivo*: targeted disruption of the FSH receptor leads to aberrant gametognesis and hormonal imbalance. *Proc. Nat. Acad. Sci., USA*, 95: 13612–13617.

Dluzen, D.E., Remirez, V.D. 1983. *In vivo* LHRH release from the median eminence of conscious, unrestrained, intact, acute castrate and long-term castrate male rats as determined with push-pull perfusion. (Abst.) *65th Ann. Mtg. Endocr. Soc. San Antonio, Tx.*, pp. 156.

Docke, F., Glaser, D. 1971. Internal feedback of luteinizing hormone in the cyclic female rats. *J. Endocrinol.*, 51: 403.

Donoso, A.O., Lopez, F.J., Negro-Vilar, A. 1992. Cross-link between excitatory and inhibitory amino acids in the regulation of luteinizing hormone-releasing hormone secretion. *Endocrinology*, 131: 1559–1561.

Drouva, S., Laplante, E., Gautron, J., Kordon, C. 1984. Effects of 7-β-estradiol on LHRH release from rat mediobasal hypothalamus slices. *Neuroendocrinology*, 38: 152–157.

Dumesic, D.A., Abbott, D.H., Eisner, J.R., Goy, R.W. 1997. Prenatal exposure of female rhesus monkeys to testosterone propionate increases serum luteinizing hormone levels in adulthood. *Fert. Steril.*, 67: 155–163.

Eagleson, C.A., Gingrich, M.B., Pastor, C.L., Arora, T.K., Burt, C.M., Evans, W.S., Marshall, J.C. 2000. Polycystic ovarian syndrome: evidence that flutamide restores sensitivity of the gonadotropin–releasing hormones pulse generator to inhibition by estradiol and progesterone. *J. Clin. Endocrinol. Metab.*, 85: 4047–4052.

Eisenfeld A.J. 1970. ^{3}H-estradiol: in the vitro binding to macromolecules from the rat hypothalamus, anterior pituitary and uterus. *Endocrinology*, 86: 1313–1318.

Emanuele, N., Lipov, S., Anderson, J., Anderson, E., Kirsteins, L., Nishimura, N., Lawrence, A.M. 1985. Sub-cellular localization of luteinizing hormone. *Neuroendocrinology*, 40: 339.

Evans, N.P., Dahl, G.E., Glover, B.H., Karsch, F.J. 1994. Central regulation of pulsalile gonadotrophin-releasing hormone (GnRH) secretion by estradiol during the period leading up to the preovulatory GnRH surge in the ewe. *Endocrinology*, 134: 1806–1811.

Farnworth, P.G. 1995. Gonadotrophin secretion revisited. How many ways can FSH leave a gonadotroph *J. Endocr.*, 145: 387–395.

Feleder, C., Jarry, H., Leonhardt, S., Wuttke, W., Moguilevsky, J.A. 1996. GABAergic control of gonadotrophin-releasing hormone secretion in male rats during sexual maturation involves effects on hypothalamic excitatory and inhibitory amino acid systems. *Neuroendocrinology*, 69: 305–312.

Ferin, M., Van vugt, D., Wardlaw, S. 1983. The hypothalamic control of the menstrual cycle and the role of endogenous opioid peptides. *Rec. Progr. Horm. Res.*, 40: 441–480.

Flugge, G., Oertel, W.H., Wutke, W. 1986. Evidence for estrogen receptive GABAergic neurons in the preoptic/anterior hypothalamic area of the rat brain. *Neuroendocrinology*, 43: 1–5.

Forman, L.J., Marquis, D.E., Stevens, R. 1985. The effect of chronic estrogen treatment on immunoreactive β-endorphin levels in intact female rats. *Proc. Soc Exp. Biol. Med.*, 179: 365–372.

Fox, S.R., Smith, M.S. 1985. Changes in pulsatile pattern of luteinizing hormone secretion during the rat estrous cycle. *Endocrinology*, 116: 1485–1492.

Funabashi, T., Diakoku, S., Shinora, K., Kimura, F. 2000. Pulsatile gonadotrophin-releasing hormone (GnRH) secretion is an inherent function of GnRH neurons, as released by the culture of medial olfactory placodes obtained from embryonic rats. *Neuroendocrinology*. 71: 138–144.

Gallo, R.V. 1981. Pulsatile LH release during periods of low-level LH secretion in the rat estrous cycle. *Biol. Reprod.*, 24: 771–77.

Gay, V.L., Midgley, R. 1969. Response of the adult rat to orchidectomy and ovariectomy as determined by radioimmunoassay assay. *Endocrinology*, 84: 1359–1364.

Gerozissis, K., Rougeot, C., Dray, F. 1986. Leukotriene C_4 is a potent stimulator of LHRH secretion. *Evr. J. Phamacol.* 121: 159–160.

Hales, T.G., Sanderson, J.M., Charles, A.C. 1994. GABA has excitatory actions on GnRH-secreting immortalized hypothalamic (GTI-7) neurons. *Neuroendocrinology*, 59: 297–308.

Han, S.K., Abraham, I.M., Herbison, A.E. 2002. Effect of GABA on GnRH neurons switches from depolarization to hyperpolarization at puberty in the female mouse. *Endocrinology*, 143: 1459–1466.

Han, S.K., Todman, M.G., Herbison, A.E. 2004. Endogenous GABA release inhibits the firing of adult gonadotropin-releasing horinone neurons. *Endocrinology*, 145: 495–499.

Harris, G.W. 1955. *Neural Control of the Pituitary Gland*. E. Arnold, London.

Hasegawa, Y., Miyamoto, R., Yazaki, C., Igrashi, M. 1981. Regulation of the second surge of FSH: effects of anti LHRH serum and pentobarbital. *Endocrinology*, 109: 130–135.

Hassing, J.M., Kletter, G.B., I'Anson H., Woods, R.L., Beitins, I.Z. Foster, D.L., Padmanabhan, V. 1993. Pulsatile administration of gonadotrophin-releasing hormone does not alter the follicle-stimulating hormone FSH isoform distribution pattern of pituitary or circulating FSH in nutritionally growth restricted ovariectomized; lambs. *Endocrinology*, 132: 1527–1536.

Herbison, A.E. 1998. Multimodal influence of estrogen upon gonadotropin-releasing hormone neurons. *Endocr. Rev.*, 19: 302–330.

Herbison, A.E., Theodorsis, D.T. 1992. Localization of estrogen receptors in preoptic neurons containing neurotensin but not tyrosine hydroxylase cholecystokinin or luteinizing hormone-releasing hormone in the male and female rat. *Neurosci.*, 50: 283–298.

Herbison, A.E. and Pape, J.R. 2001. New evidence for estrigen receptor-beta immunoreactivity in gonadotropin-releasing hormone neurons of rats and mice. 5th *Int. Cong. Neuroendocr. Bristol, UK*, pp. 336.

Herbison, A.E., Skinner, D.C., Robinson, J.E., King, I.S. 1996. Androgen receptor immunoreactive cells in ram hypothalamus: distribution and co-localization patterns with gouadotropin-releasing hormone, somatostatin and tyrosine hydroxylase. *Neuroendocrinology*, 63: 120–131.

Hoffman, A.R., and Crowley, W.F. 1982. Induction of puberty in men by long term pulsatile administration of low dose gonadotropin-releasing hormone. *N'Engl. J. Med.*, 307: 1237–1241.

Hohlweg, W., Junkman, K. 1932. Die hormonal nervose regulierung der function des hypophysenborderlappens. *Klin. Wsch.*, 11: 321–323.

Hokfelt, T., Eueritt, B.J., Theodorsson-Nerheim, E., Goldstein, M. 1984. Occurrence of neurotension-like immunorestivity in subopulations of hypothalamic, mesencephalic and medullary catecholamine neurons. *J. Comp. Neurol.*, 222: 543–549.

Hotkiss, J., Knobil, E. 1994. The menstrual cycle and its neuroendocrime control. In: *Physiology and Reproduction*, (Eds.) Knobil, E. and Neil, J.D. Ranen Press, New York, p. 711–749.

Hovarth, T.L., Pu, S., Duke, M.G., Diano, S. and Kalra, S.P. 2001. A GABAergic-neuropeptide-Y (NPY) interplay in LH release. *Peptides*, 22: 473–481.

Hrabovszky, E., Merchenthaler, I., Moenter, S.M., Liposits, Z. 2002. Comparative studies of nuclear estrogen receptor beta immunoreativity in gonadotropin-releasing hormone neurons of rats and mice. 5th *Int. Cong. Neuroendocr, Bristol, UK*, pp. 336.

Hrabovszky, E., Shughru, P.J., Merchenthaler, I., Hajszan T., Carpenter, C.D., Liposits, Z., Petersen, S.L. 2000. Detection of estrogen receptor-β messenger ribonucleic acid and 125, I estrogen binding sites in luteinizing hormone-releasing hormone neurons in the rat brain. *Endocrinology*, 141: 3506–3509.

Hrabovszky, E., Steinhauser, A., Barabas, K., Shughrne, P.J., Petersen, S.L., Merchenthaler, I., Liposits, Z. 2001. Estrogen receptor-beta immunoreactivity in luteinizing hormone releasing hormone neurons in the rat brain. *Endocrinology*, 142: 3261–3264.

Hutting, A.J., Lindgren, J.A., Hokfelt, T., Eneroth, P., Werner, S., Samuelsson, B. 1985. Leukotine C_4 as a mediator of LH release from rat anterior pituitary cells. *Proc. Natl. Acad. Sci. USA*, 52: 3834–3838.

Ichikawa, S., Sawada, T., Nakamura, Y., Morioka, H. 1974. Ovarian secretion of pregnane compounds during the estrous cycle and pregnancy in rats. *Endocrinology*, 94: 1615–1620.

Igrashi, M., McCann, S.M. 1974. A hypothalamic follicle stimulating hormone releasing factor. *Endocrinology*, 74: 446–452.

Igrashi, M., Nallar, R., McCann, S.M. 1973. Existence of FSH-RF distinct from LHRH. In: *Psychoneuro Endocrinology*, (Ed.) Hartoni, H. Karger Publ., Basel, p. 178–186.

Ikeda, Y., Swain, A., Weker, T.J., Hentges, K.E., Zanaria, E., Lalli, E., Tamai, K.T., Sassone-Corsi, P., Lovell-Badge, R. Camerino, G., Parker, K.L. 1996. Steroidogenic factor I and Dax-I co localize in multiple cell lineages: potential links in endocrine development. *Mol. Endocrinol.*, 10: 1261–1272.

Jarry, H., Perschl, A., Wutke W. 1988. Further evidence that preoptic anterior hypothalamic GABAergic neurous are part of the GnRH pulse and surge generator. *Acta Endocrinol.*, 118: 573–579.

Jensen, H.T., Cutter, C., Hardy, S., Lehman, M.N., Goodman R.L. 2003. Seasonal plasticity within the GnRH system of the ewe: changes in identified GnRH inputs and in the glial association. *Endocrinology*, 144: 3663–3676.

Kalra, P.S. 1985. Stimulation of hypoyhalamic LHRH levels and release by gonadal steroids. *J. Steroid Biochem.*, 23: 725–731.

Kalra, S.P. 1981. Neural loci involved in naloxne induced luteinizing hormone release. Effects of as norepinephrine synthesis inhibitior. *Endocrinology*, 109: 1805–10.

Kalra, S.P. 1983. Opioid peptides Inhibitory neuronal systems in regulation of gonadotropin secretion. In: *Role of Peptides and Proteins in Control of Reproduction*, (Eds.) McCann, S.M. and Dhindsa, D.S. Elsevier Biomedical, New York, p. 63–87.

Kalra, S.P. 1986. Neural circuitry involved in the control of LHRH secretion: a model for preovulatory LH release. In: *Frontiers in Neuroendocrinology*, Vol. 9, (Eds.) Ganong, W.F. and Martini, L. Raven Press, New York, p. 31–75.

Kalra, S.P., Crowley, W.R. 1984. Norepinephrine like effects of neuropeptide Y on LH release in the rat. *Life Sci.*, 35: 1173–1176.

Kalra, P.S., Crowley, W.R., Kalra, S.P. 1987a. Differential *in vitro* stimulation by naloxone of LHRH and catecholamine release from the hypothalamic of intact and castrated rats. *Endocrinology*, 120: 178–185.

Kalra, S.P., Dube, M.G., Sahu, A., Phelps, C.P. and Kalra, P.S.L. 1991. Neuropeptide Y secretion increases in the paraventri-nucleus in association with increased appetite for food. *Proc. Nat. Acad. Sci., USA*, 88: 10931–10935.

Kalra, S.P. and Hovarth, T.L. 1998. Neuroendocrine interactions between galanin, opioids and neuropetide Y in the control of reproduction and appetite. *Ann. N.Y. Acad. Sci.*, 863: 236–240.

Kalra, S.P., Leadem, C.A. 1984. Control of luteinizing hormone secretion by endogenous opioid peptides. In: *Opioid Modulation of Endocrinology Function*, (Eds.) Delitala, G. Motta, M., Serio, M. Raven Press, New York, p. 171–84.

Kalra, S.P. and Kalra, P.S. 1982. Stimulatory role of gonadal steroids on luteinizing hormone releasing hormone secretion. In: *Pitutary Hormones and Related Peptides*, (Eds.) Motta, M., Zanisi, M., Piva, F. Academic Press, New York, p. 157–170.

Kalra, S.P. and Kalra, P.S. 1983. Neural regulation of luteinizing hormone secretion in the rat. *Endocr. Rev.*, 4: 311–351.

Kalra, S.P. and Kalra, P.S. 1984. Opioids adrenergic steroid connection in regulation of luteinizing hormone secretion in the rat. *Neuroendocrinology*, 38: 418–426.

Kalra, S.P., Kalra, P.S., Sahu, A., Allen, L.G., Crowley, W.R. 1987b. The steroid–Neuropeptide connection in the control of LHRH secretion. In: *Regulation of Ovarian and Testicular Function*, (Eds.) Mahesh, V.B., Dhindsa, D.S., Anderson, E. and Kalra, S.P. Plenum Press, New York, London, p. 65–84.

Kalra, S.P., Simpkins, J.W. 1981. Evidence for noradrenergic mediation of opioid effects on luteinizing hormone secretion. *Endocrinology*, 109: 776–782.

Karsch, F.J., Cummins, J.T., Thomas, G.B., Clarks, I.J. 1987. Steroid feedback inhibition of pulsatile secretion of gonadotropin-releasing hormone in the ewe. *Biol. Reprod.*, 36: 1207–1218.

Karsch, F.J., Dahl. G.E., Evans, N.P. Manning, J.M., Mayfield, K.P., Merely, T., Poster, D.L. 1993. Seasonal changes in gonadotropin-releasing hormone in the ewe. *Biol. Reprod.*, 49: 1377–1383.

Kato, A., Hiruma, H., Kimura, F. 1994. Acute estradiol modulation of electrical activity of the LHRH pulse generator in the ovariectomized rat: restoration by naloxone. *Neuroendocrinology*, 59: 426–431.

Kaur, C. 1978. Effects of antifertility agents on hypothalamo-hypophyseal gonadal axis in male. *Ph.D. Thesis* G.N.D.U., Amritsar, Punjab, India.

Kaur, C. and Mangat, H.K. 1978. Possible effects of ovocyclin in male reproduction. *Int. J. Fert.*, 23: 229–233.

Kaur, C. and Porter, J.C. 1985. Biochemical estimation of neuronal activity of some dopaminergic systems of rat brain during estrous cycle. *Ann. Biol.*, 1: 103–109.

Kerkerian, L., Guy, J., Lefevre, G., Pelletier, G. 1985. Effects of neuropeptide Y (NPY) on the release of anterior pituitary hormones in the rat. *Peptides*, 6: 1201–1204.

Kim, K. and Ramirez, V.D. 1985. Dibutryl cyclic adenosine monophosphate stimulates *in vitro* luteinizing hormone releasing hormone release only from median eminence derived from ovariectomized, estradiol primed rat. *Brain Res.*, 342: 154–157.

Kimura, E. and Jinnai, K. 1994. Bicuculine infusions advance the timing of luteinizing hormone surge in proestrous rats comparisons with naloxone effects. *Horm. Behav.*, 28: 424–430.

Kimura, F., Kawakami, M. 1978. Reanalysis of the preoptic afferent involved in the surge of LH, FSH, and prolactin release in the proestrous rat. *Neuroendocrinology*, 27: 74–85.

King, J.C., Tai, D.W., Hanna, I.K., Pfeiffer, A., Haas, P., Ronsheim, P.M., Mitchell, S.C., Tureotte, J.C., Blaustein, J.D. 1995. A subgroup of LHRH neuron in guinea pigs with progestin receptors is centrally positioned within the total population of LHRH neurons. *Neuroendocrinology*, 61: 265–275.

King, J.C., Tobet, S.A., Snauely, F.L., Arimura, A.A. 1982. LHRH immunoreactive cells and their projections to the median eminence and organum vasculosum of Lamina Terminals. *J. Comp. Neural.*, 209: 287–300.

Korach, K.S. and Muldoon, T.G. 1973. Comparison of specific 17β-estradiol–receptor interactions in the anterior pituitary of male and female rats. *Endocrinology*, 92: 322–326.

Kusano, K., Fueshko, S., Gainer, H., Wray, S. 1995. Electrical and synaptic properties of embryonic luleinizing hormone-releasing hormone neurons in explant cultures. *Proc. Nat. Acad Sci., USA*, 92: 3918–3922.

Lamberts, R., Vijayan, E., Graf, M., Mansky, T., Wattke, W. 1983. Involvement of preoptic-anterior hypothalamic GABA neurons in the regulation of pituitary LH and prolactin release. *Exp Br. Res.*, 52: 356–362.

Lamperti, and Hill, L. 1987. The effects of anterior hypothalamic differentiation on FSH-releasing activity in the intact female hamster. *Exp. Br. Res.*, 68: 189–194.

Leadem, C.A., Crowley, W.R., Simpkins, J.W., Kalra, S.P. 1985. Effects of naloxone on catecholamines and LHRH release from perifused hypothalamus of steroid primed rat. *Neuroendocrinology*, 40: 497–500.

Leadem, C.A., Kalra, S.P. 1985. Effects of endogenous opioid peptides and opiates on luteinizing hormone and prolactin secretion in ovariectomized rats. *Neuroendocrinology*, 41: 342–352.

Lehman, M.N., Karsch, F.J. 1993. Do gonadotropin-releasing hormone, tyrosine hydroxylase and β-endorphin immunoreactive neurons contain estrogen receptors? A double label immunocytochemical study in the Suffolk ewe. *Endocrinology*, 133: 887–895.

Leipheimer, R.E., Bona-Gallo, A., Gallo, R.V. 1984. The influence of progesterone and estradiol on the acute changes in pulsatile. Luteinizing hormone release induced by ovariectomy in diestrous day I in rat. *Endocrinology*, 114: 1605–1612.

Leipheimer, R.E., Bona-Gallo, A., Gallo, R.V. 1985. Ovarian steroid regulation of pulsatile LH release during interval between mornings of diestrus 2 and proestrus in the rat. *Neuroendocrinology*, 41: 252–257.

Leipheimer, R.E., Bona-Gallo, A., Gallo, R.V. 1986. Influence of estradiol and progesterone on pulsalile LH secretion in 8-day ovariectemized rats. *Neuroendocrinology*, 37: 421–426.

Leranth, C., Shanabrough, M., Naftolin, F. 1991. Estrogen induces ultrastructural changes in progesterone receptor containing GABA neurons of the primate hypothalamus. *Neuroendocrinology*. 54: 571–579.

Levine, J.E., Bauer-Dentoin, A.C., Besecke, A.C. Besecke, L.M., Conaghan L.A., Legan, S.J., Meradith, J.M., Strobl, P.J., Urban, J.H., Vogelsong, K.M., Wolfe, A.M. 1991. Neurondocrine regulation of the luteinizing hormone-releasing hormone pulse generator in the rat. *Rec. Prog. Horm. Res.*, 47: 97–151.

Levine, J.E., Norman, R.L., Gliessman, P.M., Oyama, T.T., Bangsberg, D.R., Spies, H.G. 1985. *In vivo* gonadotropin-releasing hormone release and serum luteinizing hormone measurement in ovariectomized, estrogen-treated rhesus macaques. *Endocrinology*, 117: 711–721.

Levine, J.E., Remirez, V.D. 1982. Luteinizing hormone releasing hormone release during the rat estrous cycle and after ovariectoney as estimated with push-pull cannule. *Endocrinology*, 111: 1439–1448.

Lindgren, J.A., Hokfelt, T., Dahleu, S.E., Patron, C., Samuelsson, B. 1984. Leukotrienes in the rat central nervous system. *Proc. Nat. Acad. Sci., USA*, 81: 6212–6216.

Lindstedt, G., Nystron, E., Mathews, C., Ernest, I., Janson, P.O., Chatterjee, K. 1998. Follitropin (FSH) deficiency in an infertile male due to FSH beta gene mutation. A syndrome of normal puberty and virilization but underdeveloped testicles with azoospermia, low FSH but high lutropin and normal serum testosterone concentrations. *Clin. Chem. Lab. Med.*, 36: 663–665.

Lofstrom, A. 1997. Catcholamines turnover alterations indiscreate area of the median eminence of the 4- and 5-day cycle rat. *Brain Res.*, 120: 113–131.

Lumpkin, M.D. and McCann, S.M. 1984. Effect of destruction of the dorsal anterior hypothalamus on follicle-stimulating hormone secretion in the rat. *Endocrinology*, 115: 2473–2480.

Lumpkin, M.D., McDonald, J.K., Samson, W.K., McCann, S.M. 1989. Distruction of the dorsal anterior hypothalamic region suppresses pulsatile release of follicle stimulating hormone but not luteinizing hormone. *Neuroendocrinology*, 50: 229–235.

Lumpkin, M.D., Moltz, J.H., Yu., W.H., Samson, W.K. McCann, S.M. 1987. Purification of FSH-releasing factor: its dissimilarity from LHRH of mammalian, avian and piscian origin. *Br. Res. Bull.*, 18: 175–178.

Malven, P.V., Bossut, D.F.B., Diekman, M.A. 1984. Effects of naloxone and electroaccupuncture treatment on plasma concentrations of LH in sheep. *J. Endocrinol.*, 101: 75–80.

Marshall, J.C., Griffin, M.L. 1993. The role of changing pulse frequency in the regulation of ovulation. *Hum. Reprod.*, 8: 57–61.

Martinez de la, E.G., Choi, A.L., Weiner, R.I. 1992. Generation and synchronization of gonadotrophin releasing hormone (Gn-RH) pulses: intrinsic properties of the GT-I GnRH neuronal cell line. *Proc. Natl. Acad. Sci., USA*, 89: 1852–1855.

Marubayashi, U., Yu, W.H., McCann, S.M. 1999. Median eminence lesions reveal separate hypothalamic control of pulsatile follicle-stimulating hormone and luteinizing hormone release. *Proc. Soc. Exp. Biol. Med.*, 220: 139–146.

Mather, J.P., Woodruff, T.K., Krummon, L.A. 1992. Paracrine regulation of reproductive function by inhibin and activin. *Proc. Soc. Exp. Biol. Med.*, 201: 1–15.

Mathews, L.S., Vale, W.W. 1991. Expression cloning of an activin receptor, a predicted transmembrane serine kinase. *Cell*, 65: 973–982.

Matsuo, H., Baba, Y., Nair, R.M., Arimura A., Schally, A.V. 1971. Structure of the porcine LH and FSH releasing hormone. *Biochem. Biophys. Res. Comm.*, 43: 134–139.

McCann, S.M., Ojeda, S., Negro-Vilar, A. 1974. Sex steroid, pituitary and hypothalamic hormone during puberty in experimental animals. In: *Control of the Onset of Puberty,* (Eds.) Grumbach, M.M., Grave G.D., Mayer, F.F. John Wiley and Sons, New York, p. 1–31.

McCartney, C.R., Gingrich, M.B., Hu, Y., Evans, W.S., Marshall, J.C. 2002. Hypothalamic regulation of cyclic ovulation: evidence that the increase in gonadotrphin releasing hormone pulse frequency during the follicular phase reflects the gradual loss of the restraining effects of progesterone. *J. Clin. Endocrinol. Metab.,* 87: 2194–2201.

McCullagh, D.R. 1932. Dual endocrine activity of the testis. *Science,* 76: 19–20.

McEwen, B.S., Alves, S.E. 1999. Estrogen actions in the central nervous system. *N. Engl. J. Med.,* 340: 1801–1811.

McEwen, B.S., Alves, S.E., Bulloch, K., Weiland, N.G. 1997. Ovarian steroids and the brain: implications for cognition and aging. *Neurology,* 48: 58–515.

McNeilly, A.S. 1988. The control of FSH secretion Acta Endocrinol Suppl. *Copenh.,* 288: 31–40.

McPherson, J.C., Mahesh, V.B. 1979. Dose-related effect of a single injection of progesterone on gonadotrophin secretion and pituitary sensitivity to LHRH in estrogen primed castrated female rats. *Biol. Reperod.,* 20: 763–772.

Meites, J., Simpkins J., Brumi, J. and Advis, J. 1977. IRCS. *Med Sci.,* 5: 1.

Midgley, A.R. Jr., McFadden, K., Ghazzi, M., Karsch, P.J., Brown, M.B., Mauger, D.T. and Padmanabhan, V. 1997. Nonclassical secretary dynamics of LH revealed by hypothalamo-hypophyseal porta sampling of sheep. *Endocrine,* 6: 133–143.

Mitsushima, D., Hei, D.L., Tersawa, E. 1994. Gamma-amino butyric acid is an inhibitory neurotransmitter restricting the release of luteinizing hormone-releasing hormone before the onset of pubeaty. *Proc. Nat. Acad. Sci., USA,* 91: 395–399.

Miyake, A., Aono, T., Tamizawa, O., Kinugasa, T., Kurachi, K. 1977. Influence of chrionic gonadotropin on the response of luteinizing hormone to luteinizing hormone releasing hormone in gonadectomized women. *J. Endocrinol.*, 74: 499.

Mizunuma, H., Samsson, W.K., Lumpbin, M.D., Mottz, J.H., Kawcett, C.P., McCann, S.M. 1983. Purification of a bioactive FSH-releasing factor (FSH-RF). *Brain Res. Bull.*, 10: 623–629.

Moenter, S.M., Carrty, A., Locatelli, A., Karch, F.J. 1991. Pattern of gonadotropin-releasing hormone (GnRH) secretion leading upto ovulation in the ewe: existence of a preovulatory GnRH Snrge. *Endocrinology*, 129: 1175–1182.

Moenter, S.M., Sullivan, S.D., De Fazio, R.A. 2004. Endogenous GABA can exite GnRH neurons. In: *86th Meeting of the Endocrine Society*.

Mondschein, J.S., Hersey, R.M., Weisz, J. 1986. Purification and characterization of estrogen 2/4-hyderoxylase activity from rabbit hypothalamic peroxidase mediated cateholl estrogen formation *Endocrinology*, 119: 1105–1112.

Moore, C.R. and Price, D. 1932. Gonadal hormone functions and the reciprocal influence between gonads and hypophysis. *Am. J. Anat.*, 50: 13–17.

Mori, Y., Mishihara, M., Tanaka, T., Shimizu, T., Yamaguch, M., Takeuchi, Y., Hoshino, K. 1991. Chronic recording of electrophysiological manifestation of the hypothalamic. Gonadotropin-releasing hormone pulse generator activity in the goat. *Neuroendocrinology*, 53: 392–395.

Morrell, J., MEGinty, F. and Pfaff, D.W. 1985. A subset of β-endorphin or dynorphin-containing neurons in the medial basal hypothalamus accumulates estradiol. *Neuroendocrinology*, 14: 417–426.

Muldoon, T.G., Singh, P., Watson, G.H., Harper, S.A. 1984. Molecular determinants of pituitary responsiveness to estrogen and luteinizing hormone releasing hormone. In: *Hormone Receptors in Growth and Reproduction*, (Eds.) Sexana, B.B., Catt, K.J., Birnbaumer, L. and Martini, L. Raven Press, New York, p. 237–248.

Murphy, L.L. and Mahesh, V.B. 1985. Selective release of follicle–stimulating hormone and lutenizing hormone by 5α-dihydroxy progesterone and 3α, 5α-tetrahydro progesterone in pregnant mares serum gonadoterphin-primed immature rats exposed to light. *Biol. Reprod.*, 32: 795–803.

Negro-Vilar, A., De Paolo, L., Tesone, M., Johnston, C.A. 1982. Role of brain peptides in the regulation of gonadotropin secretion. In: *Therapy in Andrology: Pharmacological Surgical and Physiological Aspects,* (Eds.) Menchini-Febris, F., Pasini, and W. Martini L. Excepta Medica, Amsterdam, p. 115–123.

Negro-Vilar, A. and Ojeda, S.R. 1981. Hypophysiotrophic hormone of the hypothalamus. In: *Endocrine Physiology III, and International Review of Physiology,* (Ed.) McCann, S.M. University Park Press, Baltimore, p. 97–156.

Negro-Vilar, A., Ojeda, S.R., McCann, S.M. 1979. Catecholaminergic modulation of luteinizing hormone releasing hormone release by median eminence terminals *in vitro. Endocrinology,* 104: 1749–1757.

Nolin, J.M., Campbell, G.T., Nausel, D.D. and Bogdanone, E.M. 1977. Does androgen influence prolactin secretion. *Endocrinol Res. Commn.,* 4: 61–70.

Nunemaker, C.S., De fazio, R.A. and Moenter, S.M. 2002. Estradiol sensitive afferents modulate long-term episodic firing patterns of GnRH neurons. *Endocrinology,* 143: 2284–2292.

Padmanabhan, V. 2001. Neuroendocrine regulation of follicle stimulating hormone. In: *Follicular Growth, Ovulation and Fertilization: Molecular and Clinical Basis* (Eds.) A. Kumar and A.K. Mukhopadhyay. Narosa Publishing House, New Delhi.

Padmanabhan, V., Van Cleeff, J., McLeod, M.K., Nett, T.M. and Karsch, F.J. 1997a. Is there a GnRH-independent hypothalamic control of FSH secretion? *Prog. Abstr Ann Meet. Soc. Neurosci.,* St. Louis, p. 588 (Abst.).

Padmanabham, V., McFadden, K., Mauger, D.T., Karsch, F.J. and Midgley, A.R. Jr. 1997b. Neuroendocrine control of FSH secretion: I Direct evidence for separate episode and basic components of FSH secretion. *Endocrinology,* 138: 424–432.

Padmanabhan, V., McNeilly, A.S. 2001. Is there an FSH-releasing factor? *Reproduction,* 121: 21–30.

Parent, A-S., Lebrethon, M-C., Gerard, A., Bourguignon, J-P. 2005. Factor accounting for perinatal occurrence of pulsatile gonadotrophin-releasing hormone secretion *in vitro* in rats. *Biol. Reprod.,* 72: 143–149.

Pastor, C.L., Griffin-Korf, M.L., Aloi, J.A., Evans, W.S. and Marshall J.C. 1998. Polycystic ovary syndrome: evidence for reduced sensitivity to inhibition by estradiol and progesterone. *J. Clin. Endocrinol. Metab.*, 83: 582–590.

Pau, K.Y-F., Gleissman, P.M., Oyama, T. and Spies, H.G. 1991. Disruption of GnRH pulses by anti-GnRH serum and phentolamine obliterates pulsatile LH but not FSH secretion in ovarietomized rabbits. *Neuroendocrinology*, 53: 381–382.

Petraglia, F., Locatelli, V., Facchinetti F., Bergamaschi, M., Genazzani, A.R., Cocchi, D. 1986. Estrous cycle related LH responsiveness to naloxne: effect of high estrogen leaves on the activity of opioid receptors. *J. Endocrinol.*, 108: 89–94.

Petraglia, F., Locatelli, V., Penalova, A., Cocchi, D., Genazzani, A.R., Muller, E.E. 1984. Gonadal steroid nodulation of naloxone-induced LH secretion in the rat. *J. Endocrinol.*, 101: 33–39.

Pfaff, D.W. 1980. *Estrogens and Brain Function*. Springer-Verlag, New York.

Pfeiffer, D.G., Pfeiffer, A., Shimohigashi, Y., Merriam, G.R. and Loxiaux, D.L. 1983. Predominant involvement of Mu-rather than delta or Rappa–opiate receptors in LH secretion. *Peptides*, 4: 647–649.

Phillip, M., Arbelle, J.E., Segev, Y. and Parvari, R. 1998. Male hypogonadism due to a mutation in the gene for the beta sub-unit of follicle stimulating hormone. *N. Engl. J. Med.*, 338: 1729–1732.

Piva, F., Limonta, P., Maggi, R. and Martini, L. 1986. Stimulatory and inhibitory effects of the opioids on the gonadotropin secretion. *Neuroendocrinology*, 42: 504–512.

Piva, F., Maggi, R., Limonta, P., Motta, M. and Martini, L. 1985. Effect of naloxone on luteinizing hormone, follicle stimulating hormone and prolactin secretion in the different phases of the estrous cycle. *Endocrinology*, 117: 766–772.

Pompolo, S., Pereira, A., Kancko, T. and Clarke, I.J. 2003. Seasonal changes in the inputs to gonadotropin-releasing hormone neurons in the ewe brain: an assessment by conventional fluorescence and confocal microscopy. *J. Neuroendocrinol.*, 15: 538–545.

Porter, J.C., Reymond, M.J. and Kaur, C. 1983. Secretion of hypothalamic dopamine into the hyperphyseal portal vasculature. In: *Proloactin and Prolactinomas*, (Eds.) Talis *et al.* Raven Press, New York, p. 71–81.

Purnelle, G., Gerard, A., Czajkowski, V. and Bourguignon, J.P. Pulsatile secretion of gonadotrophin-releasing hormone by rat hypothalamic explants without cell bodies of GnRH neurons. *Neuroendocrinology*, 66: 305–312.

Purvis, K., Clausen, O.P.F., Olsen, A., Hang, F. and Hansen, V. 1979. Prolactin and Leydig cell responsiveness to LH/HCG in the rat. *Arch. Androl.*, 3: 219–230.

Rasmussen, D.D., Liu, J.H., Wolf, P.L. and Yen, S.S.C. 1983. Endogenous opioid regulation of gonadotropin releasing hormone release from the human fetal hypothalamus *in vitro*. *J. Clin. Endocrinol Metab.*, 57: 881–884.

Reymond, M.J., Kaur, C. and Porter, J.C. 1983. An inhibitory role for morphine on the release of dopamine into hyperphyseal portal blood and on the synthesis of dopamine in tuberoinfundibular neurons. *Brain Res.*, 262: 253–258.

Rivier, C., Rivier, J. and Vale, W. 1986. Inhibin-mediated feedback control of follicle stimulating hormone secretion in the female rat. *Science*, 234: 205–208.

Rivier, J., Spiess, J., Mclintok, R., Vanghan, J. and Vale, W. 1985. Purification and partial characterization of inhibin from follicular fluid. *Biochem. Biophys. Res. Commn.*, 133: 120–127.

Robertson, D.M. 1992. Follistatin/Activin-binding protein. *Trends Endocrinol. Metab.*, 3: 65–68.

Robertson, D.M., Devos, F.L., Foulds, L.M., Mchachlan, R.I., Burger, H.G., Morgan, F.J, Hearn, M.T.W. and deKrester, D.M. 1986. Isolation of a 31KD a form of inhibin from bovine follicular fluid. *Molec. Cell. Endocr.*, 44: 271–277.

Robinson, J.E., Forsdike, R.A. and Taylor, J.A. 1999. In utero exposure of female lambs to testosterone reduces sensitivity of the gonadotropin-releasing hormone neuronal network to inhibition by progesterone. *Endocrinology*, 140: 5797–5805.

Robinson, J.E., Kendrick, K.M. and Lambart, C.E. 1991. Changes in the release of gamma-aminobutyric acid and catecholamines in the preoptic/septal area prior to and during the preovulatory surge of luteinizing hormone in the ewe. *J. Neuroendocrinol.*, 3: 393–399.

Rush, M., Ashiru, O.A. and Blake, C.A. 1982. Effects of complete hypothalamic deafferentation on the estrous phase of FSH release in the cyclic rat. *Biol. Reprod.*, 26: 399–403.

Samuelsson, B. 1983. Leukotrienes: mediators of immediate hypersensitivity reactions and inflammation. *Science*, 220: 227–230.

Sarkar, D.K. and Yen, S.S.C. 1985. Changes in β-endorphin like immunoreactinity in pituitary portal blood during the estrous cycle and after ovarietomy in rats. *Endocrinology*, 116: 2075–2079.

Schally, A.V., Arimura, A., Redding, T.W., Dubeljuk, L., Karter, W., Dupont, A. and Vilchez-Martinez, J.A. 1976. Reexamination of poreine and bovine hypothalamic fractions for additional luteinizing hormone and follicle stimulating hormone releasing activities. *Endocrinology*, 98: 380–391.

Scharrer, B. *Amer. Zool.*, 7: 161.

Seminara, S.B., Acherman, J.C., Genel, M., Jameson, J.L. and Crowley, Jr. W.F. 1999. X-linked adrenal hypoplasia congenital; a mutation in DAXI expands the phenotype spectrum in males and females. *J. Clin. Endocr. Metab.*, 84: 4501–4509.

Seminara, S.B. and Crowley, W.F. 2001. Perspective: The importance of genetic defects in humans in elucidating the complexities of the hypothalamic-pituitary gonadal axis. *Endocrinology*, 142: 2173–2177.

Sheridan, R., Loras, B., Surardt, L., Ectors, F. and Pasteels, J.L. 1979. Autonomous secretion of follide stimulating hormone by long-term organ cultures of rat pituitaries. *Endocrinology*, 104: 198–204.

Shillo, K.K., Leshin, L.S., Kueh, D. and Jackson, G.L. 1985. Simultaneous measurement of luteinizing hormone releasing hormone and luteinizing hormone during estradiol-induced luteinizing hormone surges in ovariectonized ewe. *Biol. Reprod.*, 33: 644–652.

Shivers, B.D., Harlan, R.E., Morrell, J.I. and Pfaff, D.W. 1983. Absence of qestradiol concentration in cell nuclei of LHRM-immunoreative neurons. *Nature*, 304: 345–347.

Shull, J.D. and Gorski, J. 1984. Estrogen stimulates prolactin gene transcription by a mechanism independent of pituitary protein synthesis. *Endocrinology*, 144: 1550–1557.

Sim, J.A., Skynner, M.J., Pape, J.R. and Herbison, A.E. 2000. Late postnatal reorganization of GABA (A) receptor signaling in native GnRH neurons. *Eur. J. Neurosci.*, 12: 3497–3504.

Skinner, D.C., Caraty, A. and Allingham, R. 2001. Unmasking the progesterone receptor in the preoptic area and hypothalamus of the ewe: no colocalization with gonadotropin-releasing neurons. *Endocrinology*, 142: 573–579.

Skynner, M.J., Sim, J.A. and Herbison, A.E. 1999. Detection of estrogen receptor α and β-messenger ribonucleic acids in adult gonadotropin-releasing hormone neurons. *Endocrinology*, 140: 5195–5201.

Spergel, D.J., Kruth, U., Henley, D.F., Sperengel, R. and Seeburg, P.H. 1999. GABA and glutamate-activated channels in green fluorescent protein-tagged gonadotropin-releasing hormone neurons in transgenic mice. *J. Neurosci.*, 19: 2037–2050.

Sullivan, S.D., Defazio, R.A. and Moenter, S.M. 2003. Metabolic regulation of fertility through presynaptic and postsynaptic signaling to gonadtrophin-releasing hormone neurons. *J. Neurosci.*, 23: 8578–8585.

Sullivan, S.D. and Moenter, S.M. 2004. Gamma-aminobutyric acid neurons integrate and rapidly transmit permissive and inhibitory metabolic cues to gonadotrophin-releasing hormone neurons. *Endocrinology*, 145: 1194–1202.

Sullivan, S.D. and Moenter, S.M. 2005. GABAergic integration of progesterone and androgen feedback to gonadoterophin-releasing hormone neurons. *Biol. Reprod.*, 72: 33–41.

Suter, K.J., Song, W.J., Sampson, T.L., Wuarin, J.P., Saunders J.T., Dudek, F.E. and Moenter, S.M. 2000. Genetic targeting of green fluorescent protein to gonadotrophin-releasing hormone neurons characterization of whole cell electrophysiological properties and morphology. *Endocrinology*, 141: 412–419.

Terasawa, E., Keen, K.J., Magi, K. and Clande, P. 1999. Pulsatile release of luteinizing hormone-releasing hormone (LHRH) in cultured LHRH neurons derived from the embryonic olfactory placed of the rhesus monkey. *Endocrinology*, 140: 1432–1441.

Terasawa, E., Lushansky, L.L., Kasuya, E. and Nyberg, C.L. 1999. An increase in glutamate release follows a decrease in gamma-amino butyric acid and the pubertal increase in luteinizing hormone releasing hormone in the female rhesus monkeys. *J. Neuroendocrinol.*, 11: 275–282.

Turgeon, J.L. 1979. Estradiol-luteinizing hormone relationship during the proestrous gonadotrop in surge. *Endocrinology*, 105: 731–736.

Turgeon, J.L. and Waring, D.W. 1982. Differential changes in the rate and pattern of follicle stimulating hormone secretion from pituitaries of cyclic rats superfused *in vitro*. *Endocrinology*, 11: 66–73.

Ulloa-Aguirre, A., Midgley Jr., A.R., Beitins, I.Z. and Padmanabhan, V. 1995. Follice stimulating Isohormones; Biological characterization and physiological relevance. *Edocr. Rev.*, 16: 765–787.

Valenca, M.M., Conte, D. and Negro-Vilar, A. 1985. Diacyeglycerol and phorbol esters enhance LHRH and prostaglandin E secretion from mediam eminence terminals *in vitro*. *Brain Res. Bull.*, 15: 657–59.

Vu Hai, M.T., Logeat, F., Warenbourg, M. and Milgrom, E. 1977. Hormonal control of progesterone receptors. *Ann. N.Y. Acad. Sci.*, 286: 199–209.

Wardlow, S.C., Thoron, L. and Frantz, A.G. 1982. Effects of sex steroids on brain β-endophin. *Brain Res.*, 245: 327–331.

Watson, S.J., Khachaturian, H., Akil, H., Coy, D.H. and Goldstein, A. 1982. Comparison of the distribution of dynorphin system and enkephalin system in brain, *Science*, 218: 1134–1136.

Weiner, R.I. and Ganong. 1979. Role of brain monoamines and histamine in regulation of anterior secretion. *Physiol. Rev.*, 58: 905–906.

Weiss, J., Adams, E., Whitcomb, R.W., Crowley, W.F. and Jameson, J.L. 1991. Normal sequence of the gandotrophin-releasing hormone gene in patients with idiopathic hypogonadtropic hypogonadism. *Biol. Reprod.*, 45: 743–747.

Weiss, J., Crowley, W.F and Iameson, J.L. 1989. Normal structure of the gonadoterphin-releasing hormone (GnRH) gene in patients with GnRH deficiency and idiopathic hypogonadism. *J. Clin. Endocrinol. Metab.*, 69: 299–303.

Wetsel, W.C., Valenca, M.M., Merchenthaler, I., Lipsits, Z., Lapez, F.I., W einer, R. T., Mellon, P.L. and Negro-Vilar, A. 1992. Intrinsic pulsatile secretory activity of immortalized luteinizing hormone-releasing hormone secreting neurons. *Proc. Natl. Acad. Sci., USA*, 89: 4149–4153.

Wildt, L., Hausler, A., Marshall, G., Hutchinson I.S., Plant, T.M., Belchetz, P.E. and Knobil, E. 1981. Frequency and amplitude of gonadotrophin-releasing hormone stimulation and gonadotrophin recreation in the rhesus monkey. *Endocrinology*, 109: 376–385.

Wise, P.M., Rance, N., Barr, G.D. and Baaraclough, C.A. 1979. Further evidence that luteinizing hormone releasing hormone also is follicle-stimulating hormone releasing hormone. *Endocrinology*, 104: 940–947.

Yamanaka, C., Lebrethon, M.C., Vandersmissen, E., Gerard, A., Pumelle, G., Lemaire, M., Wilk, S. and Bourguignon, J.P. 1999. Early prepubertal outogenny of pulsatile gonadotrophin releasing hormone (GnRH) secretion: I inhibitory auto feedback control through prolyl endopeptidase degradation of GnRH. *Endocrinology*, 140: 4609–4615.

Ying, S. 1988. Inhibins, activins and follistatins: gonadal proteins modulating the secretion of follicle stimulating hormone. *Endocr. Rev.*, 9: 267–293.

Yu, W.H., Karanth, S., Walczewska, A., Sower, S.A. and McCann, S.M. 1997. A hypothalamic follicle stimulating hormone releasing decapeptide in the ram. *Proc. Nat Acad Sci., USA*, 94: 9499–9503.

Zanisi, M. and Martini, L. 1975. Differential effects of castration on LH and FSH secretion in male and female rats. *Acta Endocrinol.*, 78: 683–688.

Chapter 20

Effect of Some Antifertility Agents on the GSI in Albino Rats

Mangla Bhide* and Angel Gupta

Department of Zoology, Dr. H.S. Gour Vishwavidyalaya, Sagar – 470 003, M.P.

ABSTRACT

The antifertility agents *e.g. Carica papaya* seed extract, levonorgestrel and ethinylestradiol showed adverse effect on the GSI (MFI, FFI, UI) hence directly or indirectly affecting the fertility of experimental animals *e.g.* in albino rats in the present investigation.

The parameter of GSI plays very significant role which evaluate the reproductive potency of the concerning animal. Deviation from normal or control and increase in GSI indicating the oedematous condition or cancerous growth or hyperactivity of gonads while decrease in GSI indicating poor development, disfunctioning or inactiveness of gonads and malformation of gametes. Doses and duration of treatment dependent changes were recorded and summarized. The antifertility activity of the treated agents and these agents can used for the birth control of the animal pests having high rate of fertility.

* Corresponding Author: Dr. Mangla Bhide, 27 Vidyapuram, Makronia, Sagar –470 004, M.P., Email: manglabhide@rediffmailcom, manglabhide@yahoo.com

***Keywords:** Antifertility agents, GSI, Albino rats,* Carica papaya.

Introduction

Rats are very harmful pest of standing crops, stored food grain, fodder, vegetables, fruits and other valuable household. They not only cause damages but also act as vector of so many diseases. So to control their progeny, it is most significant to introduce them with some antifertility agents of plants origin because synthetic antifertility agents cause a lot of side effects to user but when synthetic and plants origin antifertility agents be used as synergist the antifertility effect will be prominent without any side effect.

The increase in GSI was due to the accumulation of fluid in the testes by the obstacle in initial segment of epididymis was reported by Ye *et al.* (1993) in male rats after gossypol polyacetic acid treatment. The decrease in GSI as well in testicular weight seems to be disfunctioning of the testes due to the action of antifertility agent's treatment as reported in rats by Tyagi *et al.* (1990) after *Trigonela foenum graceum* treatment, Lohiya *et al.* (1993, 1994 and 1999) and Bhide *et al.* (2002) in albino rats with *Abrus precatorius* seed extract treatment in male rats and Subhalakshmi and Reddy (1996) after *Hyosyamus niger* seed extract treatment respectively.

Antispermatogenic effect were accompanied by significant decrease in testicular weight as well as GSI in male rats was reported by Chawla *et al.* (1996); Subhalakshmi and Reddy (1996) and Lohiya *et al.* (1999) in male albino rats and male rabbits after treatment with *Carica papaya* seed extract; *Hyosyamus niger* seed extract and chloroform extract of *Carica papaya* seed extract respectively. While Tyagi and Rajalakshmi (1996) in adult Rhesus monkey after androgen exposure observed the same effect. Tyagi *et al.* (1990) reported the drastic changes in the reduction of body weight but also on the testicular weight after treating the male albino rats with *Trigonela foenum graceum* acetone extract as antifertility agent as also reported antispermatogenic effect by Sharma *et al.* (1996) in male after *Marsilea minuta* sporocarp and *Capsicum indicum* exposure; Parte *et al.* (1996) in male rats after *Xenoestrogens* exposure, Singh (1996) in male rats after *Hibiscus rosa sinensis* and Bhide *et al.* (2002) in albino rats after *Abrus precatorius* seed extract exposure respectively.

The increase or decrease in GSI is correlated with the increase or decrease in the body weight with respect to the increase or decrease in ovarian weight as observed in rats by Adhikari *et al.* (1990), Choudhari *et al.* (1991) after *Piper betle* leaf stalk extract treatment; Pandey *et al.* (1992) after *Hibiscus rosa sinensis* flower extract exposure and Sreenivasulu *et al,* (1992) after CDRI 85/287 compound treatment and stated that reduction in weight of ovaries was correlated with the anti-estrogenic activity as reported in albino rats by Gupta (2000), Bhide *et al.* (2001, 2002) and Gupta *et al.* (2003) after treatment with some extracts of plant origin.

The decrease in uterine weight was also reported in albino rats by Adhikari *et al.* (1990) and Choudhari *et al.* (1991) after *Piper betle* leaf stalk extract exposure, Pandey *et al.* (1992) after *Hibiscus rosa sinensis* flower extract exposure and by Sreenivasulu *et al.* (1992) after CDRI 85/287 compound exposure, Gupta (2000), Gupta *et al.* (2003) with *Carica papaya* seed extract respectively. While, Giand (1999) reported the increase in weight of the uterine horns in albino rats after treated female with some intermediate azasteriod was due to its estrogenic activity as indicated by the gradual increase in the ovarian weight. The present investigation is, therefore, intended to assess the antifertility activity of *Carica papaya,* levonorgestrel and ethmylestradlol on the GSI of albino rats.

Materials and Methods

The experiments were conducted in the present investigation on the live specimens of albino rats and the details of the same as follows:

Procurement and Rearing of Albino Rats

Indian Veterinary Research Institute colony-bred albino rats procured from Jabalpur (M.P.) were used in the present investigation. The rats were acclimatized for 15 days on the normal laboratory conditions prior to experiments according to the methods of Edmond (1950) and were maintained on the balanced diet (Hindustan lever Ltd.) and water *ad libitum.*

Test Animals

Albino rats of both the sexes averaging between 180 g to 200 g initial body weight and aged about 90 to 95 days were randomly bred and the male and female rats obtained from first litter were used for experiments. All the normal, control and experimental rats were housed into wooden cages at the room temperature under normal laboratory conditions and provided with balance diet and clean water *ad libitum* (Edmond, 1950).

Procurement of *Carica papaya* (C.P.) Seeds and Extraction of Antifertility Agent

Seeds of *Carica papaya* (belongs to family Caricaceae) were collected locally, washed, shadow dried, crushed mechanically into coarse pieces and weighed 100 g. The antifertility agent was extracted with petroleum ether (B.P. 40°C–60°C procured from BDH) from coarse pieces of *Carica papaya* seeds by Soxhlet method (after Sharma, 1988). The extract was concentrated and evaporated to dryness in water bath (after Sharma, 1988). The extracted antifertility agent weighed and 1 per cent stock solution was prepared in distilled water. For experimental purposes further dilution of stock solution with distilled water to prepare 0.1 per cent experimental solution.

Procurement of Levonorgestrel (N.G.) and Ethinylestradiol (E.E.) as Synthetic Antifertility Agent

In the present investigation N.G. (+) 13 ethyl-17-hydroxy-18, 19 dinor-17-pregn-4-en-20-yn-3-one) + E.E (19-nor-17 pregna-1,3,5 (10)-trien-20 yne-3, 17-diol) were procured from "Wyeth Lederie Ltd., Germany". These antifertility compounds inhibit ovulation and also used as oral contraceptive agents.

All these solutions (stock and experimental solutions) of C.P. seed extract and powdered N.G. + E.E. antifertility compounds were kept in refrigerator at 4°C in air tight glass bottles and at the time of ingestion the required amount of experimental solution and N.G. + E.E. powder were taken out and kept in separate glass tubes with corks and wait till they obtained the room temperature.

Experiments with Different Dosages of Antifertility Agents

Each groups of 5 albino rats (male or female) were treated orally with doses of antifertility agents along with synergists in triplicate as follows.

Carica papaya Seed Extract Treatment

0.2 ml C.P. seed extract/100 g body wt./dose/rat daily orally ingested up to 90 days.

Levonorgestrel and Ethinylestradiol Treatment

100 mg N.G. and 10 mg E.E./100 g body wt./dose/rat daily orally ingested up to 90 days.

Synergistic (CP + N.G. + E.E.) Treatment

Synergistic doses of 0.2 ml C.P. seed extract with 100 mg N.G. and 10 mg E.E./100 g body wt./dose/rat daily orally ingested up to 90 days.

Calculation of Gonadosomatic Index (GSI)

Mean weight of each group and their gonads and Uteri have been recorded and summarized in Tables 20.1 to 20.6 and GSI (Gonado Somatic Index) was calculated by the following formula:

$$\text{GSI (Testicular tissue) or MFI} = \frac{\text{Wet wt. of testes}}{\text{Wet wt. of male albino rat}} \times 100$$

$$\text{GSI (Ovanan tissue) or FFI} = \frac{\text{Wet wt. of ovaries}}{\text{Wet wt of female albino rat}} \times 100$$

$$\text{GSI (Uterine tissue) or UI} = \frac{\text{Wet wt. of uteri}}{\text{Wet wt. of female albino rat}} \times 100$$

Results

Tables 20.1 to 20.6 summarized the changes in body weight and genital organs weight and GSI etc. of albino rats treated with C.P and N.G. + E.E. The rats of normal and control groups did not show any remarkable changes in body weight and genital organ weight. Reduction in body wt from 28 g to 60 g in CP treatment, 20 g to 70 g body wt. in C.P + N.G. + E.E treated rats. 5 g to 55 g increase in the body weight in N.G. + E.E. treatment in male rats respectively (Tables 20.1–20.3) while body wt. gradually decreased from 20 g to 80 g in C.P seed extract treatment, from 10 g to 55 g in N.G. + E.E treatment and 20 g to 80 g in C.P. + N.G. + E.E. treated female rats respectively (Tables 20.4–20.6). The reduction in body wt. was dose and duration of treatment dependent. The testicular wt. gradually decreased from 0.950 g to 0.750 g in C.P treatment and gradual increase from 0.900 g to 1.320 g in N.G. + E.E. treatment and again decreased from 1,450 g to 0.650 g in C.P + N. G. + E.E. treated male rats respectively (Tables 20.1–20.3).

The gradual decrease in the ovarian wt. was recorded from 100 g to 0.50 g in C.P. Seed extract treatment, from 0.105 g to 0.40 g in N.G. + E.E treatment 0.200 g to 0.100 g in C.P. + N.G. + E.E treated female rats respectively (Tables 20.4–20. 6) and uterine wt. from 0.700 g to 0.450 g in C.P. treatment from 0.520 g to 0.100 g in N.G. + E.E. treatment and from 0.680 g to 0.280 g in C.P. + N.G. + E.E. treated female rats from 20 days to LTT 90 days respectively (Tables 20.4–20.6). While the GSI (in male MFI) first showed increase from 20 to 40 days of treatment and then exhibited decrease towards 90 days in CP seed extract treatment, but N.G. + E.E. treatment also showed somewhat increase in GSI from day 20th to 90th day of treatment but C.P. + N.G. + E.E. treatment showed significant decrease in the GSI in male rats suggesting the malfunctioning of the testes of the corresponding treated males (Tables 20.1–20.3).

GSI in female rats showed gradual decrease from day 40th to day 90th of C.P. seed extract treatment (Table 20.4) while GSI in N.G. + E.E. treated dams showed first increase from 20th to 40th day and afterwards decrease from 60th to 90th days (Table 20.5) but gradual decrease in GSI has been reported from 20th days to 90th days in C.P. + N.G. + E.E. treated dams (Table 20.6).

Uterine index showed gradual decrease from 40th day to 90th day of C.P. seed extract treatment (Table 20.4) while N.G. + E.E. treatment first showed an increase in uterine index from days 20th to 90th days and afterwards gradual decrease from 60th to 90th days of treatment (Table 20.5) but the uterine index of C.P. + N.G. + E.E. treated female rats showed gradual decrease from 20th to 90th days of treatment (Table 20.6) suggesting that the antifertility activity in order C.P. < N.G. + E.E. < C.P. + N.G. + E.E. and C.P. + N.G. + E.E. treatment is more effective in the present investigation.

The data from Tables 20.1–20.6 on the effect of antifertility agents treatment indicates that there is no doubts about the efficacy of *Carica papaya* seed extract as an effective male and female antifertility agent but the effect was somewhat slower than the effects of synergistic *e.g.*, N.G. + E.E. and C.P. + N.G. +E.E. respectively. The efficacy was correlated with dose and duration of antifertility agent treatment. The synergists C.P. + N. G. + E.E. showed adverse effect on body wt. gain and GSI data as evident by Tables 20.1–20.6.

Table 20.1: Effect of *Carica papaya* Seed Extract Treatment on the Body Weight and Testes Weight of Male Albino Rats upto 90 Days

Sl.No.	Experimental Group of Male Rats	Average Body Weight		Decrease or Increase in Body Wt.	Average Wt. of Testes** in Male Albino Rats	$GSI = \frac{\text{Wet Wt. of Testes}}{\text{Wet Wt. of Male Rat}} \times 100$
		IW*	FW*			
1.	Normal	200 g	220 g	20 g	1.005 g	0.502
2.	Control	200 g	218 g	18 g	1.005 g	0.502
3.	20 Days Treated	200 g	172 g	– 28 g	0.950 g	0.552
4.	40 Days Treated	200 g	164 g	– 36 g	0.942 g	0.574
5.	60 Days Treated	200 g	150 g	– 50 g	0.800 g	0.533
6.	90 Days Treated	200 g	140 g	– 60 g	0.750 g	0.535

*Body wt.: Average wt. of 5 male albino rats in each group; IW: Initial weight; FW: Final weight at autopsy.

**Wt. of testes: Average wt. of testes of 5 albino rats in each group.

Table 20.2: Effect of *Levonorgestrel* and *Ethinylestradiol* Treatment on the Body Weight and Testes Weight of Male Albino Rats upto 90 Days

Sl.No.	*Experimental Group of Male Rats*	*Average Body Weight*		*Decrease or Increase in Body Wt.*	*Average Wt. of Testes in Male Albino Rats*	$GSI = \frac{\text{Wet Wt. of Testes}}{\text{Wet Wt. of Male Rat}} \times 100$
		*IW**	*FW**			
1.	Normal	200 g	220 g	20 g	1.005 g	0.502
2.	Control	200 g	218 g	18 g	1.005 g	0.502
3.	20 Days Treated	160 g	165 g	5 g	0.900 g	0.545
4.	40 Days Treated.	160 g	180 g	20 g	1.090 g	0.605
5.	60 Days Treated	160 g	204 g	44 g	1.205 g	0.590
6.	90 Days Treated	160 g	215 g	55 g	1.320 g	0.613

Table 20.3: Effect of *Carica papaya* Seeds Extract Along with *Levonorgestrel* and *Ethinylestradiol* on the Body Weight and Testes Weight of Male Albino Rats upto 90 Days

Sl.No.	Experimental Group of Male Rats	Average Body Weight		Decrease or Increase in Body Wt.	Average Wt. of Testes in Male Albino Rats	$GSI = \frac{\text{Wet Wt. of Testes}}{\text{Wet Wt. of Male Rat}} \times 100$
		IW*	FW*			
1.	Normal	200 g	220 g,	20 g	1.005 g	0.502
2.	Control	200 g	218 g	18 g	1.005 g	0.502
3.	20 Days Treated	240 g	220 g	– 20 g	1.450 g	0.659
4.	40 Days Treated	240 g	200 g	– 40 g	1.050 g	0.502
5.	60 Days Treated	240 g	180 g	– 60 g	0.750 g	0.41.6
6.	90 Days Treated	240 g	170 g	– 70 g	0.650 g	0.382

Table 20.4 Effect of *Carica papaya* Seeds Extract Treatment on the Body Weight and Genital Organs Weight of Female Albino Rats upto 90 Days

Sl.No.	Experimental Group of Male Rats	Average Body Weight		Increase or Decrease in Body Weight	Genital Organs** Weight in Female		$GSI = \frac{\text{Wet Wt. of Ovary}}{\text{Wet Wt. of Female Rat}} \times 100$	$GSI = \frac{\text{Wet Wt. of Uterus}}{\text{Wet Wt. of Female Rat}} \times 100$
		IW*	FW*		Ovary	Uterus		
1.	Normal	180 g	240 g	60 g	1.25 g	8.20 g	0.520	3.41
2.	Control	180 g	235 g	55 g	1.20 g	8.00 g	0.510	3.33
3.	20 Days Treated	180 g	160 g	– 20 g	0.100 g	0.700 g	0.062	0.437
4.	40 Days Treated	180 g	130 g	– 50 g	0.90 g	0.690 g	0.069	0.530
5.	60 Days Treated	180 g	120 g	– 60 g	0.70 g	0.500 g	0.058	0.416
6.	90 Days Treated	180 g	100 g	– 80 g	0.50 g	0.450 g	0.050	0.450

*Body wt.: Average wt. of 5 female albino rats in each group; IW: Initial weight; FW: Final weight.

**Wt. of ovary and uterus: Average wt. of ovary and uterus of 5 female albino rats in each group.

Table 20.5 Effect of *Levonorgestrel* and *Ethinylestradiol* Treatment on the Body Weight and Genital Organs Weight of Female Albino Rats upto 90 Days

Sl.No.	Experimental Group of Male Rats	Average Body Weight		Increase or Decrease in Body Weight	Genital Organs Weight in Female		GSI = (Wet Wt. of Ovary / Wet Wt. of Female Rat) × 100	GSI = (Wet Wt. of Uterus / Wet Wt. of Female Rat) × 100
		IW	FW		Ovary	Uterus		
1.	Normal	180 g	240 g	60 g	1.25 g	8.20 g	0.520	3.41
2.	Control	180 g	235 g	55 g	1.20 g	8.00 g	0.510	3.33
3.	20 Days Treated	150 g	140 g	– 10 g	0.105 g	0.520 g	0.075	0.371
4.	40 Days Treated	150 g	120 g	– 30 g	0.100 g	0.500 g	0.083	0.416
5.	60 Days Treated	150 g	110 g	– 40 g	0.90 g	0.120 g	0.081	0.109
6.	90 Days Treated	150 g	95 g	– 55 g	0.40 g	0.100 g	0.042	0.105

Table 20.6 Effect of *Carica papaya* Seeds Extract Along with *Levonorgestrel* and *Ethinylestradiol* Treatment on the Body Weight and Genital Organs Weight of Female Albino Rats upto 90 Days

Sl.No.	Experimental Group of Male Rats	Average Body Weight		Increase or Decrease in Body Weight	Genital Organs Weight in Female		$GSI = \frac{\text{Wet Wt. of Ovary}}{\text{Wet Wt. of Female Rat}} \times 100$	$GSI = \frac{\text{Wet Wt. of Uterus}}{\text{Wet Wt. of Female Rat}} \times 100$
		IW	FW		Ovary	Uterus		
1.	Normal	180 g	240 g	60 g	1.25 g	8.20 g	0.520	3.41
2.	Control	180 g	235 g	55 g	1.20 g	8.00 g	0.510	3.33
3.	20 Days Treated	220 g	200 g	– 20 g	0.200 g	0.680 g	0.100	0.340
4.	40 Days Treated	220 g	186 g	– 36 g	0.185 g	0.590 g	0.099	0.317
5.	60 Days Treated	220 g	172 g	– 48 g	0.160 g	0.425 g	0.093	0.241
6.	90 Days Treated	220 g	140 g	– 80 g	0.100 g	0.280 g	0.071	0.200

Discussion

In this investigation onset of infertility after C.P., N.G. + E.E. and C.P. +N.G. + E.E. treatment was dose and duration of treatment dependent which is indicated by GSI (in male and female rats) as evident by Tables 20.1–20.6. The increase in GSI from 20th–40th day of C.P. or N.G. + E.E. treatment separately in male rats suggesting the initial increase was due to the accumulation of fluid in the testes due to blockage in the initial segment of epididymis (Bhide *et al.*, 2001 and Gupta *et al.*, 2003) while treated rats gained body wt. as fast as did the control was observed in male rats after gossypol polyacetic acid treatment by Ye *et al.* (1993). The decrease in GSI as well as in testicular weight seems to be disfunctioning of the testes due to intoxication of antifertility agent treatment as observed in rats by Tyagi *et al.* (1990) after *Trigonela foenum graceum* treatment, Lohiya *et al.* (1993, 1994 and 1999) with C.P. seed extract treatment. Subhalakshmi and Reddy (1996) after *Hyosyamus niger* seed extract treatment, Gupta (2000), Bhide *et al.* (2002) in albino rats after treatment with some antifertility agents respectively. Antispermatogenic effect were accompanied by significant decrease in testicular wt. as well as GSI in male albino rats (Gupta, 2000 and Bhide *et al.*, 2002), while there was no significant change in body weights of treated rats indicating normal development as reported by Chawla *et al.* (1996); Subhalakshmi and Reddy (1996) and Lohiya *et al.* (1999), in male albino rats and male rabbits after treatment with *Hyosyamus niger* seed extract and chloroform extract of *Carica papaya* seed extract. Tyagi and Rajalakshmi (1996) in adult Reshus monkey after androgen exposure respectively. Tyagi *et al.* (1990) reported the drastic reduction not only in the body weight, but also on the testicular weight after treated the male albino rats with *Trigonela foenum graceum* acetone extract as antifertility agent as also observed by Sharma *et al.* (1996) in male rats after *Marsilea minuta* sporocarp exposore; Parte *et al.* (1996) in male rats after *Xenoestrogens* exposure; Singh (1996) in male rats after *Hibiscus rosa sinensis*, Gupta (2000) and Bhide *et al.* (2002) in albino rats after treatment with *Abrus precatorius* seed extract respectively.

In the present investigation, the body weight of female showed gradual decrease (Tables 20.4–20.6) which was dose and duration of treatment dependent. While GSI in female rats showed gradual decrease from day 20th to 90th day of CP and C.P. + N.G. + E.E.

treatment (Tables 20.4 and 20.6), it showed first increase (from day 20th to 40th day) and then gradual decrease (from 60th to 90th days) in N.G. + E.E. treatment (Table 20.5). The increase or decrease in GSI is correlated with the increase or decrease in the body wt. with respect to the increase or decrease in ovarian weight as observed in the present investigation. Reduction in ovarian weight with respect to the reduction in the body wt. was observed in rats by Adhikari *et al.* (1990), Choudhari *et al.* (1991) after *Piper betle* leaf stalk treatment; Pandey *et al.* (1992) after *Hibiscus rosa sinensis* flower extract exposure and Sreenivasalu *et al.* (1992) after CDRI 85/287 compound treatment and stated that reduction wt. of ovaries was correlated with the anti-estrogenic activity as also indicated by the absence of developing, primary, Graffian follicles and corpora lutea, in the ovaries of albino rats treated with C.P., N.G. + E.E. or C.P. + N.G. + E.E. in the present investigation and also reported by Gupta (2000), Bhide *et al.* (2001) and Gupta *et al.* (2003) in albino rats after treatment with some antifertility agents of plant origin.

Uterine index showed gradual decrease in C.P. (Table 20.4) and C.P. + N.G. + E.E. (Table 20.6) treatment, while N.G + E.E. treatment first showed increase (from 20th to 40th day) and then showed gradual decrease (Table 20.5) in the present investigation. The decrease in uterine weight was also reported in albino rats by Adhikari *et al.* (1990) and Choudhary *et al.* (1991) after *Piper betle* leaf stalk extract exposure, Pandey *et al.* (1992) after *Hibiscus rosa sinensis* flower extract exposure and by Sreenivasalu *et al.* (1992) after CDRI 85/287 compound exposure, Gupta (2000) and Gupta *et al.* (2003) in albino rats after treatment of *Abrus precatorius* and *Carica papaya* seed exposure respectively while Giand (1999) reported the wt. of the uterine horns increased to 166 mg/100 g body wt. as compared to 64 mg/100 g body wt. in control group of albino rats treated with some intermediate azasteriod was due to it's estrogenic activity as indicated by the gradual increase in the ovarian weight as reported by Bhide *et al.* (2001) in albino rats after treatment with plant extract as antidote on the intoxication of *Cerbera thevetia* seed kernel extract exposure. So, from the present investigation it could be concluded that the synthetic and plant origin antifertility agents have adverse effect on the development of gonads and other genital organs *e.g.* uteri and also resulted into decrease in'o fertility of pest animals while effecting their MFI, FFI and UI.

Acknowledgement

We are very much thankful to Prof. S. Sahai, HOD, Department of Zoology and Dean of Life Sciences, Dr. H.S. Gour V. V. Sagar (M.P.) for providing the laboratory facilities and one of the authors (M. Bhide) is grateful to UGC for the financial assistance during the tenure of this research investigation.

References

Adhikari, P., Banerji, J., Choudhary, D., Das, A.K., Deb, C.C. and Mukerjee, S.R. 1990. Effect of oral administration of stalk leaves of *Piper betle* on estrous cycle and its antifertility activity in rats. *Ind. J. Physio. Appl. Sci.*, 44(3): 116–123.

Bhide, M., Gupta, A. and Modi, S. 2001. *Cerbera thevetia* toxicity remediation by placentract therapy on the reproductive physiology of albino rats. In: *Current Topics in Environmental Sciences*, (Eds.) G. Tripathi and G.C. Pandey. ABD publisher, Jaipur, India, p. 327–346.

Bhide, M., Gupta, A. and Modi, S. 2002. *Abrus precatorius* induced histopathological changes in the testes of albino rats. In: *Perspectives in Animal Ecology and Reproduction*, (Eds.) V.K. Gupta, Anil K. Verma and Jai Prakash Sharma. Daya Publishing House, Delhi, p. 293–203.

Chawla, A.S., Chinoy, N.J. and Ghosh, S. 1996. Antifertility effects of papaya seed's aqueous extract in male rats and it's reversal. *Asian J. Androl. Supple.*, PO–11.

Choudhary, D.N., Singh, J.N. and Singh, B.P. 1991. Effect of some medicinal plants on fertility of albino rats. Osooan. *J. Pharmacol.*, 23(4): 253–257.

Edmond, J. Ferris. 1950. *The Care and Breeding of Laboratory Animals.* John Wiley and Sons, New Delhi, Chapman and Hall Ltd., London.

Giand, B. 1999. Study of estrogenic effect of some intermediate azasteroids. *Ind. J. Host. Phaml.*, 8: 223–234.

Gupta, A. 2000. Synergistic effect of levonorgestrel and ethinylestradiol with some antifertility agents on the reproductive physiology of albino rats. *Ph.D. Thesis*, Dr. H.S. Gour Vishwavidyalaya, Sagar, M.P.

Gupta, A., Bhide, M. and Modi, S. 2003. Antifertility potential of *Carica papaya* in mammalian reproduction. In: *Potential of Living Resources*, (Eds.) G. Tripathi and A. Kumar. Discovery Publishing House, New Delhi, p. 43–75.

Lohiya, N.K., Goyal, R.B., Jaiprakesh, D., Sharma, S., Kumar, M. and Ansari, A.S. 1993. Induction of reversible antifertility with a crude ethanol extract of *Carica papaya* seeds in albino rats. *Int. J. Pharm.*, 30(4): 398–320.

Lohiya, N.K., Goyal, R.B., Jaiprakesh, D., Ansari A.S. and Sharma S. 1994. Antifertility effect of aqueous extract of *Carica papaya* seeds in albino rats. *Planta Medica*, 64(5): 400–404.

Lohiya, N.K., Pathak, N., Mishra, P.K. and Manivannan, B. 1999. Reversible contraception with chloroform extract of *Carica papaya* seeds in male rabbits. *Rep. Toxicol.*, 13(1): 59–66.

Pandey, D., Bisht, M. and Nabhachandra, V. 1992. The antiestrogenicity of *Hibiscus rosa sinensis* flower's extract on female rats. *Him. J. Environ. Zool.*, 6: 50–55.

Parte, P., Majdic, G., Sharpe, R.M. and Saunder, R.J.K. 1996. Investigation of the effect of neonatal exposure to xenoesterogens on gene expression in the testes and pituitary of the rats. *Ind. J. Andro. Supple.*, PO–123.

Sharma, J.D. 1988. *Forensic Science and Toxicology.* Suvidha Law House Publishers, Bhopal, M.P.

Sharma, A., Dixit, V.P. and Dhobal, M.P. 1996. Male contraceptive efficacy of *Myristica fragrans* seed extract (MeOH-fraction). *Ind. J. Androl. Supple.*, PO–35.

Singh, S.P. 1996. Antifertility effect of *Hibiscus rosa sinensis* flowers in male rats. *Ind. J. Androl. Suppl.*, PO–31.

Sreenivasulu, S., Singh, M.M., Dwivedi, A., Setty, B.S. and Kamboj, V.P. 1992. CDRI 85/287: a novel antiestrogen and anti-implantation agent. Biological profile and interaction with the estrogen receptors in immature rat uterus. *Contraception*, 45(1): 51–92.

Subhalakshmi, S. and Reddy, M.K. 1996. Effect of *Hyosyamus niger* seed extract on fertility in male albino rats. *Advance in Zool.*, 2: 241–246.

Tyagi, A. and Rajalaksmi, M. 1996. Effect of long-term administration of androgen on body weight and testicular volume in adult Rhesus monkey. *Ind. J. Androl. Supple.*, PO–120.

Tyagi, S.D., Singh, N. and Agarwal, S.C. 1990. Spermicidal effect of crude powder of *Trigonela foenum graceum* in albino rats. *Biosphere*, 2(1): 119–122.

Ye, W., Lu, G., Den, Y., Huang, Y. and Xue, S. 1993. Effect of intra-vas deferens injections of gossypol polyacetic acid on fertility and spermatogenesis in rats. *Chinese Med. Sci. J.*, 8(1): 20–24.

Chapter 21

Ciprofloxacin Acts through Oxidative Stress Generation in Impairing Spermatogenesis and Steroidogenesis in Testes of Adult Male Rats

C. Anbalagan and G. Vanithakumari
Department of Zoology, Bharathiar University,
Coimbatore – 641 046, Tamil Nadu

Introduction

The clinical utility of non-steroidal anti-inflammatory drugs (NSAIDs) to manage pain and inflammation is limited by adverse side effects (James, 1999). The testes are highly susceptible to ciprofloxacin as it crosses the blood-testis barrier and depresses spermatogenesis with a numeric reduction in spermatids and fragmentation of Sertoli cells (Dalsenter *et al.*, 1996). A few toxicological studies (Sally, 1997) have addressed the possible relationship between reproductive toxicity and exposure tc chemicals that generate reactive oxygen species (ROS). ROS are an important part of the defence mechanism against infection, but excessive generation of free oxygen radicals may damage tissues. ROS are formed in both physiological and pathological conditions in mammalian tissues. Reactive oxygen species has been shown to be involved in infertility due to defective sperm function (Aitken *et al.*, 1989). The present study was undertaken to evaluate the effect of ciprofloxacin on the testicular antioxidant system and its influence on steroidogenesis in the testes.

Materials and Methods

Materials

The drug ciprofloxacin hydrochloride T.P. manufactured by Okasa Limited, Goa, India was used for the experiment. The drug was dissolved in distilled water and was administered orally.

Animals and Treatment

Healthy adult male albino rats of Wistar strain weighing 180 g–225 g were used in the present investigation. They were housed in clean cages in a well ventilated room with 12 ± 1 hour light and 12 ± 1 hour dark schedule. They were fed with a standard balanced diet and clean drinking water was made available *ad libitum.*

Adult male rats were orally administered with ciprofloxacin at a dose of 250 and 400 mg respectively/60 kg body weight per day (low and high dose, respectively) for 30 days.

Experimental Procedure

Twenty-four hours after the last treatment the rats were killed by decapitation method. The testes were removed, washed in 0.9 per cent Saline and weighed. A 10 per cent of testicular homogenate was prepared and centrifuged at 3000 rpm. The supernatant was used for various biochemical estimations.

Chemicals and Reagents

All chemicals and reagents used for the experiments were of analytical grade and were obtained from BDH (British Drug House, England and India), E. Merck (Germany and India), Sigma Chemical Company (USA), Loba chemie (Indo. Austranol Co, India) and Qualigens fine chemicals division (Mumbai).

Statistical Analysis

All data were analysed statistically by Duncan's Multiple Range Test (DMRT) (Duncan, 1955).

Table 21.1: Effect of Ciprofloxacin on the Body Weights and Organ Weights in Male Adult Rats

Parameters	*Control*	*Low Dose*	*High Dose*
Body Weight (Grams)	$198.000^{b}\pm4.722$	$201.000^{a}\pm1.414$	$189.000^{c}\pm1.795$
Organ Weight (Grams) Testis	$1073.900^{a}\pm17.802$	$989.820^{c}\pm23.939$	$1011.460^{b}\pm11.863$

Each value is the mean ± SE.

Means followed by a common letter are not significantly different at the 5 per cent level by DMRT.

Table 21.2: Effect of Ciprofloxacin on Biochemical Parameters in Testes of Male Adult Rats

Parameters	*Control*	*Low Dose*	*High Dose*
Lipid peroxidation (n moles/mg.tissue)	$0.276^{c} \pm 0.005$	$1.556^{b} \pm 0.002$	$2.448^{a} \pm 0.001$
Superoxide dismultase (n moles/min/mg.protein)	$1.552^{a} \pm 0.018$	$0.307^{ab} \pm 0.001$	$0.287^{c} \pm 0.002$
Catalase (n moles/min/mg.protein)	$0.136^{a} \pm 0.004$	$0.035^{b} \pm 0.0003$	$0.026^{c} \pm 0.0006$
Gulathione (n moles/mg.tissue)	$0.0480^{a} \pm 0.004$	$0.0130^{b} \pm 0.0003$	$0.0090^{c} \pm 0.0002$
3β-OSHD (n moles/min/mg.protein)	$0.034^{a} \pm 0.0005$	$0.013^{b} \pm 0.0001$	$0.009^{b} \pm 0.0001$
17β-OHSD (n moles/min/mg.protein)	$0.183^{a} \pm 0.001$	$0.105^{b} \pm 0.001$	$0.090^{c} \pm 0.001$
Testosterone (ng/ml Serum)	$3.185^{a} \pm 0.004$	$0.813^{b} \pm 0.005$	$0.548^{c} \pm 0.002$
Protein (n moles/mg.tissue)	$0.185^{a} \pm 0.002$	$0.014^{bc} \pm 0.001$	$0.013^{b} \pm 0.002$

Each value is the mean ± SE.

Means followed by a common letter are not significantly different at the 5 per cent level by DMRT.

Results

The body weight of ciprofloxacin-treated rats did not show any significant change during the course of the treatment. The weights of the testis were decreased significantly from those of the control group of animals (Table 21.1).

The specific activities of superoxide dismutase, catalase and concentration of glutathione and lipid peroxidation had decreased, while the levels of hydrogen peroxide was found to be elevated in treated rats when compared to control animals. The specific activities of steroidogenic enzymes, 3β-hydroxy steroid dehydrogenase and 17β-hydroxy steroid dehydrogenase were decreased significantly. There was also a marked reduction in the levels of testosterone and testicular protein concentrations in treated group when compared to the controls (Table 21.2).

Discussion

In the present study, administration of ciprofloxacin had no marked effect on the body weight ot the animals during the long treatment schedules (4 weeks). The food intake was found to be normal during the drug treatment schedule. The drug seems to have no marked adverse effect on the body weight. On the other hand the testicular weights of ciprofloxacin treated rats were significantly decreased. The testis has been shown to be highly susceptible to ciprofloxacin as it crosses blood-testis barrier and depresses spermatogenesis (Dalsenter *et al.*, 1996). The decrease in the testicular weight of ciprofloxacin treated rats may be due to reduced tubule size, spermatogenic arrest and inhibition of steroid biosynthesis of Leydig cells (Shivanandappa *et al.*, 1983).

The antioxidant system plays an effective role in protecting testes and other biological tissues below a critical threshold of reactive oxygen species thus preventing testicular dysfunction (Oschsendorf, 1999). Antioxidant enzymes constitute a mutually supportive team of defence against reactive oxygen species (ROS). In the present study, the activities of antioxidant enzymes, superoxide dismutase, catalase and concentration of glutathione were decreased with increased lipid peroxidation in ciprofloxacin treated rats. The levels of hydrogen peroxide generation were found to be elevated which serves as a marker of increased ROS. Thus, in the present study increase in hydrogen peroxide with reduction in

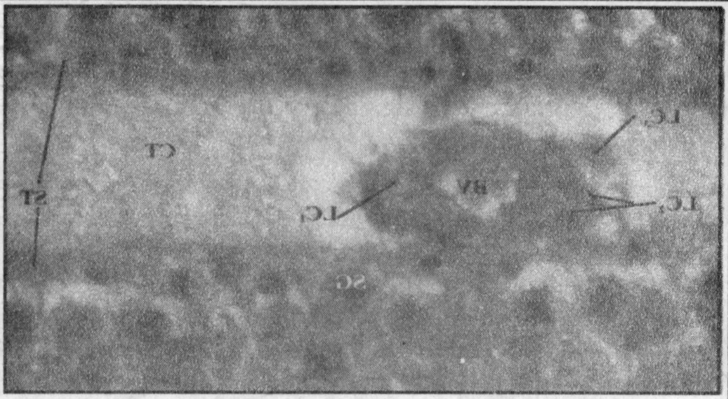

Plate 21.1: Control–H × E 400

Note: The testis of control group showed normal seminiferous tubules (ST) with active spermatogeneis. Leydig cells (LC) with three different populations.

IC: Interstitial cells; BV: Blood vessels; SC: Sertoli cells; SG: Spermatogonia; PS: Primary spermatogonia; LC_1: Small polygonal; LC_2: Polyhydrous vacuolated; LC_3: Big size; CT: Connective Tissue; PS: Primary spermatocyte; D: Debris.

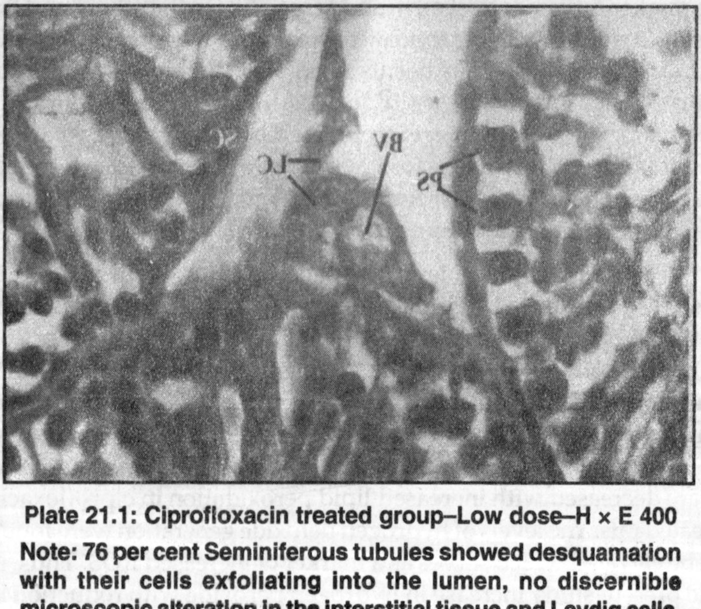

Plate 21.1: Ciprofloxacin treated group–Low dose–H × E 400

Note: 76 per cent Seminiferous tubules showed desquamation with their cells exfoliating into the lumen, no discernible microscopic alteration in the interstitial tissue and Leydig cells.

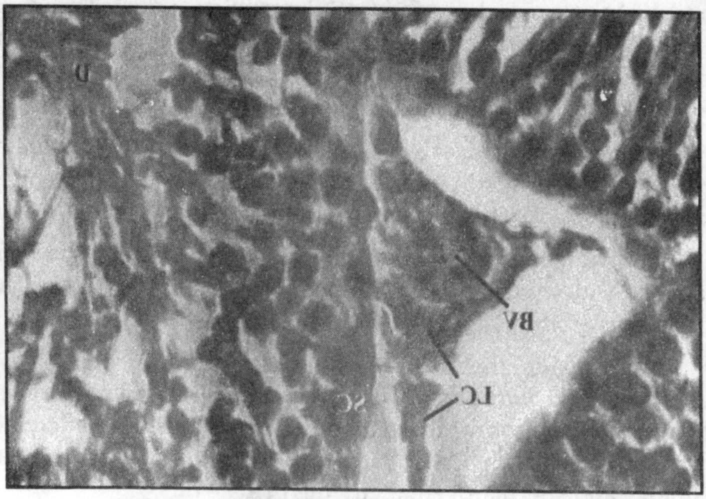

Plate 21.1: Ciprofloxacin treated group–High dose–H × E 400
Note: 83 per cent Seminiferous tubules started showing degenerative changes, the spermatogonial and primary spermatocytes were intact, but there was exfoliation of other gem cells in to the lumen, the blood vessels were congested. Leydig cell number reduced.

antioxidant enzymes indicates the oxidative stress induces by ciprofloxacin.

The decreased activities of steroidogenic enzymes 3β-Hydroxy steroid dehydrogenase and 17β-Hydroxy steroid dehydrogenase were indicative of the reduced testicular steroidogenesis. Ciprofloxacin administration had also inhibited the 17β-OSHD activity besides 3β-OHSD and this indicates the affectation of cytochrome P_{450} enzymatic activity interacting with the conversion of progesterone to androstenedione and subsequent testosterone production. Infact, a drastic reduction in testosterone levels was observed in this study confirming the above enzymatic response. The reduction of superoxide dismutase suggests that it is involved in antioxidant defence and it has been shown to act as an alternate regulatory switch in testicular steroidogenesis (Kumar *et al.*, 1990). The cytochrome P_{450} enzymes of the steroidogenic pathway are known to produce free radicals. These free radicals are produced as a result of electron leakage due to the interaction of steroid products or other pseudosubstrates with the enzymes. The inability of the pseudosubstrate to be oxygenated promotes the release of reactive oxygen species (Peltola *et al.*, 1996).

In ciprofloxacin-treated rats the decrease in the testicular contents of protein were observed, which shows that reactive oxygen species can attack vital components of the cell-like proteins.

Spermatogenesis had been abruptly arrested at early pachytene stage itself, decreasing these cells and also leading to a complete absence of secondary spermatocytes to late spermatids (Plate 21.1).

In conclusion, the present studies indicate that ciprofloxacin alters testicular functions possibly by inducing reactive oxygen species and decreasing the antioxidant enzymes, steroidogenic enzymes, lipid peroxidation, testosterone, proteins, testicular histology and spermatogenic activities in male rats.

References

Aitken, R.J., Clarkson, J.S., Fishel, S. 1989. Generation of reactive oxygen species, lipid peroxidation and human sperm function. *Biol. Reprod.*, 40: 183–197.

Dalsenter, P.R, Faqi, A.S., Webb. J., Merker. H.J. and Chahoud. 1996. Reproductive toxicity and tissue concentrations of lindane in adult male rats. *Hum. Exp. Toxicol.*, 15: 406–410.

James, D.S. 1999. The multisystem adverse effects of NSAID therapy. *J. Am. Osteopat. Assol.*, 99 (II Supple.): 51–57.

Kumar, G.P., Seerwani, N., Laloraya, M., Nivsarkar, M., Verma, S., Singh, A. 1990. Superoxide dismutase as a regulatory switch in mammalian testicular steroidogenesis. *Biochem. Biophys. Res. Commun.*, 173: 302–308.

Oschsendorf, F.R. 1999. Infections in the male genital tract and reactive oxygen species. *Human Reproduction Update*, 5: 399–420.

Petola, V., Huhtaniemi, I., Metsa-ketela, T., Ahoutapa, M. 1996. Induction of lipid peroxidation during steroidogenesis in the rat testis. *Endocrinology*, 137: 105–112.

Sally, D.P. 1997. Reproductive and Endocrine toxicology. The mature spermatozoan as a target for reproductive toxicants. In: *Comprehensive Toxicology,* (Eds.) Kim Boekelteide, Robert. E. Chapin, Patricia, B.H. and Craig Harris. Pergamon, Cambridge, UK, 10: 165–191.

Shivanandappan, T., Krishnakumari, M.K. 1983. Hexachlorocyclo-hexane-induced testicular dysfunction in rats. *Acta. Pharmacol. Toxicol.,* 52: 12–17.

Sally D.F. 1997. Reproductive and Endocrine toxicology. The mature spermatozoon as a target for reproductive toxicants. In: *Comprehensive Toxicology*, (Eds.) Kim Boekelheide, Robert E. Chapin, Patricia B.H and Craig Harris. Pergamon. Cambridge, UK. 10: 165–181

Sivanandappan, T., Kalaiselvakumari M.K. 1983. Hexachlorocyclohexane-induced testicular dysfunction in rats. *Acta Pharmacol Toxicol* 52: 12–17

Index